Connected
Geometry

Connected
Geometry

Developed by

Education Development Center, Inc.
Newton, Massachusetts

EVERYDAY LEARNING®

Chicago, Illinois

Photo Credits

cover: Photodisc™

interior: Unit 2, pp. 100-101 UPI/Corbis-Bettmann

Everyday Learning Development Staff

Editorial: Anna Belluomini, Steve Mico, Carol Zacny

Production/Design: Fran Brown, Héctor Cuadra, Annette Davis,
Jess Schaal, Norma Underwood

Additional Editorial Credits: Jody Levine, Abby Tannenbaum

 This project was supported, in part, by the National Science Foundation.
The opinions expressed are those of the authors and not necessarily those
of the Foundation.

ISBN 1-57039-589-6

Everyday Learning Corporation
P.O. Box 812960
Chicago, IL 60681
1-800-322-MATH (6284)

Visit our website at www.everydaylearning.com

1 2 3 4 5 6 7 8 9 CU 04 03 02 01 00 99

Contents

Acknowledgments

The *Connected Geometry* modules are the result of the efforts of many people besides the project team of the Education Development Center. These people provided support and encouragement throughout the writing, field testing, and final production. A group of teacher advisors and their students provided extremely valuable feedback on how the lessons worked in their classrooms. They held the first trials of the methods and ideas we use. Both teachers and students offered detailed comments and opinions about how they worked with the material. We, in turn, have incorporated their work as well as what we learned from visiting many of their classrooms over the years. **Teacher Advisors and Field Testers:** Jim Barnes, Joan Bryant, Cheri Dartnell, Larry Davidson, Paul DiNolo, Janice Enos, Kathy Erikson, Jane Gorman, Carol Haney, Betty Helm, Felisa Honeyman, Elfreda Kallock, Britt Kleiman, Phil Lewis, Barney Martinez, Carol Martingnette-Boswell, Doug McGlathery, Bill Nevin, Mary T. Nowak, Faye Ruopp, Gary Simon, Jesse Solomon, and Jennifer Takarabe.

We've also benefited from working with an Advisory Board of mathematicians, educators, college professors, and high school teachers. Our Advisory Board members brought us a wealth of resources and helped us develop ideas for topics and problems. **Advisory Board Members:** Barbara Adler, Gail Bussone, Douglas Clements, Ed Dubinsky, Joan Ferrini-Mundy, Carol Findell, Hector Hirigoyen, Paul Horwitz, James Kaput, Margaret Kenney, Harvey Keynes, Eugene Klotz, Steven Monk, Bob Moses, Barbara Scott Nelson, Jim Newton, Arthur Powell, Andee Rubin, Deborah Schifter, Marjorie Senechal, Antonia Stone, Ellen Wahl, and Bernie Zubrowski.

We are indebted to our reviewers and their students. The reviewers used drafts of the modules as course material. They pointed out omissions and errors, as well as their favorite problems and other features that they particularly enjoyed. They also shed light on new ways of looking at some of the problems and themes. **Reviewers:** Peter Baunfeld, Larry Davidson, Tommy Dreyfus, Brad Findell, Carol Findell, Wayne Harvey, Jim King, Terry Leveritch, Gary Martin, Barney Martinez, Tricia Pacelli, Libby Palmer, Judy Roitman, Faye Ruopp, Chih-Han Sah, and Hung-Hsi Wu.

We also extend thanks to the following **Special Contributors:** Tammy Jo Ruter, for her illustration expertise; Jim Sandefur, for the "120° devise" used in *Optimization*; and Jim Tattersall, for providing information on segment splitting.

Finally, we want to thank editors Barbara Janson and Eric Karnowski who provided unending encouragement, patience, and support in the development and editing of this curriculum, and our editors and publishers at **Everyday Learning Corporation:** Steve Mico, Anna Belloumini, Carol Zacny, Abby Tanenbaum, and Jody Levine.

The Connected Geometry Development Team
at Education Development Center, Inc.
55 Chapel St.
Newton, MA 02158
(617) 969-7100
email: ConnGeo@edc.org

Primary Developers: Nancy Antonellis, Al Cuoco, Pamela Frorer, E. Paul Goldenberg, Jack Janssen, Michelle Manes, June Mark, Daniel Scher

Other Project Staff: Jane Gorman, Jim Hammerman, Phil Lewis, Lawrence Morales, Patricia Pacelli, Jill Pfenning, Susan Rice, Joseph Scott, Jr., Despina Sylianou

Administrative Support: Laura Burns, Helen Lebowitz, Tammy Ruter, Albertha Walley

Preface

What is *Connected Geometry?*

Connected Geometry is a mathematics book created to help you and everyone in your classroom enjoy mathematics while you work on developing powerful mathematical ways of thinking. The activities and problems in this book are also designed to help you develop a special culture in your classroom: a culture of mathematical exploration in which you puzzle out, think through, and discuss your ideas with others. It is the authors' intention that the activities, reading, and writing you'll do will help you see some of the connections between your own experiences and various mathematical topics and their applications.

How to use this book

This book was written by a team of people. Each member of the team has a somewhat different opinion about what's important in mathematics and what it takes to learn mathematics, but there are two fundamental principles on which we all agree.

First, mathematics produces results: new facts, ideas, or ways of doing things. In other fields, any fact or idea may be very useful to one person and not at all useful to the next. Yet the *methods* mathematicians use to *find* these facts—*the mathematical ways of looking at things*—are extremely important and useful to just about everyone. This book contains plenty of facts and ideas—the ones considered most important for a high school education. But it also contains activities that are *organized* to help you develop the the mathematical ways of thinking *behind* these ideas. We sometimes refer to them as "habits of mind." So, as you read this book, keep in mind that its major focus is on clear, powerful thinking: *your* thinking.

Second, there are several very effective ways to do mathematics. Four of these ways are described below.

One of the best ways to learn something is by doing it—really doing it— yourself. Nothing else is as good as your own answer to a problem you worked on and thought hard about.

1. **Work on problems.** Most of the pages in this book contain problems for you to work on. Some are easy. Others are quite difficult. All encourage you to build mathematical ideas for yourself.

2. **Study what others have done.** Building upon what others know is an important part of learning. In training to be a plumber, artist, scientist, dancer, athlete, or architect, you work with people who already know how; you study what they do, and then in your own way *make their techniques your own*. It's the same in mathematics. Making advances in mathematics is a communal activity. Every new discovery, every famous result, comes about because someone finds a missing piece of a puzzle that has been worked on by dozens of people, often over

hundreds of years. So when you study a famous solution of a problem, you aren't "cheating." Instead, you are benefiting from the wisdom and refinement of a long evolution of ideas. In return, you and your classmates should try to add your own link to the chain; you should try to expand what people know.

3. Explain yourself. Many teachers often say they understand a topic much better after explaining it. Explaining your thoughts to others is a very important part of developing good habits of mind, and preparing to do so often brings you new insights.

When people pool ideas, what emerges is often better than what any one alone would have produced.

4. Work in Teams. It is valuable to think hard about a problem on your own, and it's also valuable to trade ideas with other people. Discussing a problem with others can provide missing pieces and inspire you to find new and creative solutions. A discussion usually helps everyone in a group build a deeper understanding of the topic at hand. While many geometry books contain the axiom, "The whole is equal to the sum of its parts," the members of a research team will tell you that the ability of the whole group is much greater than the sum of their individual abilities. Ask your teacher about good ways to work together.

This book is organized into activities. Each activity includes problems with a particular focus. You will also find these features in each lesson:

- **For Discussion** These should be worked on like any other problem, but the work involves discussing your ideas with others before you write things down or make a presentation.

- **Write and Reflect** For most problems you will probably be writing answers and results, but these particular problems ask for something like a brief essay. Sometimes these questions ask you to express your opinion; other times they ask for a carefully thought-out explanation, of the sort you might later present to the class.

- **Checkpoint** These problems are designed to help you determine or review what you've learned in the previous pages.

- **On Your Own** These problems are designed for you to do independently to reinforce ideas you may have developed during a group activity, an investigation, or a previous problem.

- **Take it Further** There are two kinds of problems in this feature. One asks you to look at extensions of ideas; the other asks you to establish results that will be useful later in the book.

Some people who reviewed this book thought the side note jokes were funny; some used the term "bad" jokes.

- **Perspectives** Mathematics is really about people and how they puzzle things out. This occasional feature includes personal stories of living mathematicians, as well as some of the history behind major mathematical results.

- **Sidenotes** These contain hints, comments from the writers, historical notes, information about words in the text, questions, and occasional jokes.

- **Theorems** Every branch of knowledge has its own way of establishing major results. In mathematics, the major results are know as theorems, which are established through deductive proof. You will see important statements that are proved in the text, statements others have proved, and statements you are asked to prove, all labeled as theorems. They are identified to help you keep track of important results.

We've had a great time writing this book. Some of the problems were brand new to us; sometimes we put aside our writing for a few days so that we could work on a problem. We hope you'll find the problems as much fun as we did.

UNIT 1

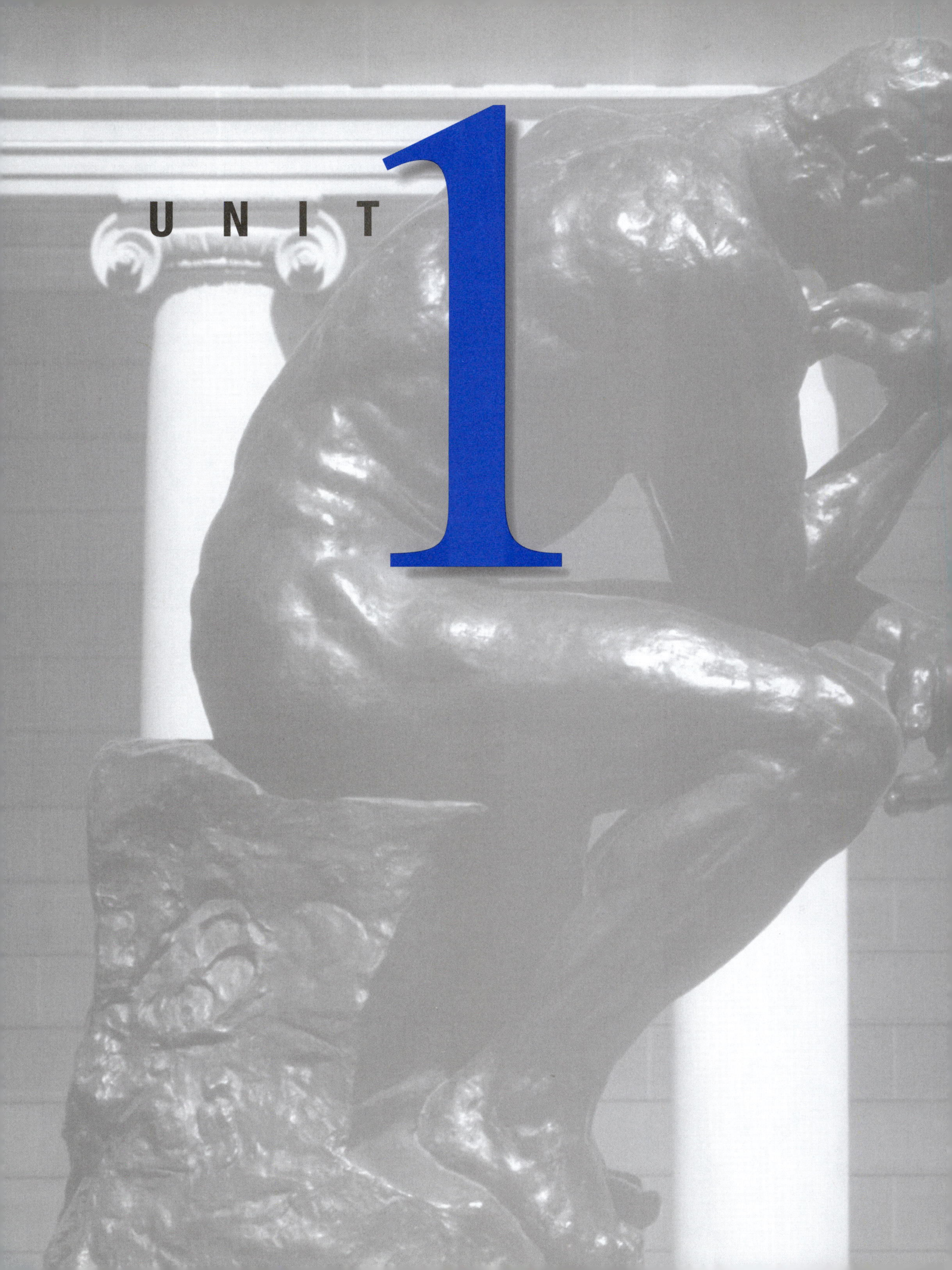

Habits of Mind
An Introduction to Geometry

1

Problem Solving in Geometry

In this lesson, you will work on many geometric problems that preview various ideas and techniques.

The problems were selected to give you a glimpse of some of the ideas, techniques, styles, and language that you will work with throughout the year. The problems are from different areas of geometry, and the connections between them may not be apparent right away. These are not 5-minute problems. Each investigation will take time.

Some problems may use unfamiliar vocabulary. Keep track of new words and ideas in your journal; find out what they mean and how they fit together.

Explore and Discuss

Think of the geometry around you.

a What shape are manhole covers? Think of a good reason for choosing *that* shape instead of all others. Explain.

b What shape are the nuts on fire hydrants? Why do you think *that* shape was selected? Write as complete an explanation as you can.

c On your own, list at least five things that are cylindrical.

d Compare all the lists of cylindrical objects in your group. You need to agree on a list of 10 objects that are cylinders. The goal is to have as many objects as possible on your list that are not on any other group's list.

ACTIVITY 1 ## Handshakes and Polygons

1. a. If everyone in your class shook hands with everyone else, how many hand-shakes would that be?

b. There are many different ways to approach this problem. Describe *your* way.

2. How many **diagonals** are in a square? In a pentagon? In a hexagon? In a heptagon? In an octagon? Write a rule that would enable you to determine the number of diagonals in *any* **regular polygon,** given the number of sides.

If penta-, hexa-, hepta-, and octa- mean 5, 6, 7, and 8, and poly- means "many," what does -gon mean?

3. Problem **2** asked about diagonals in regular polygons. Does your answer change if you consider irregular polygons, like the hexagon shown below? Explain.

4. Checkpoint How is counting handshakes similar to counting diagonals? How is it different? That is, how could knowing the answer for Problem **2** have helped you solve Problem **1**?

ACTIVITY 2

Nets

What is a face of a cube?

Imagine unfolding a cube, so all its **faces** are laid out as a set of squares attached at their edges. The resulting diagram is called a **net** for a cube. There are many nets for a cube.

Continue to unfold the cube in your mind.

5. Which of the following are nets for cubes? Explain your answer.

a.

b.

c.

d.

A tetrahedron is a 3-dimensional shape made from four triangular faces.

6. How many different nets for a cube can you make? Describe how you thought about the problem, what method you used to generate different nets, and how you checked whether or not a new one really was different.

7. Checkpoint Find all the possible nets for a regular tetrahedron.

ACTIVITY 3

The Triangle Inequality

In this activity, you will use rods of different lengths to make triangles. To begin the activity, do the following experiment.

Roll the three numbered cubes and pick three corresponding rods. For example, if you roll 5, 3, 5, pick two rods of size 5 and one of size 3. Try to make a triangle using the three rods as sides of the triangle. Some sets of three rods will work, and others won't.

8. Repeat the experiment several times. Keep a table of your results. For the combinations that do not work, write an explanation of what went wrong when you tried to make a triangle.

9. Write a rule that will tell you whether three given rods will make a triangle.

10. **Write and Reflect** Your experiments dealt only with sidelengths from 1 to 6, and not with noninteger lengths, like $4\frac{1}{2}$ or 3.14159. Write a rule that explains how you can tell if *any* three segments will actually fit together to make a triangle. Some sets of three lengths just do not work. Explain why not, and how to predict that from the lengths involved.

11. **Checkpoint** Given the three lengths, decide if they will make a triangle. (Do not construct the triangle.)

   a. 6 cm, 6 cm, 1 cm
   b. 2 cm, 4 cm, 6 cm
   c. 1 cm, 1 cm, 1 cm

   d. 4 cm, 2.1 cm, 6 cm
   e. 0.99 cm, 0.99 cm, 2 cm

ACTIVITY 4 — Angles Inscribed in Semicircles

Many important results in geometry came about because someone noticed an **invariant:** something about a situation that stays the same even as parts of the situation vary. In this activity, you will look for an invariant related to **inscribed angles.**

An angle *inscribed* in a circle has a vertex on the circle, and sides that go through two other points on the circle.

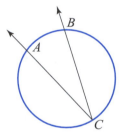

ACB is an inscribed angle.

If you have a circle, you can make a **semicircle** by drawing a **diameter** across the circle.

To make an angle inscribed in a *semicircle,* the sides need to go through the endpoints of a diameter.

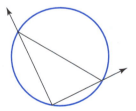

An angle inscribed in a semicircle.

How do you make sure the line segment going across is a diameter?

12. **a.** Draw a semicircle. Then inscribe an angle in that semicircle. What is the measure of that angle?

   b. Inscribe another angle in your semicircle. Measure it. What varies? What remains the same?

13. a. Draw two different diameters in one circle. Connect the endpoints of the diameters to make a four-sided figure. What kind of figure is it?

  b. Repeat the experiment with a new pair of diameters. What is invariant here? Explain.

14. A very important habit of mind is to look for connections. How are Problems **12** and **13** related? Explain.

15. Write and Reflect If you inscribed one hundred angles in a semicircle, and they all had the same measure, would that guarantee that *every* angle inscribed in a semicircle would be that size? Why or why not?

16. Checkpoint In Problem **12**, you inscribed an angle in a semicircle. What can you say about the angle inscribed in a quarter circle? In a three-quarter circle? Can you make a general rule?

∠AVB *is inscribed in an*
arc that is $\frac{3}{4}$ *of the circle.*

∠AOB *is also called the*
central angle of the arc
from A *to* B.

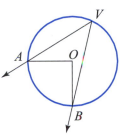

$m\angle AOB = 90°$ An angle inscribed in a three-quarter circle.

ACTIVITY 5

Cross Sections

A **cross section** is the face you get when you make one slice through an object. Problems **17–20** ask you to visualize cross sections of solid objects.

Below is a sample slice
through a cube, showing
one possible shape.

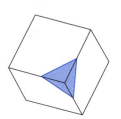

The cross section
shown is a triangular
cross section.

17. What cross sections can you make by slicing a cube? Record which of the shapes below you are able to create. Describe how you did it.

  a. a square   **f.** a hexagon

  b. an equilateral triangle   **g.** an octagon

  c. a rectangle that is not a square **h.** a parallelogram that is not a rectangle

  d. a triangle that is not equilateral **i.** a trapezoid

  e. a pentagon

18. Can you create any shapes that are not listed above? Draw and name any other cross sections you can make.

19. If you think any of the shapes on the list in Problem **17** are *impossible* to make by slicing a cube, explain what makes them impossible.

20. Checkpoint What cross sections can you get from a

  a. sphere?

  b. cylinder?

On Your Own

1. Illustrate and briefly explain each of the following terms.

inscribed   net
cross section   semicircle
diagonal   regular polygon
diameter

Bisect *means "cut in half." If two segments bisect each other, they meet at their midpoints.* Perpendicular *segments meet at 90° angles.*

2. Investigate diagonals in quadrilaterals. Classify each of the following figures by whether the diagonals are the same length, bisect each other, are perpendicular, or have no clear relationship.

Arbitrary
quadrilateral

Square

Kite: *two pairs of congruent adjacent sides.*

Kite

Rectangle

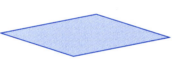

Parallelogram

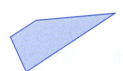

Trapezoid

Trapezoid: *one pair of parallel sides.*
Parallelogram: *two pairs of parallel sides.*

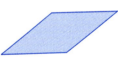

Rhombus

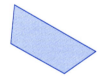

Isosceles trapezoid

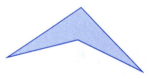

Concave kite

Isosceles trapezoid: *the nonparallel sides are the same length.*
Rhombus: *all sides are the same length.*
Concave: *at least one diagonal falls outside the polygon.*

3. Look again at the pictures from Problem **2** on page 2. In the pentagon, all of the diagonals intersect in pairs only. In the hexagon, three diagonals intersect at the center point. For which other regular polygons is there a point where more than two diagonals meet?

4. In a soccer tournament, 10 teams are competing. Each team must play every other team. How many games will there be?

5. A square pyramid has a square base and triangular faces that meet at a vertex. Find and draw all the nets for a square pyramid.

6. Find and draw a net for a cylinder. Be sure to include the top and bottom circular faces.

The figure made from eight triangular faces is an octahedron.

7. Below are several drawings of eight congruent triangles connected in some way. Decide which of these will fold up into a closed 3-dimensional figure and which will not.

a.

b.

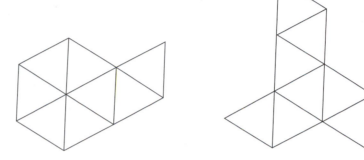

c.

d.

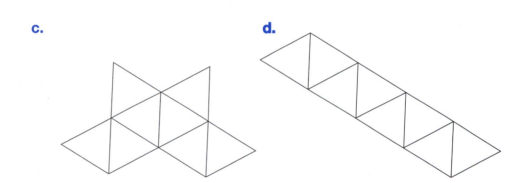

8. **Write and Reflect** "The shortest distance between two points is a straight line." Explain why this says the same thing as the Triangle Inequality.

9. Three friends on a camping trip have set up their tents so that Noriko is 100 feet from Minor, and Minor is 35 feet from Lucinda.

   a. What is the largest distance possible between Lucinda and Noriko?

   b. What is the shortest distance possible between Lucinda and Noriko?

   c. If Minor moves his tent closer to Lucinda, so that his tent is 100 feet from Noriko but only 10 feet from Lucinda, what are the maximum and minimum distances between Lucinda and Noriko?

10. A **central angle** is an angle with its vertex at the center of a circle.

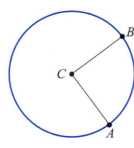 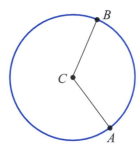

Quarter circle   One-third circle

    a. If the central angle cuts off a quarter circle, what is the measure of the angle?

    b. If the central angle cuts off a semicircle, what is the measure of the angle?

    c. If the central angle cuts off one-third of a circle, what is the measure of the angle?

    d. Can you find a general rule for central angles based on how much of the circle they cut off?

11. In the picture at the right, a central angle and an inscribed angle cut off the same part of a circle.

    a. Which angle is larger?

    b. How much larger is it?

    c. How did you make your decision?

Take It Further

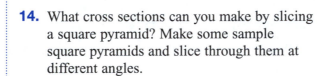

12. Draw a circle. Construct a square that is inscribed in the circle. Explain how you know it is a square.

A cone has a circular base and a point at the vertex. Cross sections of a cone are called conic sections.

13. At the right is a drawing to help you picture a **cone.** What cross sections can you make by slicing a cone? Make some sample cones. Slice through them at different angles to find as many cross sections as you can.

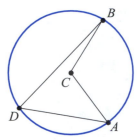

A pyramid has a polygon base and triangular sides. A square pyramid has a square for its base.

14. What cross sections can you make by slicing a square pyramid? Make some sample square pyramids and slice through them at different angles.

15. Try constructing quadrilaterals from sticks or rods of different lengths. What rules can you find about the lengths that can and cannot be used in constructing four-sided figures?

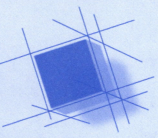

Perspective

What Is Geometry?

Geometry is the attempt to understand space, shape, and dimension. Parts of "geometry"—earth-measuring—grew out of the age of explorers to map where they had been, and of landowners to determine the boundaries of their holdings. Other parts were invented by artists, who wished to portray convincingly what they saw with their eyes or saw in their minds, and by inventors and engineers who wished to make devices that would fit together and work. Geometrical ideas have also come from the needs of architects and builders whose work needs to be both strong and beautiful, and from surveyors, planners, and workers who must be able to assure that tunnels or railroad tracks built from both ends will actually meet in the middle.

Geometric shapes are not only the simple shapes with familiar names like *square* or *pyramid*. Fractal geometry can describe tree-like shapes with surprisingly simple methods.

The following picture describes the rule for building the trees as shown after Problem **1.**

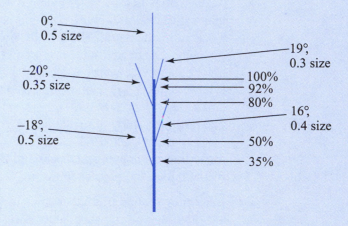

1. Decide how to interpret this pictorial "rule."

    a. How many limbs does it show growing from the trunk?

    b. How far up the trunk is the lowest limb?

    c. How long is that limb compared with the trunk?

    d. What angle does that limb make with the trunk?

    e. How far up the trunk is the highest limb?

    f. How do its size and angle compare with those of the trunk?

The picture on the left shows what happens when each limb sprouts five branches according to the same rules—the same relative distance along the branch, size relationships, and angle of growth. Apply the same rules again to grow five twigs on each branch; and the result looks like the middle picture. The last picture, with five leaves per twig, is very tree-like.

Mathematics is sometimes pictured as if it were a field unto itself, unrelated to other classes, fields, or hobbies. And geometry is often presented as unrelated even to other kinds of mathematics. Historically, none of this was the case. Within mathematics, geometry has been a source of great insight into other mathematical domains, like algebra, and vice versa. Some of the most important mathematicians were also artists, scientists, inventors, clerics, or a blend of these careers. Art, mathematics, science, and social sciences are still closely related. One of the fastest growing scientific fields combines the study of mathematics, psychology, computer science, and biology. Artist M.C. Escher (1898–1972) was fascinated with mathematical ideas and incorporated many into his drawings. Dancer and choreographer Michael Moschen uses ideas from mathematics and physics to inspire his remarkable art.

Examples abound of mathematics influencing other fields and vice versa, but these are often neglected when subjects are taught separately. The *Connected Geometry* units are designed to help you find these connections or invent them for yourself, building links with your other interests and studies, and within mathematics itself.

How has geometry changed in the last 3000 years?

Euclid's series of books was titled The Elements.

The most famous geometry books of all time, *The Elements,* were written in Egypt about 2300 years ago. Their author, Euclid, compiled and systematized in them his own ideas and all the geometrical ideas that he had known, many of which had been developed by the mathematicians who lived before him.

But mathematics, including geometry, is not all thousands of years old, all known, or all finished with nothing new to discover. It should be no surprise that a lot of new geometry has been invented since Euclid. After all, 2300 years is a long time. Art, architecture, mechanical design, clothing design, navigation, communication, and other fields have all changed and have raised new geometrical questions.

In fact, geometry has changed enormously in the last twenty years, and it continues to change rapidly today. New applications, especially involving computers, have changed the scope of what is possible to explore in geometry.

Computer graphics and animation have created new jobs and demanded new research. New geometrical techniques have been developed to solve problems of optimizing paths—for example, finding the most efficient routes for snow plows, the best routes for airlines, the least expensive network of phone wires, or the smallest microchip. The mathematical study of knots, which started in support of a defunct theory held by chemists a century ago, has recently become a powerful tool in modern physics and biochemistry. Astrophysicists have new notions of the shape of space and have invented new mathematical tools like optical geometry. The advent of the computer age has made possible mathematical modeling of complex natural phenomena like weather, and has given rise to a new branch of mathematics called dynamical systems. And there is combinatorial geometry, algebraic geometry, differential geometry, and so on. This is one reason why such different problems as the ones you have just done can all be considered geometry.

Some of the geometry you will learn predates Euclid. But you will also learn about geometry that is being researched and discovered today. What ties Euclid to modern geometers, and high school students to professional mathematicians, is their use of the same powerful mathematical habits of mind. These habits will become a central focus of your work in this course as you learn to visualize new situations, to experiment with fresh ideas, and to prove your own conjectures.

Picturing and Drawing

In this lesson, you will learn to visualize pictures and develop drawing techniques.

Many problems are solved or made easier by making pictures. The pictures can be on paper, on a computer, or in your head.

Even when your goal is to make a picture on paper, mental pictures are important. Practice in making clear and detailed pictures in your head can help you to draw better.

Explore and Discuss

Sometimes you make mental pictures because the things that you need to see are not where you can see them. Answer these questions without moving from where you are.

a How many windows are in your home?

b Is the bathroom doorknob on the right or left as you enter?

c Where did you find your shoes this morning?

d Think of someone you saw this morning but can't see now. What color clothing was he or she wearing?

e How many outside doors are in your school?

Sometimes you make mental pictures because the things that you need to see are abstract ideas. Picture a point hanging in midair somewhere in the room. Picture a second point somewhere else.

f How many different straight lines pass through those two points?

g How many different circles pass through those two points?

h Are there squares that can be built using those two points as corners? How many can you think of?

i How many different sized cubes are there which would use the two points as corners?

When you need to see things with your mind, it often helps to close your eyes, so that you aren't distracted by what you see with your eyes.

ACTIVITY 1 | ## Shadows

For artists to make convincing drawings of scenes that they have pictured in their imagination, they must be able to draw the correct shape of a shadow.

1. Imagine a square casting a shadow on a flat floor or wall. Can the shadow be non-square? Nonrectangular? That is, can the angles in the shadow ever vary from 90°?

2. Which of the shapes on the following page could cast a square shadow and which could not? Explain.

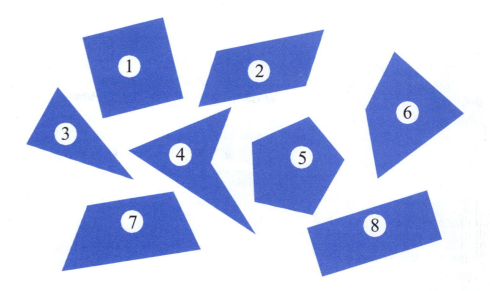

3. What shadow-shapes can a circle cast?

4. [Checkpoint] What shadow-shapes can an **equilateral triangle** cast?

ACTIVITY 2 **Drawing Pictures**

Pictures drawn on paper or by computer can help when the images with which you are experimenting become too complicated to keep in your head. For Problem **5,** try to notice when you prefer mental pictures and when you prefer paper ones.

T can look like T or T or t. You can decide how your letters will look, but keep them simple.

5. Some letters of the alphabet can be cut into two parts that are pretty much the same. Some cannot.

    a. The letter T can be cut into matching parts in two different ways. A slice can separate the top from the stem, making two straight "sticks" of roughly the same size. Or, a vertical slice through the stem can separate the T into two upside-down L-shaped pieces. Draw pictures to illustrate these two different methods.

    b. Which other letters can be cut into two matching parts by a line?

Analyzing complex scenes geometrically can often help you see and sketch them more clearly. For example, here are two sets of *guidelines* for a portrait. The first set shows no details; the construction lines simply show the rough sizes and placement of the features. The second set adds some detail.

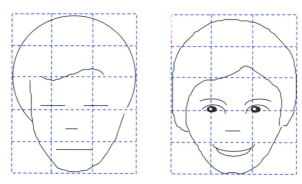

Plans for the layout of a face.

6. Describe the size and placement relationships that these plans show for this face. For example, the mouth is roughly one third the width of the head.

7. Draw guidelines, like the left-hand sketch on page 13, for a friend's face. Notice where the corners of the mouth are with respect to the eyes; get the mouth and nose guidelines roughly the right fraction of the way from chin to eyes. Attend closely to proportions, but don't worry about fractions of an inch. For example, are your friend's eyes roughly halfway from top to bottom, or more like two thirds of the way?

8. **Checkpoint** Consider these problems.

   a. Write a word so that it can be cut into two matching parts with a line. Draw the line that will do it.

   b. Find a picture of a famous person, like Eleanor Roosevelt, Albert Einstein, Martin Luther King, or the Mona Lisa. Draw guidelines only for that face. Do not include shading or other fancy details. Do your guidelines make the face recognizable?

Analyzing and Drawing Three-Dimensional Figures

It is especially important to pay attention to the geometry when you are trying to draw 3-dimensional figures. Some of the figures you will draw in this activity are impossible figures, pictures that confuse the eye. Why impossible figures? Why not just ordinary things? Partly for fun, but mostly because these strange pictures do confuse the eye. In order to draw them, you must take them apart in your mind to see how they are made up. And when you do, you will understand what about their geometry makes them impossible. You will see them differently and be able to draw them.

9. Look at the drawing at the right. It is a strange thing. Or is it a thing at all?

   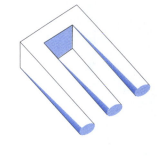

   a. What do you see? Explain why the drawing looks "off."

   b. When the drawing is partially covered (see below), what is left does not look nearly so peculiar. Why does covering part of the drawing help?

   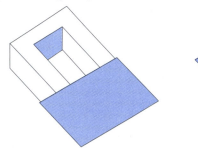

   c. Sketch your own copy of the original drawing.

   d. Now, close the book and do not look at your copy of the picture. Try to draw this thing from memory. Explain how you drew the picture.

It is not *just* because this is a trick picture that it is hard to draw. Part of what makes it hard is that even though it is drawn on flat paper, you see it as if it were 3-dimensional.

You have probably at some time tried to draw your name or someone else's in letters that have a 3-dimensional look.

When the flat shape is a polygon, then the solid created by translating that base shape into the third dimension is called a prism.

Below is a recipe for turning *flat* letters like these:

A E F

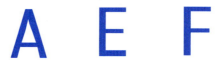

into *solid* letters like these:

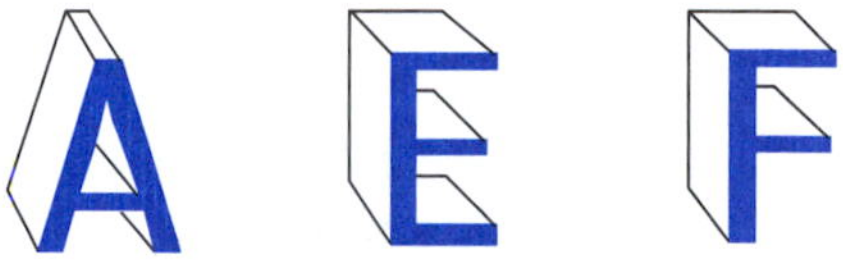

Follow the steps as they are written. This recipe uses special terms—like **prism, parallel,** and **line segment**—so that it will apply to any drawing you make.

Sample Recipe

Step 1 Choose a letter or other shape and draw it in your notebook. The shape you have drawn is called the *base* of the prism that you will draw. The example here shows an L-shaped hexagon and a house-shaped pentagon.

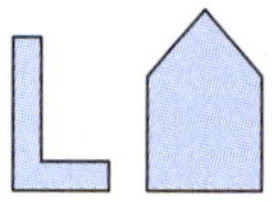

Step 2 *Optional step*. If you like, you can rotate the flat shape into the third dimension. Vertical lines stay vertical. Parallel lines stay parallel. Other changes you "make by eye."

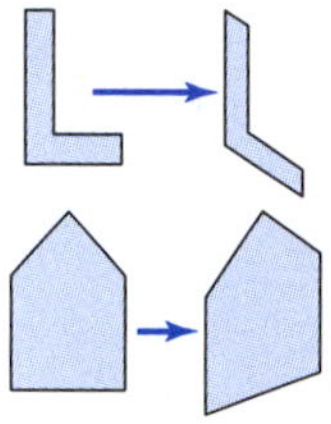

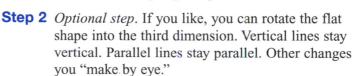

Step 3 A prism needs two parallel bases. So, draw a second copy of your base shape near the first. Take care to make the line segments of the second base parallel to, and the same size as, the corresponding line segments in the first base.

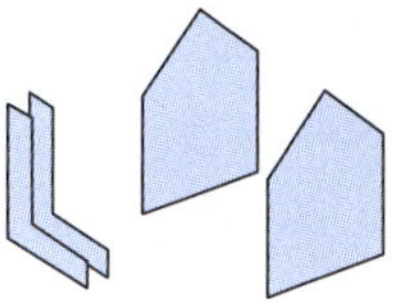

Step 4 Connect the corresponding corners of the two bases. This is called a "wire-frame drawing."

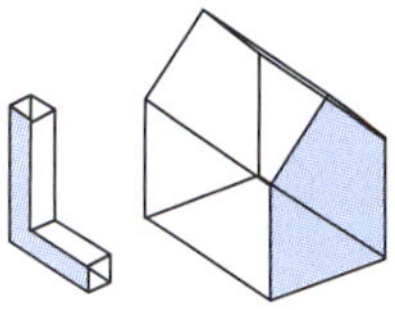

Step 5 Wire-frame pictures can be visually confusing. Hiding (erasing) the back lines helps the eye make sense out of the pictures. It is easier to start with the wire frame and then erase than to draw the correct view from scratch. Hide the appropriate lines in your picture.

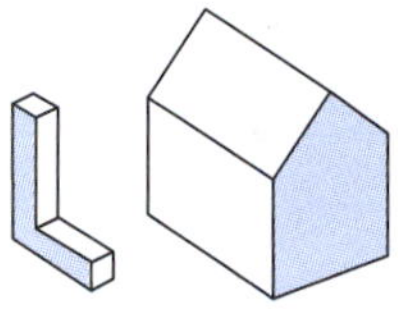

Step 6 Shading also helps the eye make sense of a picture. Faces that are parallel to each other should, in general, be shaded the same way.

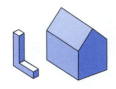

Now that the pentagonal prism looks like a house, the word "base" is especially confusing. The *house*'s "base" is a rectangle sitting on the ground. The *prism* is "based on" a pentagon. Its bases are the front and back of the *house*.

10. Let your imagination go. Add some details, but keep true to the rules so that the picture continues to look right.

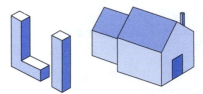

11. **Write and Reflect** The lines that connect the two bases in this kind of wireframe drawing will be parallel to each other. First, convince yourself that this is true. Then find a convincing way to explain why.

12. **a.** In the letters or shapes you drew, which faces are *parallel faces?*

    b. Why do you think *parallel faces* would normally be shaded the same way? Under what conditions would they be shaded differently?

13. Another *impossible* 3-dimensional figure to draw is shown at the right. Study the figure. Cover parts if that helps. Then try to draw the figure. Include the shading. What information does the shading give?

14. Nothing appears *wrong* about the three figures below. Yet each is the figure above with a single bar covered. On the basis of these figures or using some other insight of your own, explain what is *wrong* with the original figure shown in Problem **13.**

15. You have seen these figures before—almost. Draw them, first from the sketches, and then from memory. The task is somewhat different from the task of drawing their impossible cousin. Be sure to include the shading.

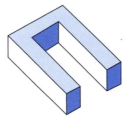
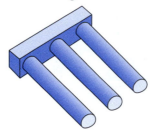

16. **Checkpoint** Consider these problems.

  a. Figures that can be seen in more than one way play tricks on the eye, similar to the tricks the impossible figures play. Try drawing these. What confuses the eye this time?

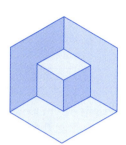

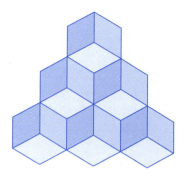

  b. Choose a letter. Draw a 2-inch tall block version of it using the technique shown on pages 15 and 16. Shade in a base of the prism. Explain why that shape is a *base*.

ACTIVITY 4 # Using Pictures to Explain Ideas

Although pictures are shapes in space, they do not have to be about shape or space. Sometimes pictures help us *visualize* quantities, or relationships among quantities—things that are not really visual at all.

17. Figure out and explain in words, what the pictures below can tell you about the multiplication of binomials.

$\leftarrow a \rightarrow \! \leftarrow b \rightarrow$	
a^2	ab
ab	b^2

$(a + b)^2 =$
$a^2 + 2ab + b^2$

$\leftarrow c \rightarrow \! \leftarrow d \rightarrow$	
ac	ad
bc	bd

$(a + b)(c + d) =$
$ac + ad + bc + bd$

18. Explain how the picture at the right says the same thing that the algebraic equation does.

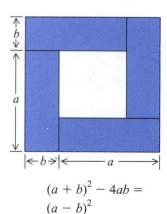

$(a + b)^2 - 4ab =$
$(a - b)^2$

19. A mathematics teacher once sketched the graph at the right to describe her feelings about tests and examinations. Explain what her picture is saying.

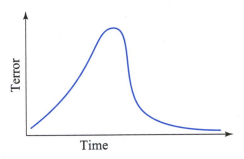

Terror

Time

20. Checkpoint Sketch a graph of your own, such as your energy over the course of a day, how hungry you feel as the hours pass, and so on. Explain what your picture is saying.

On Your Own

1. Look back through your work, the reading, and problems. List the mathematical terms. Find a definition for any that you are not certain about.

2. If a light shines directly above a triangle held parallel to the floor, what properties of the shadow are the same as the original triangle? What changes?

3. If you are looking at shadows cast by a regular hexagon, which shapes are impossible to get? Why?

a. a segment

d. a circle

b. a regular hexagon

e. an irregular hexagon

c. a triangle

f. a square

When a shape or a letter can be folded in half so that the two halves fit exactly on top of each other, the shape is **symmetric,** and the line containing that fold is a **line of symmetry.** For example, a vertical slice through the letter T shows a line of symmetry, but dividing it into two matching pieces—the top and bottom—does not show symmetry.

4. Classify the quadrilaterals below by their symmetries. Look for horizontal, vertical, and diagonal lines of symmetry.

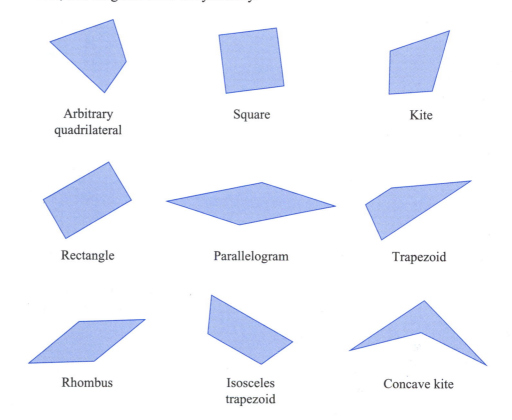

Arbitrary quadrilateral	Square	Kite
Rectangle	Parallelogram	Trapezoid
Rhombus	Isosceles trapezoid	Concave kite

5. What are the lines of symmetry of a circle?

6. Three-dimensional figures can also have symmetry. A plane must divide them into two identical pieces. If you replace the plane with a mirror, the figure should look the same. Find five different objects around your house, such as tissue boxes, cans of soup, and so on. Describe the planes of symmetry. You may include drawings with your descriptions.

7. Find a drawing of a prism in a book, magazine, or newspaper. Does it seem to be drawn using the method shown on pages 15 and 16? Describe the similarities and differences between the printed prism and the one drawn by the method given in this lesson.

8. Make a sketch that illustrates the equation $d(c + f) = dc + df$.

9. Below are two different pictures showing $a^2 - b^2$ and their matching equation. Explain each step in the algebraic equation and how it relates to the pictures.

a. $a^2 - b^2 = a(a - b) + b(a - b)$

$\qquad\quad = (a + b)(a - b)$

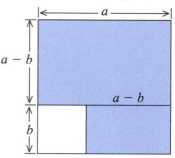

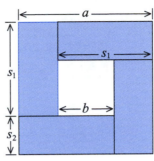

$$2s_1 = a + b$$

$$s_1 = \frac{a + b}{2}$$

$$s_1 + s_2 = a$$

$$s_2 = a - s_1$$

$$= a - \frac{a + b}{2}$$

$$= \frac{a - b}{2}$$

$$a^2 - b^2 = 4\left(\frac{a + b}{2}\right)\left(\frac{a - b}{2}\right)$$

$$= (a + b)(a - b)$$

10. On your drawing of the *impossible* pronged figure, identify as many of the following as you can.

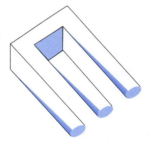

plane	face
line segment	edge
line	vertex
cylinder	midpoint
circle	right angle
perpendicular planes	perpendicular segments
intersection of planes	intersection of segments

11. How many different planes are pictured in the drawing below? (Hint: How many planes are hidden?)

What does infinite *mean?*

12. A plane is infinite. One infinite line on it will divide it into two regions.

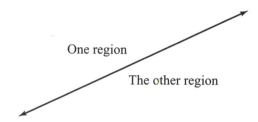

One region

The other region

Two lines may cut the plane into three or four regions, depending how they are placed.

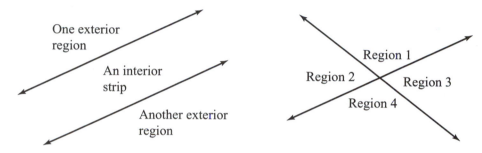

One exterior region

An interior strip

Another exterior region

Region 1

Region 2

Region 3

Region 4

What is the maximum number of plane regions that you can produce with five lines?

This is the 3-dimensional version of Problem 12, but it is hard to picture. The authors have known people to puzzle over this for weeks.

13. Space is infinite. One infinite plane through it will divide it into two regions. Two planes may cut space into three or four regions, depending how they are placed. With three planes, it is possible to cut space into as few as four and as many as eight separate regions. What is the maximum number of regions that you can produce with four planes? Five planes?

14. There is also a 1-dimensional version of Problem **12**. A line is infinite. One point on it will divide it into two regions.

a. Into how many regions will two points divide it?

b. Is there more than one possible answer?

c. What is the maximum number of regions that you can produce with five points?

Ways to Think About It

Compare dividing a line (Problem **14**) to dividing a plane (Problem **12**). What was the same? What changed? How can you use this reasoning to help you extend the idea to dividing space (Problem **13**)?

15. How could you use guidelines to make a good enlargement or reduction of a picture of a face? Take one of your pictures from Problem **7** or **8b** on page 14, and make a face that is twice as large. Write a paragraph to explain how you did it.

16. Take a guideline plan of a face and sketch it onto a stretched grid that has been stretched in one direction. Compare the stretched version with the original.

17. **Project** Research the lives and work of various artists who have used and contributed to geometric ideas. For example, learn about M.C. Eschers *impossible* pictures, and why and how he drew them. Attempts to explore *impossible* geometry have not been limited to drawing and painting. Photographers have also contributed their ideas and skills by creating *impossible* photographs. Many artists, like Victor Vasarely and Piet Mondrian, have used geometric ideas as the basis for much of their art. Salvaldor Dali often played with the perspective or other aspects of the geometry or topology of his subjects. The less well-known artist Pavel Tchelitchew gradually added more and more geometric abstraction to his drawings and paintings.

Topology is a field of mathematics in which one studies the properties of geometric figures that are unaffected when the figure is distorted (bent, stretched, or squashed) but not torn.

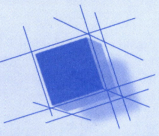

A Mathematical Memoir,

by Marion Walter

This section has been shortened and adapted from "A Mathematical Memoir," originally published in Mathematics Teaching #117, December 1986, a journal of the Association of Teachers of Mathematics, Derby, England. Printed with permission from the ATM.

Professor Walter earned her master's degree in mathematics at New York University and her doctorate in mathematics education at Harvard. She is internationally known in the field, perhaps especially for her work in informal geometry and in helping students become mathematically creative—helping them learn how to invent and investigate new problems. She has taught extensively and has written numerous books—for children as well as adults—and many more articles that combine her life-long interests in mathematics, creativity, art, and design. In this essay, she tells about some of the childhood experiences that moved her towards mathematics and teaching.

My father was in the bead business and I know I had boxes full of beads to play with. I must have done much threading, counting, and pattern making. Could my interest in symmetry stem from trying to make balanced necklaces? I also had sets of parquet blocks and I loved to make the designs provided on sheets. Is that how I got my intuitive feelings for geometry?

Of course, there were also classroom activities that influenced me. We learned with an abacus (10 rows of 10 beads each). I am pretty sure of the importance to me of learning with an abacus as I never had to "learn" my number facts—I just knew them from my experience with the abacus. Numbers that added to 10 were very easy as were ones that added to less than 10. Then sums like $8 + 7$ were easy; even today, as soon as someone says 8, I *feel* the missing 2, and the 7 obliges by "breaking up" into $2 + 5$. I often wonder why research is not done on *how* people cope with numbers and with what model they learned. I often ask my students to add two numbers, such as $27 + 15$, in their head and then to tell how they did it. Some are "breaker-uppers" but many are column-adders. I am definitely a breaker-upper and I am sure it is because I learned with an abacus. I might add $27 + 15$ by thinking $20 + 7 + 15 = 35 + 7 = 42$ or $27 + 15 = 30 + 15 - 3$ or $27 + 15 = 27 + 10 + 5$. I couldn't possibly do it as the column-adders do it: "$5 + 7 = 12$, so write down a 2 and carry the 1," and so on. My memory is far too poor to deal in my head with "carrying" and recalling digits.

After attending a public school, I went to a private, very progressive, Jewish boarding school which had strong emphasis on arts and crafts and the outdoors. I remember only a few things about academic school work there. One of the things I do recall is saying our times tables forwards and backwards; we made a game of it 7, 14, 21, 28, 35, …, 70, 63, 56…. We did not say $1 \times 7 = 7$, $2 \times 7 = 14$, …. I seem to recall it was obvious that 35 was 5×7 and not 4×7 or 6×7. Because I *talk* very fast, I was the quickest in saying these tables. It was during this time that I must have begun to believe that mathematics was safe. $7 + 17 = 24$, no matter what, and it was safe to say it.

What does Dr. Walter mean by "the missing 2"?

How do you do such a problem?

My sister and I arrived in England on 16 March 1939, and at an English boarding school in Eastbourne the very next day. Not a word of English did I know. Knowing that we spoke no English, the headmistress immediately put us into a mental-arithmetic class, since numerals are language-free. Dozens of problems were written in a booklet and one put up one's hand when a problem was completed. My sister and I always finished before anyone else! A few days later during regular classes, I found myself in what must have been beginning algebra. Here, not speaking English may have been an advantage! Since I could not understand what the teacher was explaining, I had to find my own reasons for things. I found solving linear equations no problem at all—perhaps we should more often let students figure things out for themselves. (However, I was disappointed when, on my first test—on solving those very same equations—I lost three points for spelling "where" wrong three times!)

We had geometry along with algebra and arithmetic, and did much work with straightedge and compass. When I heard, in class, that one could not trisect an angle with straightedge and compass, I simply didn't believe it! My teacher diligently checked my many—and often very long—"proofs" of how one *could* trisect any arbitrary angle, and pointed out my errors. It was not until college that I believed that the task was truly impossible. That may have been a good thing, for I learned a lot of geometry trying to show how one *could* trisect angles.

School Certificate is a graduation exam. If you pass, you graduate.

In 1944, at the standard age of 16, I took School Certificate. Because I got a distinction in mathematics, the headmistress called me during the holidays to tell me that the mathematics teacher had resigned and that there was no way that the school could get a new one because of the war shortage. Since she knew I had not decided what to do, she asked me if I would come back and teach mathematics! I went back for two terms to teach in upper school. I still had to sleep in the student dormitory, but I was now allowed in the staff room, had an afternoon off, and was paid a small amount. In the summer, the headmistress asked me if I wanted to teach. If I did, I would have to go to college.

Because I had not been planning to go to college, I hadn't spent my school days worrying about passing tests to get in. I feel sorry now for children who do worry so much about things like that. Luckily, the headmaster took a chance on me. He knew that I had no physics or chemistry in school. Though those courses were a prerequisite, he accepted me provided I studied some over the summer. (He did not know that I was afraid to light a match—which made my physics and chemistry classes a bit difficult!) In later years, I was able to visit him and thank him for taking a chance. I did not find college easy, but I took and passed the Intermediate Bachelor of Science degree before moving to New York. Just crossing the Atlantic made me a better student! I was amazed to find that in America one took 15-week courses and was examined on each course separately, and directly afterwards. This made exam-taking much easier, though learning less thorough. I spent several hours each day tutoring high school students. After two more years, I obtained a degree from Hunter College in New York and a teaching license.

I have been influenced by many great teachers including the mathematician and teacher of heuristics, George Polya, with whom I was fortunate to have a course during a summer institute at Stanford. One of the most valuable courses for *teaching* mathematics I ever took was a three-dimensional design course I had at Harvard University. In a sense it was the most thinking course I ever took. After all, in mathematics, and certainly in science, one is given some techniques and one has ways of checking if one is right. In the design course, we had to find out on our own the quality of paper, wood, and so on; and we had to *find* ways to check the validity of our designs—no answer to look up, no substituting a value to check. The practical experience in this design course greatly influenced my work in informal geometry throughout my career.

What might Dr. Walter mean by "mathematics requires no memory"?

It was partly because I felt mathematics requires *no* memory that I liked it best. Of course, there was also the feeling that it was safe—one could not get into trouble for making a statement about mathematics. Beside my strong interest in mathematics and teaching, I've always felt a strong connection between mathematics and art. (Lately, I have been teaching a college course on "Links Between Mathematics and the Visual Arts.") I still like to work on mathematics, and most of the time it gives me pleasure. I often feel that the pleasure a pleasing piece of art gives—and the pleasure of indulging in some art—is the same as is given by a pleasing piece of mathematics or by working on a piece of mathematics.

LESSON

3

Drawing and Describing Shapes

In this lesson, you will describe shapes by their names, features, and recipes.

It is often quite important to be able to describe shape and other spatial information accurately with words. There are many ways that shapes can be described.

Names Some shapes have special names, like *circle* or *square.* A name may be enough to describe your shape. When shapes are *like* things you can name, you may need some extra words, such as "like an upside down L," "like a house lying on its side," "saddle-shape," and so on.

Features Often the shape is not known, and must be figured out from some set of features. Scientists face this situation when they try to deduce the shape of a molecule from what they know about the atoms that make it up, or from the way it scatters light or x-rays.

Recipes Sometimes saying what a picture looks like is not as helpful as saying how to make it. For example, when giving traveling directions, you would be much more likely to say "walk two blocks, turn left, and walk another block" than to describe the path as "an upside-down L." The recipes we use in mathematics are often called *algorithms* or *constructions.*

As you describe or draw pictures from descriptions in this activity, try to notice when you are thinking about *names, features,* and *recipes.*

Direct sunlight on a clear day generally casts the most-distinct and least-distorted shadows.

Explore and Discuss

For most objects, the shape of a shadow depends on how the object is lit. The next few problems ask you to think about shapes and the shadows they may cast, and to *deduce* properties of a shape based on its shadow.

What shapes are possible for the shadow of a soup can? A cube?

Shadows and Directions

1. A solid object casts a circular shadow on the floor. When it is lit from the front, it casts a square shadow on the back wall.

    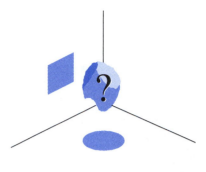

    a. What shape might the object be? Try to make one out of clay, sponge, dough, or other material.

    b. Describe the shape as well as you can.

    c. Draw a picture of the shape.

2. A solid object casts a circular shadow on the floor. When it is lit from the left, it casts a triangular shadow on the right wall.

    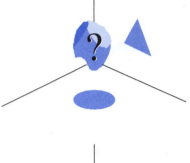

    a. What shape might the object be? Try to make one out of clay, sponge, dough, or other material.

    b. Describe the shape.

    c. Draw a picture of the shape.

3. Suppose the object cast a circular shadow on the floor, a triangular shadow when lit from the left, and a square shadow when lit from the front.

    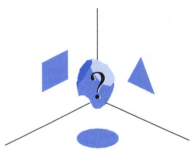

    a. What shape might the object be? Try to make one out of clay, sponge, dough, or other material.

    b. Describe the shape.

    c. Draw a picture of the shape.

4. Read the following directions.

    Face North. Walk four feet. Turn right. Walk six feet. Turn right again. Walk four feet. Turn right again, walk six feet. Turn right again.

    a. If you followed these directions, what shape would your path be?

    b. What direction would you be facing at the end?

    c. "Turn right" doesn't say how far to turn, and yet you probably have made an assumption about it. What was your assumption? What makes it seem reasonable?

5. **Checkpoint** In the last two problems, when were you thinking *Names*? *Features*? *Recipes*?

1. Pick a simple shape and describe it

   a. by its name.

   b. by a recipe to draw it.

2. Read the following recipe.

   Draw two perpendicular segments that meet at their midpoints. Connect the four endpoints in order.

   a. Draw a shape described by the recipe above.

   b. Does the recipe describe only one shape? Why?

3. Read the two recipes below.

   Recipe 1: Draw two perpendicular segments that share an endpoint; make one 3 cm long and the other 6 cm long. Connect the other endpoints.

   Recipe 2: Draw a right triangle with legs 3 cm and 6 cm.

   a. Do the two recipes describe the same shape?

   b. Draw the shapes.

4. A quadrilateral has horizontal, vertical, and diagonal lines of symmetry.

   a. Draw it.

   b. Is there only one possible shape that fits the description? Explain.

5. For this 3-dimensional shape, every possible cross section is a circle. Name the 3-dimensional shape.

Drawing from a Recipe

In this lesson, you will follow a recipe to create a drawing and create a recipe that describes a drawing.

As you translate words into drawings and drawings into words, you will encounter ideas and terminology needed for talking about geometry.

Tangent means just touching.

The two figures are tangent here, but cross here.

A geometric line is infinite. You cannot draw it, so you draw a stand-in for what you imagine.

Explore and Discuss

A picture is worth 192 words. Here's proof!

a Carefully read and follow the recipe below.

Step 1 Draw a horizontal line segment.

Step 2 Draw two same-size circles above that line segment and **tangent** to it. Leave some space between the two circles—roughly as much space as the diameter of the circles.

Step 3 Draw a line segment above the two circles and *tangent* to them. It should be just long enough to extend slightly beyond the two circles. Label this segment's left endpoint *L* and the right endpoint *R*.

Step 4 At *L*, draw a segment about half the length of $\overline{LR}$, extending *up* from *L* and perpendicular to $\overline{LR}$. Label its top endpoint *B*. Draw an identical vertical segment up from *R*. Label its top endpoint *F*.

Step 5 Draw $\overline{BF}$.

Step 6 With a very light pencil line (which you will later erase), extend $\overline{BF}$ about two thirds of its length to the right. Label the new endpoint *X*.

Step 7 Lightly sketch a perpendicular down from *X*, making it roughly the same length as $\overline{FR}$. Find the **midpoint** of this new segment and label it *M*.

Step 8 Draw $\overline{MR}$. Erase the construction lines that you drew in the last two steps.

b What is your picture?

Recipes

1. Carefully and precisely, draw a large triangle. Find the midpoint of each side. Connect each midpoint to the opposite vertex. Label the place where all three connecting lines cross.

The next three problems give geometric directions for drawing letters of the alphabet.

Equi-, 'equal.'
Lateral, 'side.'
Equilateral, 'Equal sides.'

2. Draw an equilateral triangle, two inches on a side, with a horizontal base. Find and connect the midpoints of the two nonhorizontal sides. Erase the base of the triangle. What letter did you draw?

What meaning do the words quadrant, quadrilateral, quarter, *and the Spanish word* cuatro *have in common?*

3. Draw a circle with a half-inch radius. Draw a slightly larger circle directly below the first and tangent to it. Lightly sketch a vertical construction line through the centers of both circles. Lightly sketch two horizontal construction lines, one through the center of each circle. In the top circle, erase the 90° **arc** in the bottom right **quadrant.** In the bottom circle erase the 90° arc in the top left quadrant. Then erase the construction lines. What letter did you draw?

Arcs of the circle are measured by angle. The whole circle is 360° of arc, so a 90° arc is a quarter of the circle.

4. **Checkpoint** Make a circle. Divide it into 90° arcs with two diameters that are about 45° from vertical and perpendicular to each other. Erase the arc on the right side of the circle. Then erase the diameters. What letter did you draw?

On Your Own

1. Use precise language to write directions for drawing each of your initials. Some letters are complicated, so take advantage of any terms from geometry that will help make your directions clear.

2. Write careful directions for walking from the door of your math classroom to your school's main office.

3. Write directions for drawing the figure below. Then give the directions to a classmate to see if he or she can draw the shape.

4. a. Carefully read and follow this recipe.

Step 1   Draw a circle. Label the center of the circle *A*.

Step 2   Draw a radius of the circle. Label the point where it touches the circle *B*.

Step 3   Draw a segment tangent to the circle at *B*. The segment should be longer than the diameter of the circle.

Step 4   Draw a second radius of the circle, perpendicular to $\overline{AB}$. Label the point where it touches the circle *D*.

Step 5   Draw a segment tangent to the circle at *D*. The segment should be longer than the diameter of the circle.

Step 6   Label the point where the tangents intersect *C*.

b. What kind of quadrilateral is *ABCD*?

c. Make a conjecture: In a circle, what is the angle formed between a radius and a line tangent to the circle at the endpoint of the radius?

Take It Further

5. Use the pictures provided by your teacher.

a. For each picture you are given, figure out how to draw it. Then write directions so someone else could draw the original picture without seeing it.

b. Exchange your directions with a partner. Draw nothing more or less than what is written in the directions. Do you get the same pictures that your partner started with? Did your partner get the same pictures that you started with?

5

Constructing from Features: Problem Solving

In this lesson, you will learn the difference between a construction and a drawing, and learn some construction techniques.

In the previous activity, you were given recipes for a drawing but you were not told what the drawing was. Life more often presents problems the other way around: you know what is needed, but not how to make it. Such problems have always inspired creative, inventive thinking. The solution—the how-to-make-it part—always depends on what tools are available.

Explore and Discuss

For example, suppose you are allowed one ruler, one piece of paper, and one pencil. Your task is to draw a triangle whose sides are 3 inches, 4 inches, and 6 inches.

a Using only one ruler, one piece of paper, one pencil, and no other tools or aids, make a picture of the triangle.

b Describe the construction method you used to make the triangle.

ACTIVITY 1

Constructing Triangles

*Drawings **are aids to problem-solving;** constructions **are solutions to problems.***

"It is said that geometry is the art of applying good reasoning to bad drawings."
Henri Poincaré

Geometers distinguish between a *drawing* and a *construction*. Drawings are intended to aid memory, thinking, or communication, and they don't have to be much more than rough sketches to serve this purpose quite well. The essential element of a construction is that it is a kind of guaranteed recipe—it shows how a figure can be accurately drawn with a specified set of tools. Your method for the *Explore and Discuss* above is a construction. The picture that you make just illustrates your construction.

Hand Construction Tools

A ruler is a straightedge with special marks on it.

A *compass* is any device—even a suitably rigged piece of string—that allows you to move a pencil around any point you pick, and keeps the pencil at a fixed distance from that point.

Any *straight edge*—even the edge of a piece of paper—can help you draw a straight line segment. When something is called a *straightedge,* it generally means it is unmarked, and cannot be used to measure distances.

A ruler and a protractor are *measuring devices.* A ruler can measure the length of a segment or the distance between two points. A protractor or goniometer measures angles.

Paper is not just a place to write and draw. The symmetries created by folding it can be used in very creative ways to construct geometric figures. Dissection—cutting paper figures and rearranging the parts—can be a powerful aid to reasoning.

String and tacks help you build devices to construct circles, ellipses, spirals, and other curves.

For each problem below, use whatever hand construction tools seem best. Keep track not only of your answer, but *how* you solved the problem—what tools you used, and what you did with them.

1. If possible, construct a triangle with sides of the given lengths.

   a. 3", 5", 7"   c. 3", 8", 4"

   b. 3", 5", 4"   d. 2", 3", 3"

2. For each triangle you constructed in Problem 1:

   a. Measure the angles of your triangle.

   b. Compare your results with someone else's for the same triangle. Are the two triangles identical? Do the angles of the two triangles match exactly?

   c. Summarize and explain what you observed.

3. Sum the three angles in each triangle you drew for Problem 1. Is the result invariant?

4. You may have found an invariant in Problem 3, but you only tested a few triangles.

   a. Do you believe that your invariant will hold for all triangles? For only some triangles? Which ones?

   b. What would it take to convince you that no matter what kind of triangle you have, the sum of the angles will be the same?

5. If possible, construct a triangle with the given angles.

   a. 40°, 60°, 80°   d. 90°, 90°, 90°

   b. 60°, 70°, 80°   e. 120°, 30°, 30°

   c. 30°, 60°, 90°

In one of the triangles you made, two sides were the same length. What do you think caused that?

6. For each triangle you constructed in Problem 5:

   a. Measure the sides, in inches or centimeters.

   b. Take the ratio of the longest side to the shortest side.

   c. Compare your results with someone else's for the same triangle. Are the two triangles identical? Are the ratios identical?

   d. Summarize and explain what you observed.

7. In Problems **6c** and **6d,** you compared triangles that different people constructed with three given angles. In Problems **2b** and **2c,** you did the same sort of thing for triangles constructed from three given sides. Compare the two experiments. In what ways are the results different?

8. Take one of your triangles from Problem **1.** Without measuring anything, make a new triangle with sides half as long as the sides of your original triangle.

9. [Checkpoint] In Problems **1** and **5,** some of the triangles were impossible to construct. List the sets of sides and angles you could not construct. Explain what went wrong as you tried to make these triangles.

A C T I V I T Y   2 **Compasses, Angles, and Circles**

Here is a method that students have used to draw a triangle when all three side-lengths are known. It will be illustrated using the sidelengths 3 inches, 4 inches, and 5 inches.

Draw a segment to have one of the given lengths. In the picture below, the marked segment is the 5-inch segment. The ruler that measured it is still showing. Then, using two rulers ..., and so on.

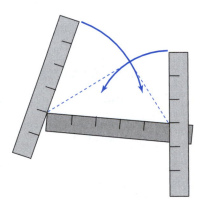

10. Look carefully at the picture above. Figure out what the students did to construct the 3-inch, 4-inch, 5-inch triangle. Then, complete the description of their method, describing carefully and precisely how each ruler was used.

11. Describe how to adapt this method to work with only one ruler.

12. This section is titled "Compasses, Angles, and Circles," but they have not been mentioned yet.

  a. What does this ruler trick have to do with compasses?

  b. How can compasses help you in geometric constructions?

13. a. Construct an **obtuse triangle**—a triangle that has one angle greater than 90°. Label the vertex at the obtuse angle C. Label the other two vertices B and A. Construct a circle whose diameter is $\overline{AB}$. The circle will pass through vertices A and B. Is vertex C inside, outside, or on the circle?

  b. Construct an **acute triangle**—a triangle in which all angles are less than 90°. Label its vertices D, E and F. Construct a circle whose diameter is $\overline{DE}$. Is vertex F inside, outside, or on the circle?

c. Construct a **right triangle**—a triangle that has one angle exactly 90°. Label the vertex at its right angle *G*. Label the other two vertices *H* and *I*. Construct a circle with $\overline{HI}$ as its diameter. Is vertex *G* inside, outside, or on the circle?

14. Compare your results for Problem **13** with those of other classmates. Summarize your observations.

    a. What tentative conclusions can you make?

    b. What other problems or facts have you encountered in this book that relate to this one?

15. **Checkpoint** Use a compass to construct two circles that pass through each other's centers. Call the center of one circle *A*, the center of the other circle *B*, and one point of intersection *C*. Then draw $\overline{AB}$, $\overline{BC}$, and $\overline{AC}$. What kind of triangle is formed? Will this always happen, or does it depend on the circles drawn? Explain.

A tentative conclusion is called a conjecture.

ACTIVITY 3 ▶ **More Construction Problems**

For the following construction problems, draw the figure and describe how you made the constructions. Use any hand construction tools you want, *except* rulers. No measurement allowed. Do not forget paper-folding. Some problems are harder than others. If you have difficulties with one problem, try another. Later, when you return to the hard one, you will have more experience and knowledge.

16. Draw a line segment. Without measuring anything, construct its midpoint.

17. Draw a line. Then construct a line that is

    a. perpendicular to it.

    b. parallel to it.

18. Start with an ordinary sheet of paper $\left(8\frac{1}{2}" \times 11"\right)$. Using just folding and scissors, create the largest square you can.

19. Start with the largest square you can construct from an $8\frac{1}{2}" \times 11"$ piece of paper.

    a. Construct a square with exactly one fourth the area of your original square.

    b. Construct a square with exactly one half the area of your original square.

20. Draw an angle. Construct its bisector. An **angle bisector** is a ray that cuts the angle exactly in half, making two equal angles.

21. For each construction in parts **a–d**, start with a freshly drawn segment. Then construct the shape.

    a. An isosceles triangle with your segment as one of the two *equal* sides.

    b. An isosceles triangle whose base is your segment.

    c. An equilateral triangle based on your segment.

    d. A square based on your segment.

A line goes on forever, so you can just draw a segment.

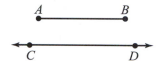

$\overleftrightarrow{CD}$ *is the line and* $\overline{AB}$ *is the segment.*

22. Starting with an equilateral triangle, construct a circle that passes through all three of the triangle's vertices.

23. Illustrate each of these definitions with a sketch. The first one is done for you as an example.

a. A triangle has three **altitudes,** one from each vertex. An altitude is a perpendicular segment from a vertex to the line containing the opposite side.

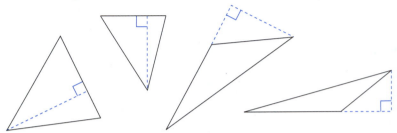

b. A triangle has three **medians.** A median is a segment connecting any vertex to the midpoint of the opposite side.

c. A triangle has three **midlines.** A midline connects two consecutive midpoints.

24. Draw four triangles. Use one for each construction below.

a. Construct the three medians of the first triangle.

b. Construct the three midlines of the second triangle.

c. Construct the three angle bisectors of the third triangle.

d. **Challenge** Construct all altitudes of the fourth triangle.

25. Compare your solutions to Problem **24** to other students' solutions. What are the similarities and differences in your results? Write any conjectures you have.

26. Start with a square. Then construct its diagonals. Study the diagonals—how their lengths compare, what angles they make with each other, how they divide the square's area, and so on. Record your observations.

27. Checkpoint Consider these problems.

a. Without measuring, create two rectangles that have the same area.

b. Construct a picture of triangles within triangles like the one at the right. (Hint: Each smaller triangle is formed by connecting the midpoints of its "parent" triangle.)

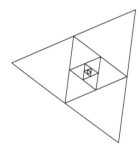

Group Thinking

Each member of your group will receive from your teacher one or at most two cards, each containing a clue to a drawing.

Rules

Do not show your clue card to anyone else. You may read your clue out loud, draw a sketch based on your clue, or discuss the clues and drawings. But you cannot *show* your clue.

Together, your group must make and agree on a drawing that fits *all* the clues the group has. Sometimes more than one drawing will fit the clues.

28. **Checkpoint** Your teacher will provide you a card which describes a geometric term or construction. Some of them are easy to make; a few of them may be impossible without the correct tools.

   a. Make the construction on your card. Explain *how* you created it, especially if you come up with a clever method.

   b. Figure out how to make the construction with *only* compass and straightedge. Some may be impossible with only those tools. If this seems to be the case, think about what other tools you would need. Does a ruler make it possible? For which constructions do you need a protractor?

   c. Now that you know how to make these objects, write directions for someone who may not know the definitions of many of these geometric terms.

   A restriction: Any word that is part of the name of an object cannot be used in your instructions for drawing that object. For instance, the word *perpendicular* cannot be used to describe how to make a perpendicular bisector.

1. Which of these four-sided figures are possible?

   a. A rectangle in which a diagonal is twice the length of one of the sides.

   b. A rectangle in which a diagonal is the same length as one of the sides.

   c. A quadrilateral whose diagonals are perpendicular to each other.

   d. A rectangle whose diagonals are perpendicular to each other.

2. You have been introduced to many mathematical terms that have special meanings. Look back through the lessons and list any terms you are still uncertain about.

3. Write what you know about each of these terms. Explain what they mean as precisely as you can.

line   invariant
perpendicular   line segment
vertex   tangent
altitude   bisector
midline   median
diagonal

4. Can you make a triangle with the following angles? (Try not to construct the triangle.)

   a. 50°, 50°, 50°

   b. 60°, 60°, 60°

   c. 45°, 45°, 90°

   d. 72°, 72°, 36°

   e. Can you make a triangle with two 90° angles?

5. You have a rule—the Triangle Inequality—that tells you whether three given lengths can form a triangle. Write a new rule that will tell you whether three given angle measurements will form a triangle.

6. Below is a segment.

   Use a straightedge and a compass to create an isosceles triangle with two sides the same length as this segment.

7. Use a compass to create two circles, one with a radius the same length as the other's diameter.

8. Construct a quadrilateral with at least one 60° angle and sides that are all the same length.

9. Draw several different triangles. For each triangle, construct a circle that passes through all three vertices. For what kinds of triangles will the circle's center be

   a. inside the triangle?

   b. on the triangle?

   c. outside the triangle?

10. Salim has planted three new saplings in the schoolyard. The school is going to put in a new rotating sprinkler to water all three. Where should the sprinkler go?

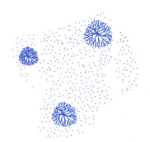

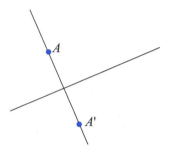

   a. Trace the trees onto your paper.

   b. Show where the sprinkler should go.

   c. Explain how you found that spot.

11. To reflect a point over a line,

- construct a perpendicular from the point to the line that crosses the line.

- mark off an equal distance along the perpendicular on the other side of the line.

   a. On your own, draw a point and a line.

   b. Follow the directions to reflect your point over your line. Think of the line as a mirror. A' is the reflection of A in that mirror; it is the same distance from the mirror and in the same "relative position."

Take It Further _____

12. Write a set of directions like those from this lesson, for a group to construct some figure.

13. Start with a square. Using folding, construct a square that has three fourths the area of your original square.

14. **Project** Start with a square. Using folding, for what fractions $\frac{m}{n}$ can you make a square with that fraction of the original square's area?

15. Construct a square. Without measuring lengths or areas, construct a second square that has *exactly* twice the area of your first square. How did you do it?

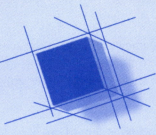

Perspective

Impossible Constructions

Problem **15** on page 39 is difficult, but it can be done. Well over 2000 years ago, scholars knew that if you doubled the length of the sides of a square, the area would be quadrupled. But if you built a square on the diagonal, the area would be doubled.

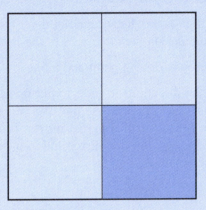
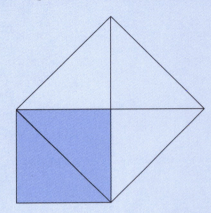

The Greeks limited themselves to a compass and straightedge. With these limitations, the following three constructions have been proved *impossible*. With a ruler, however, some of them are possible.

Pi (π) is the ratio of the circumference of a circle to its diameter.

There were other constructions, however, that the ancient Greek mathematicians were never able to do with just a compass and straightedge. Three famous constructions—doubling the cube, squaring the circle, and trisecting an angle—have since been *proved* impossible to do if one is allowed to use only a compass and straightedge. Two of the constructions were not proved until the 1800s. There is no evidence that the algebraic techniques used in the proofs were known by the ancient mathematicians.

Many people, when they hear about the "impossible constructions," take that as a challenge, and set out to construct them. But there is a big difference between something that has been proved impossible and something that simply hasn't been solved. The proofs show what kinds of line segments *can* be constructed with a straightedge and compass, and a length of π (which would be needed for squaring a circle) is not one of those possible lengths.

Doubling the Cube There are two stories about the origin of the problem of doubling the cube. In one story, King Minos had a cubical tomb constructed for his son. When it was complete, he thought the tomb was too small, so he ordered the builders to double its size by doubling the length of each edge.

1. **a.** Was the King right or wrong about how to double the tomb?

    b. If each edge of a cube were doubled, how would the final volume compare with the original?

2. A new cube is built so that each edge matches the diagonal of a face of the original cube. How does the new volume compare to the original? Is it doubled? More than doubled?

3. Suppose you have a cube with sidelength 1 inch. What length segment would you need to construct to make a cube that has double the volume?

In the second story, the gods sent a plague on Athens because they were unhappy with the cubical altar to Apollo. If the Greeks doubled the size of the altar, they were told, then the plague would stop. First they tried to build an altar with each edge twice the length of the original edge, but that wasn't what the gods had asked for, so the plague continued. You may wonder how the plague ever stopped, since there was no way that the Greeks could have constructed—with just compass and straight-edge—the appropriate length!

Squaring the Circle The problem of "squaring the circle"—constructing a square with the same area as a given circle—is very old. The Greeks knew about it by 400 B.C. and similar problems were raised in Ancient Egypt. If a circle had a one-unit radius, its area would be π units. A square of the same area would need to have a side of length $\sqrt{\pi}$. That length has been shown impossible to construct with a compass and straightedge.

Remarkably, this construction sometimes looks like it is *not* impossible. One student suggested the following construction for squaring the circle:

Start with your circle.

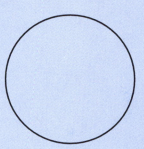

Construct a square inside the circle, with all of its vertices on the circle.

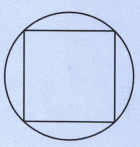

Next, construct a square outside the circle. Each side is tangent to the circle, and all the sides are parallel to the first square.

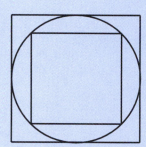

The first square is clearly smaller than the circle because it fits inside the circle. The second square is larger because the circle fits inside of it. But maybe the square halfway between them has the same area as the circle.

4. How could you construct the square that's "halfway between," as shown in the picture at the right?

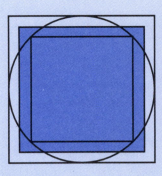

5. This is a good problem to investigate with geometry software because the software can calculate the area ratio for you. If the ratio is 1, the areas are equal. If the ratio is not 1, but is invariant as you change the circle, then there's some other relationship. Use geometry software to build and investigate this conjecture. Did the student "square the circle"?

6. How are the square and the circle related in area? Geometry software can give you a decimal approximation for the ratio of their areas, but can you find the exact relationship? What is a formula for the square's area in terms of the circle's radius? How does it compare with the circle's area?

Trisecting an Angle In compass and straightedge constructions, any angle can be bisected easily. Also, any line segment can be divided up into any number of pieces. It seems quite reasonable to believe, then, that angles can be subdivided into any number of pieces, too. Well, it turns out that it is impossible, except for some special angles like 90°, to divide an angle into *three* equal parts.

7. Write and Reflect Many people have claimed that they have found ways to trisect arbitrary angles, but flaws have always been found with their constructions. Either they work only for special angles, or they require more than just a straightedge and compass. Find out about one of these constructions and write how it is done in your own words. Include pictures!

Constructing from Features: Moving Pictures

In this lesson, you will be introduced to geometry software and use it to solve some challenging construction problems.

Below is a triangle.

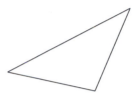

To find out what is true about *this particular triangle* is easy enough. You just study it, measure it, think about it, and list what you find.

But to find out what's true about *triangles* in general is quite another matter. Too many triangles, too little time. One needs another way of investigating.

If only you could stretch and pull and change the diagram of a triangle so that it represented not just one shape but all shapes of triangles. Then, studying that squirming figure, you might begin to figure out what seems to be true of *all* triangles, regardless of their shape.

In this lesson, you will learn to use geometry software. The software lets you create moving pictures, bringing your figures to life. With it, you can construct pictures with the *required features* built in, and yet with other features variable in all the ways you need for your experiments. You can build squares that stay square because you *built in* the features of perpendicular and equal sidelengths. The squares can be rotated or resized at will. You can even animate your pictures. For example, you can invent a carnival ride and set it in motion on your screen!

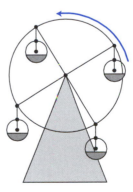

A Ferris wheel

ACTIVITY 1

Geometry Software's Basic Tools

Every geometry software program has tools for creating the basic elements of geometric constructions, for labeling these elements, and then for moving them around. Most programs also have ways of hiding "construction lines," lines that help you *make* the drawing, but that are not really part of the finished drawing.

Follow the steps below to complete the first construction.

Sample Construction

Step 1 Use the point tool to place two points on your screen roughly like this:

Step 2 Then, using *only* the circle tool, construct the picture below. Make certain that your picture *never* contains more than four points.

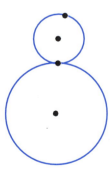

1. Move each point around. Describe the effect on the drawing. You may want to use a labeled sketch to help you identify the points you are describing.

2. Create a triangle that has two vertices that can be moved about freely, and one that can be moved only on a circle.

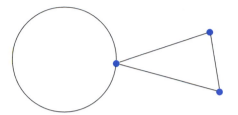

3. Draw two lines. Construct a triangle that has two vertices that can be moved, but only along one of the lines, and one vertex that lies on the other line.

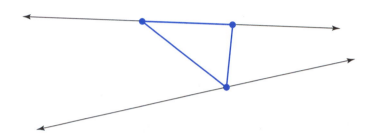

4. **Checkpoint** Consider these problems.

   a. Create a quadrilateral whose vertices stay on a circle.

   b. Create a pentagon whose vertices stay on a circle.

ACTIVITY 2 # Drawings vs. Constructions

To make moving pictures that stay the way you want them, you must *build in* the necessary features. Points that are to be *on* a line or circle cannot be created first and then adjusted to look right. They must be *created on* the line or circle. Lines that are required to pass through a point must *use* the point as part of their definition. The rest of the definition might be a second point or a slope, or perpendicularity to another line, and so on. Squares that are to stay square must have perpendicular sides, and at least one other property, built into them.

As you construct a figure, you can tell the software to build in the properties that you want. Then, if you drag around one part of your figure, all the other parts will adjust accordingly. Parallel lines will remain parallel, midpoints will remain midpoints, and so on.

Follow the directions below to make a moveable windmill. Be sure to save your finished sketch. You will need it for Problem **5.**

Labels may not appear
or may not be the same
as the examples shown
here. Many software
tools allow you to
change the labels any
way you like.

Constructing a Windmill

Step 1 Get a New sketch page on your computer.

Step 2 Make point *A* and $\overline{BC}$.

Your screen should now show four objects: three points *A*, *B*, and *C*, and $\overline{BC}$.

Step 3 Use point *A* and $\overline{BC}$ to make these three constructions:

- Use the appropriate tool to *construct* a line through point *A* and perpendicular to $\overline{BC}$, that will remain perpendicular to $\overline{BC}$, no matter how points *A*, *B*, or *C* are moved.

- *Construct* a line through point *A* and parallel to $\overline{BC}$.

- *Construct* a circle that has point *A* as its center, and has its radius defined by $\overline{BC}$. Stretching or shrinking $\overline{BC}$ should cause the size of the circle to stretch or shrink to keep its radius equal to $\overline{BC}$. Again, you will need a special tool for this construction.

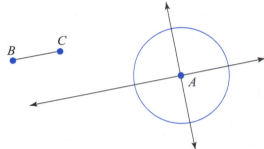

All three constructions

5. Use the selection tool to drag *A*, *B*, and *C* around. Describe what happens.

Sometimes you need to use a line or circle to build a figure, but you do not want to see these construction lines in the finished product. Recall when you used construction lines in making hand drawings, you later erased them. With computer sketching, you must not erase construction lines, but you can hide them. Find out how to hide parts of a figure with your software.

Continue the windmill sketch. Be sure to save your sketch. You will need it for Problem **7.**

Windmill Sketch

Step 1 Place points where the circle intersects the two lines.

Step 2 Construct segments from the center of the circle to each of the intersection points.

Step 3 Hide, but do not delete, the circle and lines. Do not hide the segments.

Steps 1–3 completed.

6. **a.** Describe what happens to the segments when you rotate $\overline{BC}$.

   b. Describe what happens to the segments when you stretch or shrink $\overline{BC}$.

7. Most geometry software provides a way to **Trace** the position of objects as they move. Turn the **Trace** feature on for the four points at the end of your windmill. Then move point B about, and watch what happens. Describe the effect.

8. Use geometry software to draw two intersecting segments. Adjust the segments until they look about the same length and are perpendicular to each other. Move around one of the endpoints. How does this construction behave compared to the one you made above?

What properties must you build in to guarantee that your figure remains a rectangle?

9. **Checkpoint** Construct a guaranteed rectangle. You should be able to change its height, width, or orientation when a point or side is dragged, but *it must remain rectangular.*

ACTIVITY 3 ▶ **Drawing UnMessUpable Figures**

A way to think about the difference between a construction and a drawing is to ask what properties are guaranteed to remain when points or other parts are moved. A square that you *draw*, not construct, is guaranteed to remain a quadrilateral. It may happen, at the moment, to look like a square, but it can be changed into a nonsquare by dragging a vertex or side. Its squareness is not guaranteed.

In each of the constructions below, your mission is to construct pictures that are guaranteed unmessupable. Nobody should be able to mess your picture up by dragging a point or a segment around. Work to create the figure that is specified, not just something that looks like it.

10. Make a square that remains a square.

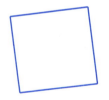

11. Make a figure that can be changed into all sorts of parallelograms, but *only* parallelograms.

12. Draw two circles through each other's centers. As you change the size of one, the other should change with it.

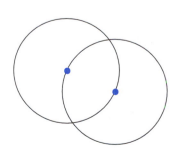

Can you explain why the three points shown in this figure are exactly the same distance from each other?

13. Draw three circles, each through the center of the other two. Changing the size of one should change all three.

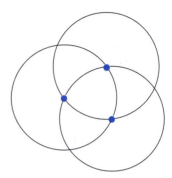

14. Build an equilateral triangle that can be adjusted in size and orientation, but it remains an equilateral.

Can you make your **T** so it always stays upright?

15. Draw the letter **T**. As the points are dragged, the top must remain perpendicular to the stem and centered on it.

16. a. Make two rectangles with their sides lined up and their diagonals along the same ray. Make sure the rectangle vertices are built to be on the ray, no matter what you drag around.

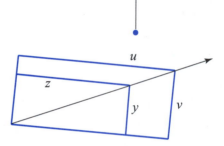

   b. Your software provides a way to compute the ratio of the lengths of two segments. Use that feature to compare the two ratios $\frac{z}{y}$ and $\frac{u}{v}$.

17. Checkpoint Do the two constructions described.

   a. Construct an isosceles triangle so that when you change the length of the base, the lengths of the two legs stay equal, but not necessarily constant.

   b. Find two different ways to construct a guaranteed unmessupable rhombus with geometry software. Write clear directions for each construction. If either construction is specialized so that it makes only *certain* rhombic shapes but not others, explain how you did that.

A Scavenger Hunt

Problems **9** through **16** challenged you to construct figures that maintain certain specified geometric relationships. All of them could be solved with parallels, perpendiculars, and circles. But many other wonderful tools exist in your geometry software.

You will have to look at all the tools, the different options for each tool, and the different menus to find all the items.

18. See how many tools you can find and use. Describe what tool to use and what objects you must select to build each item. Describe a situation where you might use each one.

   a. Construct the midpoint of a segment.

   b. Construct the perpendicular bisector of a segment.

   c. Construct an angle bisector.

   d. Create a polygon or polygon interior.

   e. Measure the distance between two points.

   f. Measure the area of a polygon. (In most software, you must first designate the region as a polygon, or construct the polygon's interior.)

   g. Measure the slope of a line, ray, or segment.

   h. Calculate the sum of two measures.

   i. Construct a copy of an object rotated by a fixed angle of 45° around a given point.

   j. Construct a copy of an object rotated by a variable angle around a point.

   k. Calculate the ratio of two lengths.

   l. Construct a copy of an object reflected through a given line.

19. **Checkpoint** Find one other tool and investigate it. Describe how to use it and what it does.

On Your Own

   1. For Problems **1–3** on page 45, write directions for making the sketches. Include how you get a point to "stick" to a line or circle.

   2. While working on Problem **3** on page 45, Buddy created a sketch that looked like this:

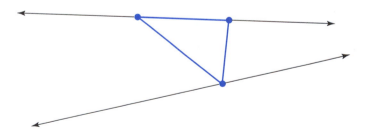

But when his teacher came to check it, she selected a point and moved it to a new position.

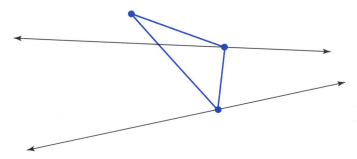

    a. What did Buddy do wrong?

    b. How can he fix his mistake?

3. Does a line perpendicular to $\overline{BC}$ have to touch $\overline{BC}$?

4. What is the difference between "a circle with radius $\overline{BC}$" and "a circle whose radius is $\overline{BC}$"?

5. Write and Reflect While you were learning how to use the software, you also did some geometric thinking. List some geometric ideas, terminology, or techniques that you learned, relearned, polished up, or invented.

6. Write and Reflect Compare the tools used in hand constructions (paperfolding, compass, straightedge) and the "basic tools" of geometry software. How are they similar? How are they different? Do you think you can do more with one set of tools than with the other? Explain.

7. In Step 3 of the windmill construction on page 46, why was it necessary to construct the line parallel to $\overline{BC}$ rather than *drawn* parallel to $\overline{BC}$?

8. Why did you need to use the circle in the construction of the windmill if it was hidden once you completed the figure?

9. a. What is the definition of a rhombus?

    b. Is there more than one way to define it? Explain.

10. No geometry software allows you to construct a perpendicular to a line or a segment without identifying both a line *and* a point. Why is that a sensible restriction?

11. Write and Reflect Select one of the unmessupable figures you constructed. Write a detailed description of the construction process. Someone should be able to follow your directions, step-by-step, and have just the figure on the screen. To test your directions, exchange them with a partner. Did you both get the predicted results?

12. Look back at Problem **14** on page 48. How did you make sure that the triangle you built was an equilateral? Describe what features of the construction or the resulting figure guarantee that the triangle has all sides of equal length.

Can you set up both
pairs of opposite sides to
be equal in length but
not parallel?

13. **Challenge** Invent a kind of quadrilateral that can be deformed in all sorts of ways, but *always* has one pair of opposite sides equal in length but not necessarily parallel.

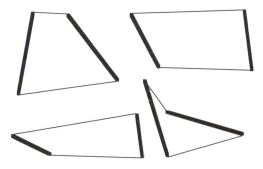

The heavy lines are the opposite sides that are equal in length.

14. Construct a picture of a house so that it looks like it is in perspective. You can build it so that as the points on the "near end" of the house are adjusted to make it look less crooked, the corresponding points on the far end adjust automatically. The point marked Drag should be able to move around, changing the house to look like it is being seen from different perspectives.

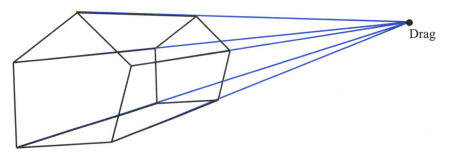

Drag

Ways to Think About It

When you drew a house in Lesson 2 on pages 15 and 16, the two "bases" were identical, their corresponding edges were parallel, and the segments that connected corresponding vertices were all parallel to each other. Here, the connecting segments are not parallel, but instead converge to a point. The bases are no longer identical, but they are the *same shape*. Their corresponding edges are still parallel.

This fact suggests one of many strategies for constructing the picture. Sketch the near end of the house; create the Drag point (called a *vanishing point* in both mathematics and art). Connect the front vertices to the vanishing point. Then, construct the far end of the house making sure that each segment of it is parallel to the corresponding segment of the near end.

The Optimization unit is
all about geometric ways
of thinking about
problems like these.

15. A good bit of mathematics is devoted to *optimization* problems: finding the shortest path, the lowest cost, the greatest benefit, and so on. Here is an optimization problem that you can investigate with geometry software.

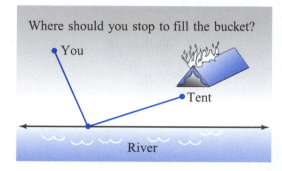

You are on a camping trip. While walking back from a hike, you see that your tent is on fire. Luckily, you are near a river and holding a bucket. Where along the river should you get the water to minimize the *travel* it takes to get back to your tent? Justify your answer.

Ways to Think About It

By drawing the situation, measuring and adding together the segments of your walk, and experimenting, you can find where the best spot is. But what could you possibly *say* about it? Is there a pattern that allows you to predict where the best spot will be? Any fact that is invariably true? Or does each new placement of you, the tent, and the river give a new solution with no relation to the previous one?

Here are three ways that good thinkers use when they are trying to solve difficult problems.

- *Think about "ballpark" solutions rather than worrying about precision.* For a given placement of the tent and you, is the best spot along the river closer to the tent or closer to you?

- *Think about how the exact solution depends on one feature of the problem.* In this problem, for example, you might think about how the solution changes as, say, the tent's distance from the river changes. If the tent were very close to the river, how would that affect the best place to run? If the tent were very far from the river, how would *that* affect the best place to run?

- *Try to make an analogy with a related situation that you understand well.* In this case, that might mean thinking about what you know about "shortest distances." For example, one thing that we all assume is that the shortest distance between two points is a straight line. You might try to find some way of looking at this problem that makes use of that idea.

16. Use geometry software to draw a picture of a person, an animal, a carnival ride, or a machine. Find a way to get the tail to wag, or the eyes to roll, or the ride to move as some part of the picture is moved or animated.

17. Use geometry software to construct a triangle that has sides that are adjustable, but that have unvarying proportions: one side is 2 units long, the other two sides are 3 and 4 units long.

Numerical Invariants

In this lesson, you will learn what an invariant is and practice looking for one in different mathematical situations.

"What kinds of *invariants* should I be looking for in geometry? What strategies will help me find them?" This part of the book begins to answer these questions, and gives you some practice.

Geometry software can often be a big help. By stretching and shrinking various parts of a figure, you can get a feel for how the parts *work* together, what patterns exist, and what stays the same as the details change.

The *invariants* you find by experimentation in this section might, when proven, become useful theorems. Visualize, draw pictures, make calculations, do whatever helps you make an educated guess. For now, do not worry about proof.

Invariants over a set are things that are the same about each member of the set.

Explore and Discuss

Something that is true of each thing in a collection is an *invariant* for the collection. Below are three different collections to study.

One set contains squares of numbers that end in 5.

a Perform the calculations to see what these numbers look like.

b What invariants do you find?

Another set contains *pairs* of square numbers. Within each pair, the numbers, before squaring, have unit digits whose sum is 10.

c What invariants do you find in the set of pairs of squares?

The third set contains the numbers 1, 4, 7, 10, 13, ..., and so on.

d Decide whether or not 301 is in the set. Give a reason for your decision.

e What seems to be true of the product of any two numbers chosen from the set?

f What seems to be true of the sum of any four numbers from the set?

g What about the difference of any two numbers?

15^2

35^2 615^2

5^2 ...

$(27^2, 103^2)$

$(31^2, 39^2)$

$(14^2, 6^2)$

$(34^2, 6^2)$ ...

$(16^2, 254^2)$

55   1   22
34   37   10   28
43   58   88   25   49
64   46   13   31   52   ...
67   70   7   91   4   61   16
76   19   40
100

Positions of points, intersections of lines, lengths of segments, angle measures, or even sums or ratios of these measurements may be invariant.

Approach these problems as experiments. Draw and measure the objects with geometry software. Drag parts of the figure; watch what changes and what remains the same. Make conjectures that seem likely; find a way to test them. Organize and record your results.

1. Using geometry software, create a circle and its diameter. Measure the circumference, the diameter, and the area. Also calculate ratios of each pair of measurements. What seems invariant as you change the size of the circle?

2. Create two parallel lines a fixed distance apart. Create $\triangle EFG$ with points E, F, and G on the lines as shown below.

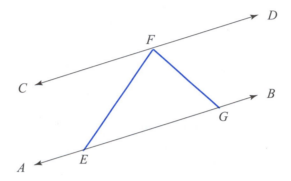

a. What seems invariant in $\triangle EFG$ as F moves along $\overleftrightarrow{CD}$?

   i. the measure of $\angle EFG$

   ii. $FE + FG$

   iii. the perimeter of $\triangle EFG$

   iv. the area of triangle $\triangle EFG$

   v. the sum of the measures of $\angle FEG + \angle FGE + \angle EFG$

b. Can you find any invariants that are not listed above?

3. Study the figure as F moves along $\overleftrightarrow{CD}$.

a. Find a pair of angles that have equal measures no matter where F is placed on the line.

b. Find pairs or groups of angles that have an invariant sum of 180°.

c. Is there a pair of angles where the measure of one is always greater than the measure of the other?

What is a median?

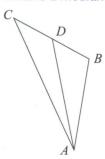

4. Draw $\triangle ABC$. Construct midpoint D of side $\overline{BC}$. Draw median $\overline{AD}$.

a. As you stretch and distort $\triangle ABC$, what remains invariant? Be sure that point D remains a midpoint.

b. Find segments whose lengths are in constant ratio.

c. Are there any invariant areas? Ratios of areas?

d. Find at least one other invariant. Provide a chart or table of measurements and sketches to demonstrate the measures or ratios that have not changed.

5. Use geometry software to construct a circle and one of its diameters. Place a new point on the circle. Complete the triangle as shown below.

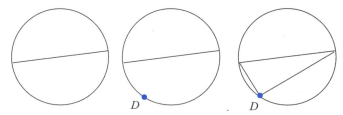

    D   *D*

    a. Try moving point *D* around the circle. What measures or relationships are invariant as *D* moves? Look at angles, lengths, sums, and ratios.

    b. Try leaving *D* in one place and stretching the circle. What measures or relationships remain invariant as the size of the circle changes?

6. **Checkpoint** Construct rectangle *ABCD* that can be stretched to any length or width. Which of the following are invariants?

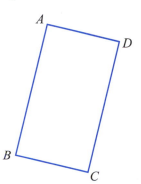

    a. The length-to-width ratio: $\frac{AB}{BC}$.

    b. The ratio of the lengths of the opposite sides: $\frac{AB}{CD}$.

    c. The perimeter of rectangle *ABCD*.

    d. The ratio of the lengths of the diagonals: $\frac{AC}{BD}$.

ACTIVITY 2 > Searching for Patterns

Each table below contains four pairs of values.

c	d	e	f	x	y
2	8	2	8	2	$2\frac{1}{2}$
4	10	3	7	4	$1\frac{1}{4}$
5	11	5	5	5	1
7	13	6	4	10	$\frac{1}{2}$

w	z	g	h	m	n
2	8	2	8	10	25
3	$5\frac{1}{3}$	4	16	36	90
4	4	6	24	40	100
5	$3\frac{1}{5}$	9	36	50	125

7. In each table, there is an invariant relationship between the numbers in the pairs. That is, for each pair of numbers you can compute a third number that is the same for all four pairs in the table. In these tables, the invariant relationship will be a constant sum, difference, product, or ratio. Find an invariant relationship for each table.

8. In Problem **7,** you found very specific invariants. More generally, there are only two patterns:

   - The values change in the *same direction:* both increase or both decrease.

   - The values change in the *opposite direction:* one increases as the other decreases.

   a. For each table, find which pattern fits.

   b. For each pattern, which of the operations, $+$, $-$, $\times$, or $\div$, helped you compute an invariant? Explain why one should expect those particular operations to fit with those particular patterns.

9. Choose one type of numerical invariant (constant sum, difference, product, or ratio) and create a table that follows the *same direction* pattern.

10. Create an *opposite direction* table that does not seem to have a numerical invariant.

11. **Challenge** Below is a table that does not use a simple sum, difference, product, or ratio relationship. The numbers are deliberately messy, too. Find the invariant relationship.

r	s
45	114.0
60	151.5
72	181.5
108	271.5

12. **Checkpoint** Create three different tables, each with an invariant relationship between the two variables. Each table should use a different operation. In each case, the invariant value—the number you get that stays constant—should be 8.

Constant Sum and Difference in Geometry

Use geometry software to draw a line containing points *A* and *B*. Place *C* on the line, between *A* and *B*. Measure segments *CA* and *CB*. Keep track of their lengths in a table like the one below.

CA	CB
⋮	⋮

13. While keeping points A and B fixed, move point C back and forth between A and B. Do the measured lengths change in the same or opposite directions? Compute a numerical invariant. To what does this number correspond?

14. Now experiment with locations for C that do not lie between A and B. Do the measured lengths change in the same or opposite directions? Make a new table and inspect your data. What is the invariant now?

You may already be convinced that the sum of the angle measures in a triangle is invariably 180°. What about the sums of angle measures for other polygons? Will they *all* be 180°? Or might each type—trapezoids, parallelograms, rectangles, pentagons, hexagons—have its own special fixed number? If so, is there some way to predict, given a type of polygon, what the sum of the angle measures will be?

15. Use geometry software to draw and measure angles of quadrilaterals, pentagons, hexagons, and so on. Be sure to draw both regular and irregular shapes. Which groups of polygons have constant sums of angle measures?

16. Checkpoint In the picture below, point C is fixed on $\overline{AB}$, but point D can move around. Find two sums that stay constant as D moves, and as C moves on $\overline{AB}$.

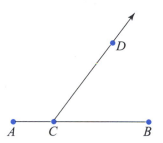

A C T I V I T Y   4   ## Constant Product and Ratio in Geometry

A segment whose endpoints are on a circle is called a **chord**.

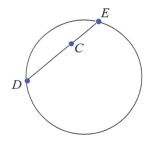

Using geometry software, make a circle and a point C anywhere inside it. Place point D on the circle. Construct a line through D and C to create a new intersection E. Mark that intersection, hide the line, and construct $\overline{CD}$, $\overline{CE}$, and $\overline{DE}$. As you move D along the circle, $\overline{DE}$ will pivot about C.

17. Measure $\overline{CE}$ and $\overline{CD}$. As the chord pivots about C, note how $\overline{CE}$ and $\overline{CD}$ vary. Do these measurements change in the same or opposite directions? Use that information to guide you in computing a numerical invariant.

18. The number you found did not depend on the location of D—you could move D, and this number remained fixed. But the number does *not* remain fixed as C is moved. For what location of C inside the circle is this number largest? Why?

Use geometry software to draw a triangle. Construct and connect the midpoints of two sides. Your construction will look something like the one pictured at the right, where D is the midpoint of $\overline{AC}$ and E is the midpoint of $\overline{AB}$.

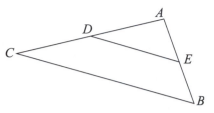

19. Change the triangle's shape by moving one of its vertices. As you change the triangle's shape, what stays the same? List at least three invariants.

20. Measure the lengths of $\overline{DE}$ and $\overline{CB}$. Compare these lengths as you change the shape of the triangle. Do the measured lengths change in the *same* or *opposite* direction? Use calculations to conjecture the exact relationship between these two segments.

Draw new $\triangle ABC$. Place point D arbitrarily on $\overline{AC}$. Through D, construct a parallel to side $\overline{CB}$. Use that line to construct $\overline{DE}$. Then hide the line. Your construction resembles the earlier one, but this time D and E are movable points rather than midpoints.

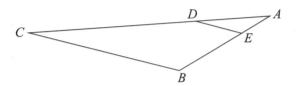

21. As D moves along $\overline{AC}$, $\overline{DE}$ moves with it. Look at lengths and areas. Try to find some invariant relationships. Record your conjectures and appropriate supporting evidence.

Ways to Think About It

Use the familiar strategy of looking for "same-direction changers" and seeing if their relationship is constant. There might also be interesting "opposite-direction changers."

This is a good example of slightly changing a problem. From Problem **19,** you already have some ideas about a special case—when D was located at the midpoint. Some ratios were invariant in that case. If D is in a different *fixed* location and the vertices A, B, or C are moved, are those ratios still invariant, or did their invariance depend on D being precisely at the halfway mark?

As D moves, do the ratios remain constant? If so, you have an invariant again. If not, perhaps there is a relationship between two or more of the ratios.

22. **Checkpoint** Construct a triangle that has one right angle, even if you alter other parts of it. Construct the midpoint of the **hypotenuse**—the side opposite the right angle. Measure the distances from the midpoint of the hypotenuse to each vertex. Look for an invariant ratio and describe it.

On Your Own

1. Find five cylindrical objects. For each object, measure the diameter of the top and the circumference. What is the ratio, $\frac{\text{circumference}}{\text{diameter}}$, for each object?

2. Tennis balls are sold in cans of three. Which is greater, the height of the can or its circumference?

3. How could you make a right triangle with the same area as $\triangle EFG$?

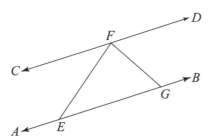

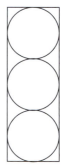

4. For each table below, do the following.

   a. Find the invariant relationship. In other words, find an operation you can do to the two numbers in the table to get a third number that remains constant.

   b. Find three more pairs of numbers that fit this invariant relationship.

   c. In which tables could the number 0 appear? For the tables where it can, what would the value in the other column be? For the tables where 0 cannot appear, explain why not.

p	q	a	b	x	y	m	n
15	105	45	135	1	1	5	5
30	120	72	108	$\frac{3}{2}$	$\frac{2}{3}$	0.25	$\frac{1}{4}$
45	135	94	86	10	0.1	1.5	$\frac{3}{2}$
72	162	144	36	4	0.25	2	$\frac{4}{2}$

5. In the table at the right, there is an invariant relationship between the numbers in the pairs.

   a. Find a computation that shows the invariant relationship.

   b. Do the numbers in the table change in the same direction or in the opposite direction?

   c. Change the statement about "only two patterns" from Problem 8 on page 56 to account for this situation.

a	b
1	−2
2	−4
5	−10
12	−24

6. Make tables of values to fit the following invariant relationships.

    a. a and b values, where $a + 2b$ is invariant

    b. x and y values, where $x(x + y)$ is invariant

Every polygon with n sides can be cut up into $n - 2$ triangles.

How can one possibly show that this is really true for every polygon? Can you find a convincing argument?

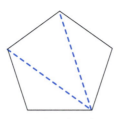
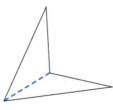

Three triangles from   Two triangles from   Six triangles from
a pentagon   a quadrilateral   an octagon

7. Assume that the angle sum in a triangle is invariant. Use that fact to write an argument that for n-sided polygons, the angle sum is also invariant. Find a rule that will tell you the angle sum if you know the number of sides.

8. Which of these will have a constant ratio, even if the picture is stretched. Explain how you make your decisions.

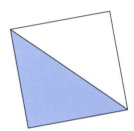

    a. The square remains square. Will the ratio of the areas $\frac{\text{triangle}}{\text{square}}$ be invariant? If yes, what will it be?

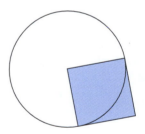

    b. The circle remains a circle, the square remains a square, and the sides of the square are fixed to the circle's radius. Will the ratio of the areas, $\frac{\text{circle}}{\text{square}}$, be invariant? If yes, what will it be?

    c. The triangle remains a triangle, but can be reshaped in any way. Will the ratio, $\frac{\text{perimeter}}{\text{area}}$, be invariant? If yes, what will it be?

Take It Further

9. On paper or on the computer, draw points *A* and *B*, which never move. Make the distance between them no longer than 6 units. Imagine a point *P* that can move along various paths. Each problem below describes one of these paths in terms of *P*'s relationship to *A* and *B*. Draw and describe what *P*'s path *looks* like in each case.

   a. As *P* moves along this path, *PA* always equals *PB*. What is the shape of the path?

   b. This time *P*'s path keeps $PA = 5$.

   c. As *P* moves along this path, $PA + PB = 6$.

   d. $m\angle APB = 90°$, no matter where *P* is along this path.

   e. $m\angle APB = 30°$, no matter where *P* is along this path.

   f. $m\angle ABP = 30°$, no matter where *P* is along this path.

This value, known as the golden ratio, is well-known in mathematics. The golden rectangle, shown here, is said to be the most pleasing rectangle to look at. What do you think?

10. The picture below shows a special kind of invariant ratio in geometry. In these two rectangles, the ratio $\frac{\text{long side}}{\text{short side}}$ is the same. In fact, you could continue constructing squares inside, using the length of the shorter side of the rectangle. Each new rectangle you get would have the same ratio $\frac{\text{long side}}{\text{short side}}$. What is the numerical value of this ratio?

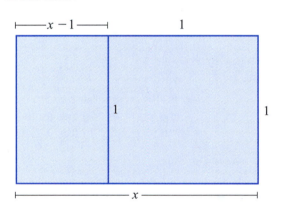

LESSON 8

Spatial Invariants

In this lesson, you will learn about three geometric invariants: shape, concurrence, and collinearity.

Explore and Discuss

The picture below shows a quadrilateral and a triangle. Every edge of each polygon is divided into thirds. Each vertex is connected to two of these "third-points" at the widest possible angle. The connecting lines surround a region. From the picture, it would appear that when the outside shape has four sides, the inside shape has eight; and when the outside shape has three sides, the inside shape has six.

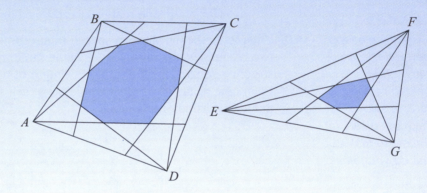

Is this a reliable pattern? That is, does the inside shape always have twice the number of sides as the outside shape when vertices are connected to "third-points" in this way? Explain.

To make the angle as wide as possible, connect the vertex to the opposite sides and to the closest of the two third-points.

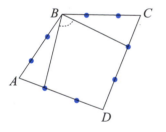

ACTIVITY 1

Shape: A Geometric Invariant

If you want to be precise, you can learn or make up a way of subdividing a segment accurately into thirds. For this construction, however, it is okay to subdivide the segment "by eye" or by measurement.

Geometry software is designed to preserve proportions along a segment as you stretch or shrink the segment. So, your "approximate thirds" will stay as accurate or inaccurate throughout the experiment as they were at the start, unless you deliberately change them.

1. Experiment with the constructions described above. Look at special cases; for example, only triangles, only regular outside polygons, and so on. Also look at general cases. When, if ever, is there a *shape* invariant? Explain.

2. Stating the limits of what can happen is often as useful as saying what must happen.

a. If the outside polygon has *n* sides, can the inside polygon ever have *more* than 2*n*? If so, what is the largest number it can have?

b. Can the inside polygon ever have fewer than 2*n* sides? If so, what is the smallest number?

c. Can the inside polygon ever be *regular?* Explain.

In other words, is there anything besides the number of sides that is invariant here?

There was a shape invariant in the previous experiments, though not as strong a one as the pictures might have suggested. The invariant only had to do with the *number* of sides, and not with the relationship of the sides to one another.

3. **Checkpoint** Draw a quadrilateral. Construct the midpoints of its sides. Then connect those midpoints in order.

a. Explain why the inside figure *must* be a quadrilateral.

b. Can that inside shape be *any* kind of quadrilateral, or are certain kinds not possible? Explain.

ACTIVITY 2   Concurrence: A Geometric Invariant

Below is a picture that might have come from one of the experiments you performed in Problems **1** and **2.**

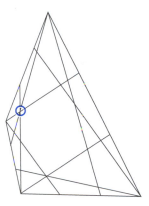

At each vertex of the outside quadrilateral, four lines "run together," or **concur.** That's no surprise; it was intentional.

Three or more lines that meet or intersect at a single point are called concurrent.

But it also happens, in this particular picture, that three *inside* lines run together. In fact, this figure shows two such concurrences. These concurrences were *not* deliberately built in; they are something of a surprise.

4. a. Look at the figure above. One concurrence is circled. Find the other.

b. Are the concurrences *invariants* for this construction? That is, if the vertices of the quadrilateral are moved, will there *always* be two internal points at which three lines meet? Explain.

When concurrence happens *reliably* in a figure—that is, when it is an invariant for that figure—it signals that something special is going on. As you work through Problems **5–14,** keep careful track of your observations and conjectures because you will need them for the last problem in this activity, which asks you to organize your observations into a report.

5. Your teacher will provide you a page with pictures of several different regular polygons. For each polygon, draw all its diagonals.

   a. For which polygons are there concurrences among the diagonals?

   b. What conjectures can you make for other regular polygons?

6. Use geometry software to place five points. Connect them with segments to create an arbitrary convex pentagon. Construct the **perpendicular bisector** of each side of your pentagon.

   a. Is there a point at which three or more perpendicular bisectors are concurrent?

   b. If not, is it possible to adjust the vertices of the pentagon to make three bisectors concur at a single point?

   c. Is it possible to adjust the vertices of the pentagon to make all five bisectors concur at a single point?

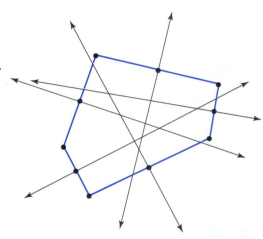

No concurrence.

What is an angle bisector?

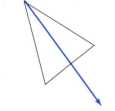

7. Try the same kind of experiment with angle bisectors. Start with an arbitrary pentagon and construct an angle bisector at each vertex. Is it possible to adjust the vertices of the pentagon to make *all five* angle bisectors concur at a single point?

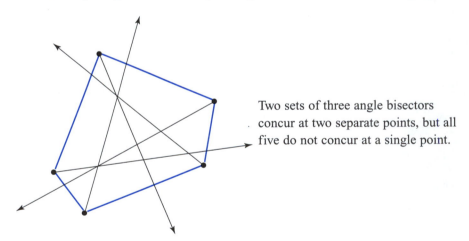

Two sets of three angle bisectors concur at two separate points, but all five do not concur at a single point.

8. Try the same kinds of experiments with triangles. Make a triangle. Construct the perpendicular bisectors of all three sides. Can you adjust the triangle so that all three perpendicular bisectors are concurrent?

9. Hide the perpendicular bisectors. Construct angle bisectors at all three of your triangle's vertices. Can you adjust the triangle so that all three angle bisectors are concurrent?

10. Under what special circumstances might it be possible for all three perpendicular bisectors and all three angle bisectors to concur at the same point?

Starting an investigation with special cases is often a good idea. It simplifies what one has to look at, and often leads more quickly to fruitful conjectures. Among polygons, the triangle is special—it is the simplest. The following experiments suggest other special cases.

This pentagon is said to be inscribed *in the circle.*

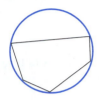

This pentagon is circumscribed *about the circle.*

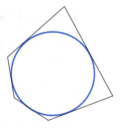

11. Regularity is a very special case. Is "concurrency of perpendicular bisectors" an invariant for *regular* polygons? Experiment. Be sure to try, at least, regular quadrilaterals (squares), regular pentagons, and regular hexagons. What do your experiments show? Explain.

12. Is "concurrency of angle bisectors" an invariant for *regular* polygons? Again, experiment, describe a conjecture, and explain your result.

13. Use geometry software to construct a circle. Place five points on it. Connect the points to form an irregular pentagon. Check angle and perpendicular bisectors for concurrence. Do you observe any invariants?

14. Construct a circle. Build an irregular polygon outside of it, carefully adjusting so that all of its sides are tangent to the circle. Perform the two concurrence experiments again. Do you observe any invariants?

It may have been surprising that the perpendicular bisectors in triangles are concurrent. If you analyze the situation, though, it becomes less surprising.

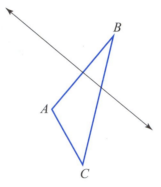

All the points on the perpendicular bisector of $\overline{AB}$ are the same distance from A and B.

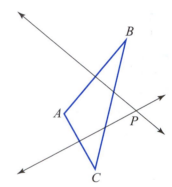

All the points on the perpendicular bisector of $\overline{AC}$ are the same distance from A and C.

15. Write and Reflect Use the drawings and explanations above to describe point P, where the two perpendicular bisectors meet. Why *must* the perpendicular bisector of $\overline{BC}$ also go through point P?

A similar argument can be made for the angle bisectors.

The points on the angle bisector of $\angle ABC$ are the same distance from sides $\overline{AB}$ and $\overline{BC}$, ..., and so on.

16. Write and Reflect Finish the argument above. Explain why all three angle bisectors in a triangle must meet at the same point.

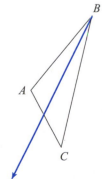

17. **Checkpoint** Review the results of Problems **5** through **14**. Write your observations, conjectures, and reasoning in your journal.

  a. For what polygons will the angle bisectors show concurrency?

  b. For what polygons will the perpendicular bisectors show concurrency?

ACTIVITY 3 # Collinearity

Three or more points that fall on the same line are called collinear.

A disc *is a circle and its interior.*

Challenge How can you accurately find the center of a circle if you trace it from a jar or can?

Just as it is very special when three lines intersect at the same point, it is noteworthy when three apparently unrelated points lie on the same line. Perform this experiment.

Experiment

Step 1 Trace a circle on a sheet of paper.

Step 2 Locate the exact center and poke a small hole through the paper at that point.

Step 3 Carefully cut out your disc.

Step 4 Draw two points on a large sheet of paper. Place them close enough together that your smallest disc can touch both.

Step 5 Working with two or three classmates, take one of your discs and lay it down so that both points lie exactly on the disc's edge (the circumference of the circle).

Step 6 While your disc is just touching the two points, make a tiny mark through its center onto the paper.

Step 7 Remove that disc and do the same thing with the next one.

18. **a.** When you have used all of your discs, look at the marks that show where their centers were located. What pattern is there to the arrangement of the centers?

  b. Draw two new points. Without using your discs, draw the pattern along which their centers would lie.

19. Construct trapezoid *ABCD* whose vertices and sides can be dragged around.

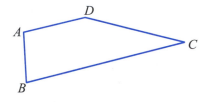

  a. Construct the diagonals and their intersection, and the midpoints of the two parallel sides.

  b. Find two intentionally built-in collinearities. Find a collinearity that was not intentional. Experiment to see if that collinearity is invariant.

20. Using a piece of paper or geometry software, build an arbitrary quadrilateral. On one edge place an arbitrary point. Connect the point with the two opposite vertices. Do the same with the remaining two vertices and the opposite edge.

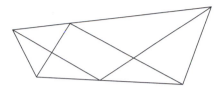

Finally, draw in the diagonals of the quadrilateral. Find two nonsurprising collinearities. Find a surprising one.

21. **Checkpoint** The picture below shows the diagonals of a regular pentagon. Points *A, B, C,* and *D* are collinear. Why is this *not* a surprising collinearity?

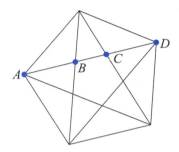

On Your Own

1. You have learned many new terms. Look through your work and make a list of them. Write definitions or explanations for each new term. If there are words that you are still uncertain about, list them separately.

2. Write, as precisely as you can, what you know about each of these terms.

diagonal	intersection	inscribed
median	concurrent	circumscribed
collinear	angle bisector	invariant

For now, each of these rules will be considered tentative. Many of them will be proven formally later in the year.

3. **Write and Reflect** For each situation you investigated, choose an invariant that you found. State it as clearly as you can in the form of a general rule. Write each rule in a special section of your notebook that contains results that are waiting for further verification.

4. Draw each of the following quadrilaterals and their medians. Do you find any concurrence? Do you find any collinearity?

A median connects a midpoint to a vertex. For a quadrilateral, there will be eight of them.

   a. square

   b. kite

   c. parallelogram

   d. rectangle

   e. trapezoid

   f. isosceles trapezoid

   g. rhombus

   h. arbitrary quadrilateral

5. Draw several different triangles. Construct their medians. Describe any concurrence or collinearity you find.

6. How many medians does a triangle have? A quadrilateral? A pentagon? A hexagon? Find a general rule, based on the number of sides of a polygon, for finding the number of medians.

7. Consider the following statement: "In *any* hexagon with all diagonals drawn in, there can be *at most* one concurrence of three diagonals." Do you think this statement is true or false? Explain your reasoning.

Parallel Lines

In this lesson, you will look for invariants in a specific geometric content: parallel lines cut by a transversal.

You have seen **parallel lines** all your life. They may even have been a part of your mathematical studies. By looking for invariants in situations with parallel lines, you will understand more about how they work.

Another way to say "everywhere equidistant" is "the same distance apart, no matter where you measure."

Explore and Discuss

One definition for parallel lines says, "Parallel lines are lines in the same plane that do not intersect." You might also have seen another definition: "Parallel lines are everywhere equidistant."

A line that intersects two or more parallel lines is called a **transversal.** The angles that are formed when transversals intersect parallel lines also have special names.

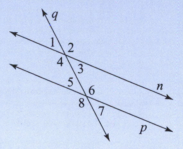

Pairs of angles like ∠3 and ∠5, or ∠4 and ∠6 are called **alternate interior angles.**

Pairs of angles like ∠1 and ∠5, or ∠4 and ∠8 are called **corresponding angles.**

Pairs of angles like ∠2 and ∠4, or ∠5 and ∠7 are called **vertical angles.**

The angles that are on the same side of the transversal and between the lines, for example, ∠3 and ∠6, are sometimes called same-side interior angles.

The same terms are used when the lines are not parallel. In the figure below, lines j and k intersect. So, ∠3 and ∠5 are alternate interior angles, and m is a transversal.

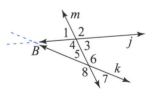

- **a** Explain the name *alternate interior angles*. Why *alternate*? Why *interior*?

- **b** Figure out what **alternate exterior angles** means. Name a pair of alternate exterior angles.

- **c** Name all the pairs of corresponding angles in the figure.

- **d** Explain what *corresponds* about corresponding angles.

- **e** What name could describe an angle pair like ∠2 and ∠7?

Explorations

1. Use geometry software to construct a pair of intersecting lines along with a transversal. Measure the angles in your figure.

   a. Move the transversal while the other lines remain fixed. What invariants can you find? Look especially for pairs of angles that remain equal and for pairs that have a constant sum.

   b. Try moving one of the intersecting lines while the transversal stays fixed. Record and explain what you find.

2. Construct a pair of parallel lines with a transversal.

   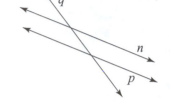

   a. Which angles stay equal when the transversal moves, or one of the lines moves while remaining parallel?

   b. What angle sums are invariant? Compare these results to your findings in Problem **1**.

3. Construct a figure with two parallel lines (like *a* and *b)* and two transversals (like *c* and *d*).

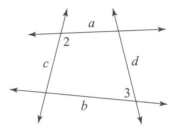

   Angles 2 and 3 are not necessarily equal in measure. In the picture above, they are certainly not equal. Move the lines around to make $m\angle 2 = m\angle 3$.

   a. When $m\angle 2 = m\angle 3$, do lines *c* and *d* have any special relationship? Is that relationship invariant over changes in the equal measure of $\angle 2$ and $\angle 3$?

   b. If lines *c* and *d* are parallel, do the measures of $\angle 2$ and $\angle 3$ have an invariant relationship?

   c. How would your answers to the previous two parts of this problem be different if lines *a* and *b* were not required to be parallel?

4. Construct a pair of parallel lines with a movable point *P* between them. Draw $\overline{PA}$ and $\overline{PB}$ to connect the point to the parallel lines.

   a. What invariants can you find in this situation?

   b. In the figure below, what would you describe as transversals, alternate interior angles, and corresponding angles?

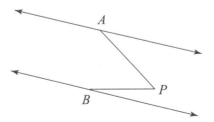

5. Construct a figure like the one below, with two intersecting lines and points on those lines that when connected form three segments. Experiment with the figure by moving various parts until you have made $\overline{BC}$, $\overline{CD}$, and $\overline{DE}$ all the same length.

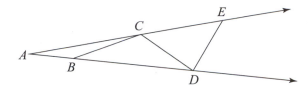

   a. If $BC = CD = DE$, what conclusions can you draw about the angles in the figure? Describe any invariant relationships you find.

   b. If $BC = CD = DE$, is it possible to make $\overline{BC}$ and $\overline{DE}$ parallel? Why or why not?

6. Checkpoint In a parallelogram, opposite sides are parallel. Draw conclusions about the angles in a parallelogram.

On Your Own

You already have overwhelming evidence that the sum of the angles in a triangle is 180°. You have used it to support other arguments, and you may know a few different ways to argue that it's true. Now you can write a reasoned argument of this fact, using what you know about parallel lines, transversals, and alternate interior angles.

1. Write an argument that the sum of the angles in *any* triangle is 180°. The picture may give you some ideas.

2. Use the figure below to answer the questions.

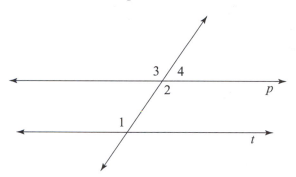

Lines p and t are parallel.

   a. Why does $m\angle 1 = m\angle 2$?

   b. Why does $m\angle 1 = m\angle 3$?

   c. Why does $m\angle 2 = m\angle 3$?

3. Does $m\angle 2 = m\angle 3$ in the picture below? Use Problem **2** or angle 1 to support your answer.

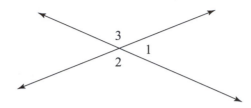

4. a. Lines n and m are parallel. Find the measures of all the numbered angles in the figure below.

b. Explain how you found each angle measure.

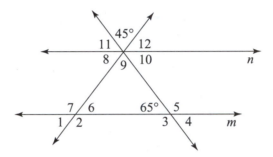

Take It Further

Many mechanical devices use parallel rods to make them work properly. Windshield wipers on some cars and many buses are often made as pairs of parallel rods driving the wiper blades on the windshield of the car. The rods sweep back and forth in a circular motion, but they remain parallel. Trucks and buses sometimes have a wiper system that uses parallel parts in a different way.

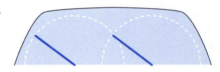

5. Study the way parallel parts are used in windshield wipers by looking at several vehicles. Then construct physical or computer models of what you saw.

a. What seem to be the advantages of each type?

b. Is it necessary for the wipers to remain parallel?

c. What shapes do the different wipers sweep clean?

d. Which design sweeps a larger area on the windshield?

e. What factors might make one type of wiper better for a particular vehicle?

6. Use your knowledge of parallel lines to decide which of the following constructions are *impossible*. Explain what makes them impossible.

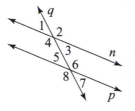 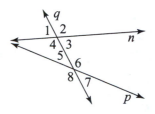

a. Lines *n* and *p* are not parallel and
$m\angle 3 + m\angle 6 = 180°$.

b. Line *n* is parallel to line *p* and $m\angle 4 = m\angle 6$.

c. Line *n* is parallel to line *p* and $m\angle 2 = m\angle 5$.

d. Measures of $\angle 4 + \angle 5$ is greater than
$m\angle 2 + m\angle 7$.

e. In the figure below, $\overleftrightarrow{AB}$ and $\overleftrightarrow{CD}$ are parallel, $\overline{GH}$ and $\overline{JH}$ are angle bisectors, and $m\angle GHJ < 90°$.

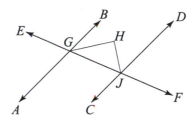

Investigations of Geometric Invariants

In this lesson, you will learn more about invariants by looking for them in some constructions and by building them in to other constructions.

Any mathematical investigation is a search for invariants and an attempt to explain them. These next activities suggest five geometric investigations.

Explore and Discuss

Here are some general guidelines for mathematical research.

Experiment Use hand or computer drawings to help you visualize the situation, explore it, and gather data.

Record your experiment Describe carefully what you did, and what happened as a result. Explain as well as you can why things behaved the way they did. If you have unanswered questions at the end of the investigation, record them, too.

Summarize your work Write up a brief, clear presentation, describing the situation and results, including whatever drawings you need. List your conjectures, any important terms, theorems, rules or ideas that came up in the investigation. Include any questions that require further exploration.

Make a list of some of the invariants you should look for. The list is started below.

- constant measure
- constant sum
- concurrence

Midlines and Marion Walter's Theorem

A midline connects the midpoints of two sides of a triangle.

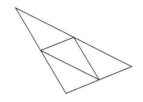

1. Use geometry software to construct a triangle and all three medians. Change the shape of the triangle by dragging around one vertex or side. Look for invariants as the triangle changes. Consider angles, lengths, and areas, as well as anything else you might find. List your conjectures.

2. Construct a triangle and all three midlines. Again, consider angles, lengths, and areas. List your observations.

   a. From the length of one midline, what can you determine about the triangle?

   b. If you know the area of the original triangle, what can you say about the areas of other parts of the figure?

Medians and midlines are results of cutting sides in half. What happens when a triangle's sides are cut into thirds?

Dividing segments is not always easy. If you are not sure how to do it, your teacher should be able to provide some tips.

3. *Situation 1:* Construct a triangle and cut each side into thirds. Connect each vertex to the first trisection point (clockwise) on the opposite side. Change the triangle by dragging vertices and segments. Look for invariants. Record your observations.

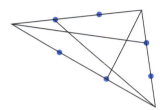

Marion Walter's Theorem states a relationship between the areas of the inner hexagon and the outer triangle.

4. *Situation 2:* Connect the remaining trisection points to the opposite vertices. Change the triangle by dragging vertices and segments. Look for invariants. Record your observations.

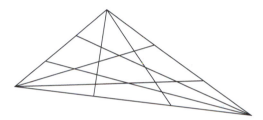

5. Organize your ideas about cutting-in-half and cutting-in-thirds in a way that allows comparison. For example, all your results about area might be placed together. Look for ways to generalize the results.

6. **a.** Choose one conjecture from your lists. Explain how it changed as the situation changed from cutting-in-half to cutting-in-thirds.

   b. Predict how that conjecture would change if you tried subdividing the triangle's sides into fourths, fifths, or some other number of pieces. Test your predictions with an experiment.

7. **Checkpoint** Use geometry software to design a triangle with all six trisection points. Working clockwise, connect the first trisection point on one side to the first trisection point on the next side. Then connect the second trisection point on one side to the second trisection point on the next side (still working clockwise).

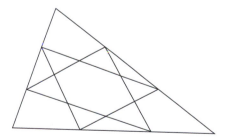

As you change the triangle by moving vertices or sides, what invariants can you find?

Folding Investigations

Mark a point somewhere on a rectangular sheet of paper. For example:

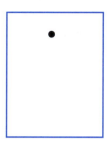 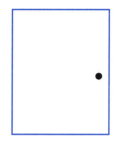 

One of these should work.

Bring one corner of the paper directly to the point you have marked and crease carefully. For example:

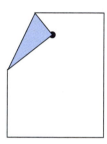 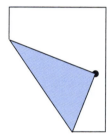 

Next, unfold the paper to see the crease you have made. For example:

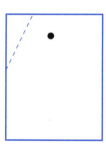 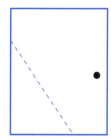 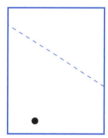

Connecting the point to the corner with a line segment might help you see the relationship.

8. Describe the geometric relationship of the crease line to your choice of point and corner.

Now start with a *square* piece of paper. Mark an interior point *P*.

One at a time, and unfolding in between, fold *each* corner in to touch point *P*. Crease carefully each time.

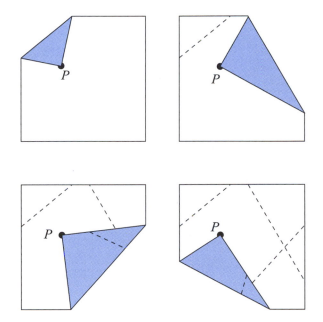

When you are done, the creases will divide the square into several polygonal region as shown at the right. In this example, the region containing the original point is a hexagon. The sides of the hexagon can be either creases or sides of your paper.

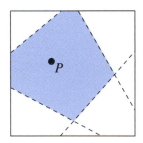

9. Perform the investigation several more times. Does the region containing *P* always have six sides, no matter where you place *P* at the start? If you think it does, give a good reason why that should happen. If you do not think so, give at least one counterexample.

10. Imagine starting with a square, picking some location for *P*, and folding each corner to *P* as described above. Find a good reason why the region that contains *P* must have.

   a. fewer than nine sides.

   b. more than three sides.

11. Problem **10** asked you to show that the region that contains *P* cannot have less than four sides, and cannot have more than eight. Perform some investigations that help you narrow the range even more. What are the minimum and maximum number of sides you can find?

12. **Challenge** Create a model of this investigation using geometry software. You should be able to move point *P* and see where the creases would be. How can you make the *creases*?

13. Checkpoint How does the placement of *P* determine how many sides the polygon will have? Where can you place *P* to get the minimum and maximum number of sides? Why?

Circle Intersections

You can construct two intersecting circles using geometry software.

Sample Construction

Step 1   Construct $\overline{AB}$.

Step 2   Place point C near the middle of $\overline{AB}$.

Step 3   Construct $\overline{AC}$ and $\overline{CB}$.

Both $\overline{AC}$ and $\overline{CB}$ have been constructed
as segments on top of $\overline{AB}$.

Step 4   Place points D and E so that their distance from each other is less than $\overline{AB}$.

$DE < AB$

Step 5   With D as center and $\overline{AC}$ as radius, construct a circle.

Step 6   With E as center and $\overline{BC}$ as radius, construct a second circle.

Step 7   Find the intersections of these two circles.

Step 8   Label your entire construction as illustrated below.

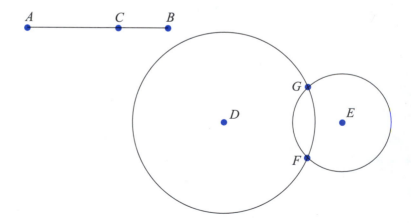

What lengths in the diagram determine the lengths of the sides of triangle DGE?

14. As you move C back and forth on $\overline{AB}$, notice that the circles do not *always* intersect. Use the Triangle Inequality to explain why the circles sometimes intersect and sometimes do not.

Trace the intersections of these two circles. Then experiment with the diagram by dragging A, B, C, D, and E. The patterns that you see as you move these points are kinds of invariants.

15. Trace the intersection points F and G as shown in the picture on page 78. Describe what you see when you move A around.

Any aspect of the pattern that does not depend on the placement of C is an invariant.

   a. What shapes do the points trace out as A moves?

   b. Does the pattern you see as you drag A depend on where C started out along $\overline{AB}$? For example, does it matter if C starts out near to A, far from A, or close to the middle of the segment? What is the invariant here?

   c. Make a reasonable argument for or against this statement: "Whatever pattern (invariant) I see when I move A, I should see exactly the same one when I move B."

16. Leave A and B fixed. Move C along $\overline{AB}$. What pattern do you find in the intersection points of the circles?

17. Move D or E around while leaving everything else fixed. Describe what happens.

18. | Checkpoint | An **ellipse** is all the points that are the same total distance from two points. When C is moved, you get an ellipse. Why?

ACTIVITY 4

Constructing Invariants

This activity is an "invariant search" in reverse. Up until now, geometric figures were described in the unit. Your job was to build them, study them, and find invariants. In the problems below, you are given an invariant and asked to construct a geometric figure which produces it. Be creative. There are many different correct solutions to most of the problems.

For Problems **19–25**, use geometry software to construct a figure that has the specified invariant. Write a set of directions for making that figure. Be prepared to explain your solution.

19. Construct two circles that have an invariant 2:1 ratio of circumferences. That is, the circumference of one, even as it grows and shrinks, is always twice the circumference of the other.

20 Construct a triangle and a rectangle whose areas remain in a fixed ratio when either figure changes size.

21. Construct a square and a circle that have any invariant ratio between their areas—a ratio that remains fixed as either shape grows or shrinks.

These four problems are really asking you to construct constant sums (perimeters) and constant products (areas).

22. Construct a triangle whose area can change but whose perimeter is invariant.

23. Construct a triangle whose perimeter can change but whose area is invariant.

24. Construct a rectangle whose area can change but whose perimeter remains fixed.

25. Construct a rectangle whose perimeter can change but whose area is invariant.

For Discussion

Discuss the different strategies that you used to work on the invariant problems in this activity. Explain why you chose to use a particular strategy. Describe which methods were useful and which were not.

26. Checkpoint Consider these problems.

  a. Points *C* and *D* are movable, but only on $\overline{AB}$. Would this help you make a constant sum or product? Explain.

  b. Point *O* is the midpoint of $\overline{MN}$. Would this help you make a constant product or ratio? Explain.

On Your Own

1. Look at this new way to connect trisection points.

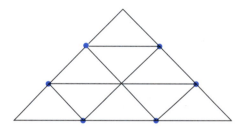

Make some conjectures.

  a. What ratios might be invariant based on your work with midlines?

  b. What might the value of the ratios be?

  c. What area relationships would you investigate? What do you expect them to be?

  d. Can you generalize these conjectures for cutting the sides in fourths, fifths, and so on?

2. Start with a circular piece of white paper. Mark the center point. Fold the edge in to just touch the center and make a crease. Do this at least 30 times for different points on the edge of the circle.

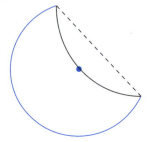

a. Draw the shape you end up with.

   b. What shape do the intersections of the creases make?

   c. When you fold the edge into the center you construct a perpendicular bisector between the center of the circle and a point on the circle. Can you explain why the folds make the shape they do?

3. Start with another circular piece of white paper. Fold it as you did in Problem **2.** But instead of using the center point, use a different point inside the circle. Draw the shape you end up with. What shape is this?

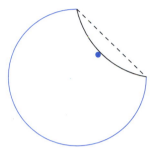

4. Look back in this lesson to help you define these terms.

   midline
   ellipse

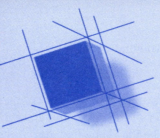

Perspective

A High School Student Extends the Theorem

Marion Walter is a professor at the University of Oregon (see Perspective on page 23). The theorem that bears her name states a relationship between the areas of the inner hexagon and the outer triangle in the trisection construction you made in Problem **4** on page 75.

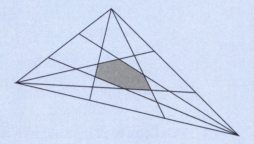

Ryan Morgan was a sophomore in high school when his geometry teacher asked his class to prove Marion Walter's Theorem. Ryan used computer software to help him investigate the problem. During his explorations, he made an important discovery that was reported in several newpapers. Ryan gave a presentation about his discovery to a faculty seminar for the Towson University mathematics department. Here is Ryan's description of his experience working on Marion Walter's Theorem.

"One day, my geometry teacher took our class across the hall to our school's computer lab. He wanted us to get familiar with the use of the computers. He did this by having us 'discover' Marion Walter's Theorem on our own. He told us how to draw the triangle, trisect the sides, and draw the hexagon in the middle. It was our job to find something 'neat' about the measurements of the shapes. When I found Marion Walter's Theorem, I got curious, and wanted to see if the same thing worked with any shapes other than triangles.

"This is where all the hard work began. Not having a computer of my own, I was forced to use my school's computers after school. The first thing I did, once I became familiar with the software, was to experiment with squares. I attempted to trisect the sides of the square, and see if there was any special relationship there. At first I thought there was. I can't remember exactly what the value was, but it did seem that there was a constant ratio between the square and the octagon that was in the middle of the square. At that point, I really thought I was on to something.

"So, the next thing I did was try the same thing with a pentagon. I trisected the sides, compared areas between the pentagon and the now 10-sided figure inside ... there was no constant ratio. I was upset, because at this point I had spent maybe a week or two, every day after school, working on this thing, and now I had hit a dead end. But I didn't give up."

Ryan first tried to extend the theorem by using figures other than triangles, but that didn't produce any invariant. So he returned to triangles and found a different way to extend the theorem.

"This time I concentrated only on triangles, and no other shape. Trisecting it was the whole basis behind Marion Walter's Theorem... [so] I started 5-secting, 7-secting..., the triangle." These new ways of "secting" the triangle, as Ryan called it, also produced hexagons in the center.

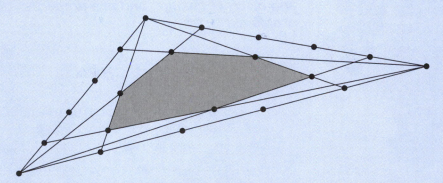

"When I compared the area of the triangle and area of the internal hexagon, I noticed a constant ratio between the two. And for every example I tried (odd number secting per side only), I was able to find a ratio that was constant no matter how the triangle was altered in size.

"The next step was to find a relationship between the number of sections per side and the ratio between the areas of the triangle and hexagon. Using the regression functions on a regular scientific calculator, I was able to do just that, and came up with the formula

$$y = \frac{9}{8}n^2 - \frac{1}{8},$$

where n is the number of sections per side, and y is how many times bigger the area of the triangle is than the area of the hexagon.

"It took a lot of time and effort, but it was worth it. And before I can take credit for my theorem, I must thank Marion for creating her theorem to begin with; without hers, mine never would've come around."

LESSON 11

Reasoning by Continuity

In this lesson, you will learn about the difference between continuous and discrete change, and use properties of continuity to solve geometric problems.

Explore and Discuss

Newborns start out at different birthweights. You now weigh much more than you did as a baby. Nonetheless, it can be said for certain that there was a time when you weighed exactly 30 pounds.

How could such a thing be known *for certain?* There was a time when you weighed less than 30 pounds and now you weigh more. So, no matter how your weight may have increased or decreased in between, there must have been at least *one* time, however brief, when you weighed exactly 30 pounds.

a Look at the graph at the right. Did the person ever weigh exactly 30 pounds?

b What are some other examples where you know something *for certain* based on knowing how it started and how it is now?

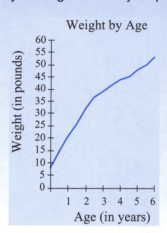

Weight by Age

Weight (in pounds) vs Age (in years)

ACTIVITY 1 — Continuous Change

1. At 6:00 one morning, the temperature in Boston was 64°F. At 2:00 that afternoon, the temperature was 86°F. Can you be certain that there was *some* time that day when the temperature was exactly 71.5°F? Can you tell what that time was? Explain.

2. In the 1950s, the town of Sudbury had a population of roughly 5,000. By 1990, the population was well over 40,000. Can you be certain that there was *some* date at which the population was exactly 10,000? Can you tell when that date was? Explain.

3. Some car companies advertise how quickly their cars can go from 0 to 60 mph. If a car goes from 0 to 60 mph, is there a time when it is traveling exactly 32 mph? Explain.

4. **Write and Reflect** Problems **1** and **2** look nearly the same except for the numbers. But they are profoundly different and have very different answers. Explain why.

5. **Write and Reflect** Some people find Problems **1** and **3** essentially the same, while others find them quite different. Where do you stand? Why?

This problem is, in some ways, the most important problem in this entire activity.

Situations that change like temperature or speed are said to be **continuous.** Situations that change by steps, like population, are said to change **discretely.**

6. Starting at dawn, you hike up a mountain, arriving at the summit roughly at dusk. You fix a delicious dinner, camp the night, and sleep quite late the next morning. After enjoying the beautiful view and a great breakfast, you hurry down the mountain following the same path you took up (stopping briefly in the middle to catch your breath), and arrive at the bottom while the day is still bright.

   a. Is it *possible* that you passed some point along the path at exactly the same time of day on both ascent and descent?

   b. Is it *certain* that there is such a point? Could there be exactly two such points? Explain.

7. **Checkpoint** Which of the following situations change *continuously,* and which change in *discrete* steps?

   a. Number of students in the cafeteria at lunchtime.

   b. Amount of water in a tub as you fill it up.

   c. Height of a tree.

   d. The cost of postage for letters of various weights.

ACTIVITY 2 **The Box Problem**

By cutting identical squares out of each corner of a 5" × 8" index card, you can create a shape that can be folded into an open box.

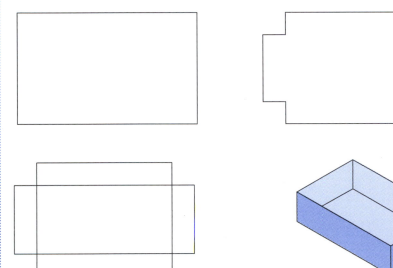

8. How does the box's **volume** change as the size of the cutout increases? Perform some experiments to see how the volume depends on the size of the cutouts. For now, do *not* bother with formulas or numbers. Notice as the size of the cutout increases whether the volume always increases, always decreases, stays the same, or does something else.

9. a. Is there at least one cutout that makes a box with the greatest volume? How can you be sure?

    b. Is there a cutout for a box with the smallest volume? How can you be sure?

These two questions do not ask whether you can find such a cutout. You are only to state whether such a cutout exists.

10. Make the box with the largest volume that you can. Describe what thinking led you to that box. What reasoning got you to change the cutouts in the ways you did?

11. Find a way to test which box in your class has the most volume. What cuts were made for that box? Can you be certain that no bigger box is possible?

12. Checkpoint One class graphed cutouts versus volume. The class decided that point *A* gave the largest volume possible. Do you agree? Why or why not?

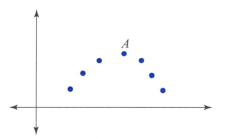

ACTIVITY 3 · **The Ham Sandwich Problem**

For some figures, it is easy to find a line that can cut them into two pieces of equal area.

   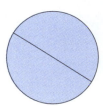

13. For each figure above, how do you know the two pieces have the same area?

14. For some shapes, it is not as obvious how to cut them into two pieces of equal area with a straight line. For some shapes, it may even seem quite impossible to do.

    a. Trace the shapes that follow. Try to cut them into pieces of equal area with a straight line. How do you know if you succeeded?

    b. Perhaps for one or another of these shapes there really is no straight line that cuts it into two equal areas. How can you be sure whether or not there is such a line? For any of these, could there be more than one such line? Explain.

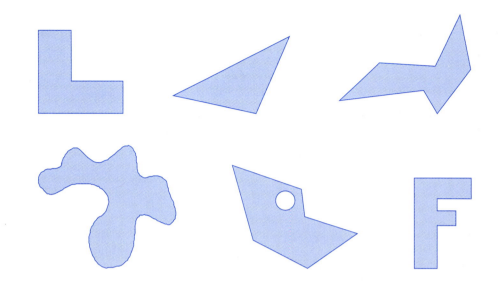

15. a. Try to create a shape whose area absolutely *cannot* be bisected by a straight line.

   b. Try to create a shape whose area can be bisected by only *one* particular straight line and no other.

Even if it is hard to find a line that will cut some shapes in half, it is possible to convince yourself that such a line exists. The argument requires reasoning by continuity. Take one of the shapes from Problem **14,** for example. Copy the figure. Draw a line somewhere completely outside of the figure, but never intersecting it. Then think of slowly passing the line over the figure, until it slides completely off again, no longer intersecting it.

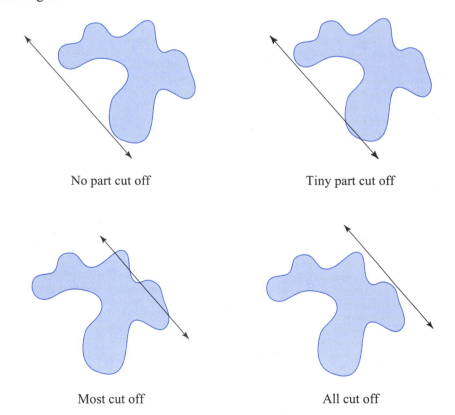

No part cut off   Tiny part cut off

Most cut off   All cut off

16. Use the preceding pictures to write an argument that there is some line that cuts the figure exactly in half. Does knowing this help you to find the line?

17. Show, with pictures and an explanation, that it is always possible to bisect the *combined* area of three arbitrary shapes with a single straight line.

18. Show, with pictures and an explanation, that it is not always possible to find a single straight line that will simultaneously bisect the area of each one of three arbitrary shapes.

19. Challenge Is it always possible to find a single straight line that will simultaneously bisect the area of each one of *two* arbitrary shapes?

This problem, bisecting irregular figures, has sometimes been dubbed the ham sandwich problem because of its 3-dimensional version, which can be stated like this:

> Imagine a ham sandwich made up of two slices of bread and a slice of ham. You do not know exactly how everything is arranged. For example, the ham may not be the same size and shape as the bread, and it may be off-center. Is there a plane that cuts *all three parts* of your sandwich exactly in half?

20. Write and Reflect Work on the ham sandwich problem. Describe your strategies. Do you believe that the statement that "a single plane can bisect three distinct volumes" is true? Explain.

> ## Ways to Think About It
>
> Recall the strategy "Change the Problem." The ham sandwich problem is certainly hard enough to be worth simplifying in some way.
>
> How can you simplify this problem? You can simplify the problem by
>
> - reducing the number of objects from three to two or one.
> - reducing the dimensions from three to two.
> - thinking about bisecting the *combined* volume instead of bisecting each object individually.
> - thinking about simple, regular 3-dimensional shapes like spheres or cubes rather than bread and ham.

21. **Checkpoint** For a single figure in a plane:

a. Is there necessarily a line that bisects the area of the figure?

b. Can you always find that line?

c. For what kinds of figures is it easy to find the line?

In fact, there are infinitely many lines that work. Why?

1. Define the following terms in your own words.

   continuous   discrete   volume
   area   bisect

2. Label the following situations as continuous change or discrete steps.

   a. Number of airplanes in the sky.

   b. Number of leaves on a tree.

   c. Amount of candy in a piñata.

   d. The temperature of a glass of water as it freezes.

3. Give your own example of something that changes continuously and something that changes in discrete steps.

4. Below are examples of graphs of continuous and discrete functions.

Continuous functions

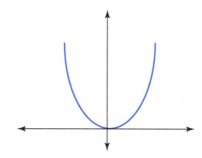 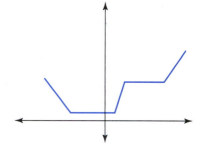

Discrete functions

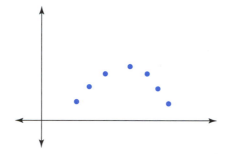 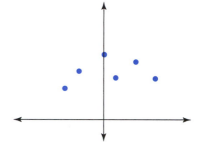

a. Draw a graph that would represent the following situations.

    i. Number of students in the cafeteria at lunchtime.

    ii. Amount of water in a tub as you fill it up.

    iii. Height of a tree.

    iv. The cost of postage for letters of various weights.

b. Write and Reflect Using your own words, write a definition for a

    i. continuous function.

    ii. discrete function.

12

Definitions and Systems

In this lesson, you will define the word straight and explore two different geometry systems.

Jeff Weeks' book *The Shape of Space* is a wonderful study of geometry and topology at both high school and college levels. The book begins with a story based on a much older book, Edwin Abbott's *Flatland.* All the characters in Weeks's story are 2-dimensional creatures, and they all assumed their world was flat, a plane. All, that is, except one—A Square, by name—who had the theory that their world was actually a *hypercircle* (a circle of one higher dimension, what we would call a sphere). Now, the creatures live in their surface-world, not on top of a surface as we do, so they are unable to escape into the third dimension to view the surface from afar. How could these creatures check out A Square's theory?

A Square and
two friends on
a plane

... and on a
hypercircle.

A Square thought that a journey might settle the issue:

> "He reasoned that if he were willing to spend a month tromping eastward through the woods, he might just have a shot at coming back from the west."

> "He was delighted when two friends volunteered to go with him. The friends didn't believe any of A Square's theories—they just wanted to keep him out of trouble. They insisted that A Square buy up all the red thread he could find in Flatsburgh. The idea was that they would lay out a trail of red thread behind them, so that after they had travelled for a month and given up, they could then find their way back to Flatsburgh."

> "As it turned out, the thread was unnecessary. Much to A Square's delight—and the friends' relief—they returned from the west after three weeks of travel. Not that this convinced anyone of anything. His friends thought that they must have veered slightly to one side or the other, bending their route into a giant circle in the plane of Flatland." (Weeks, 1985, pp. 5, 6)

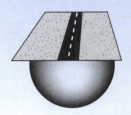

ACTIVITY 1

Life on a Sphere

What is the shortest path between two points? Is there more than one?

What is the shortest path between these two points? Is there more than one?

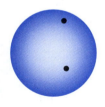

1. Ignoring details like mountains and oceans, picture two straight roads in *our* world, both starting at right angles to the equator and about a block apart, and extending straight north for 6000 miles or so.

   a. Do they stay a block apart, or does the distance between them change?

   b. Can two 6000+ mile-long roads on earth remain the same distance apart and still both be straight?

2. **a.** Mark two points on a plane and draw the shortest path between them. Take that path as the definition of *straight* on the plane. Extend that path in a natural way at both ends. What figure is that extended path?

   b. Is it possible to find some arrangement of two points on the plane so that there is *more than one* shortest route from one to the other?

3. **a.** Now mark two points on a sphere and draw the shortest path between them. The path must be *on the sphere,* not going through the interior of the sphere. Take that path as the definition of *straight* on the sphere. Extend that path in a natural way at both ends. What figure is that extended path?

   b. Is it possible to find some arrangement of two points on the sphere so that there is *more than one* shortest route from one to the other?

Great circle is the name for a line on a sphere. Great circles are considered the equivalent of lines because they represent *straight* paths, or the shortest distance between two points. Any great circle divides a sphere into two equal parts, or hemispheres.

The triangle's sides must be line segments, so they must be segments of great circles, not just any path between two points.

4. Not all the constructions that can be made on a plane can be made on a sphere. Can a triangle be drawn on a sphere? If so, tell how the figure you draw fits the definition of a triangle. If not, explain why not.

5. What about a square? Can A Square exist on a sphere? Explain.

6. On a plane, the sum of the angles in a triangle is 180°. What can you say about the sum of the angles of a triangle on a sphere?

   a. Draw a triangle on a sphere. Measure each of the angles. What is the sum of the angles for your triangle?

   b. Draw another triangle on your sphere, either much bigger or much smaller than the one you drew in part **a.** Measure and sum the angles.

   c. Is it possible to draw a triangle with three right angles on your sphere?

7. What can you say about the sum of angles in a triangle on a sphere? Is the sum constant, as it is in a plane? Is the sum bounded by any limits? Test several more triangles and make a conjecture.

8. Checkpoint Is there a triangle inequality on a sphere? Explain.

A C T I V I T Y   2

Parallel Lines and Spheres

When you are working on a sphere, the first definition should really be changed to read "Parallel lines lie on the same sphere, and never intersect."

Geometry on the plane makes a lot of use of parallel lines. Can you imagine a world *without* parallel lines? Would you believe that you live on one?

There are two common definitions for parallel lines. You can work with whichever definition you prefer.

   Parallel lines lie in the same plane, and never intersect.

   Parallel lines are everywhere equidistant.

9. Why is the shorter statement, "Parallel lines never intersect," not a good enough definition? Describe two lines that never intersect, but are not parallel by the second definition. These nonparallel, nonintersecting lines are called *skew lines.*

10. You have read that you cannot draw parallel lines on a sphere. Why not? Use your model of a sphere to try to draw two great circles that never intersect. Write a description, including pictures, of what goes wrong.

11. If there are no parallel lines on the surface of the earth, can there be perpendicular lines? If not, why not? If so, how are you defining perpendicularity on a sphere?

One rule that is different is that the sum of the angles of a triangle on a sphere is always greater than 180°.

What you have been doing in the last few problems is exploring a new system: the geometry of a sphere. This system is different from the geometry of a plane: it starts from different assumptions and, as a result, different rules or theorems arise.

A good way to learn about a new system is to see how it is the same as, and how it is different from, a system you already know. You have already looked at squares, triangles, perpendicular lines, and parallel lines.

Things that are true regardless of what surface is being investigated are "invariant under a change of surface."

12. Checkpoint Think of some familiar properties or shapes in planar geometry. Explore related ideas or shapes on your sphere. Describe your spherical geometry findings. Discuss what remains true of *both* systems.

Building Up from Rules

Use wires and beads to design *one object* that follows *all four* of these rules:

- Each pair of wires has exactly one bead in common.

- Each pair of beads has exactly one wire in common.

- Every bead is on exactly three wires.

- Every wire contains exactly three beads.

13. Compare all of the objects built in your class. In what ways are they different? In what ways are they alike? You might look at attributes like shape, number of beads, number of wires, or number of places where two wires join at an end-point. What else might be useful to compare?

Draw a picture of your object. If you call the beads *points* and the wires *lines,* then you have created another new system: a *seven-point geometry.* You can draw conclusions about your seven-point geometry based on the models constructed in class.

14. How many lines must you have in a seven-point geometry?

15. Triangles are still figures with three sides. Draw pictures of some of the triangles you can find in your seven-point geometry.

16. Checkpoint List two other conclusions you can draw or objects you can define in your seven-point geometry.

On Your Own

1. a. If Flatland were really flat and A Square told his friend to walk east, how far away from his starting point could he get?

b. If Flatland were a sphere and A Square told his friend to walk east, how far away from his starting point could he get?

2. Two Flatlanders are arguing about the best route from one point to another.

a. If Flatland is flat, can there be two best routes? Why?

b. If Flatland is a sphere, can there be two best routes? Why?

3. Here is another finite geometry:

- Every line has three points.

- Every line intersects another line at one point.

- Every point is on exactly two lines.

a. Draw a picture of this finite geometry.

b. How many points are there?

c. How many lines are there?

d. Draw some triangles using this finite geometry.

Remember: In finite geometry a line might not be straight.

4. If you switch "point" with "line" and "line" with "point" you get a new finite geometry:

  • Every point is on three lines.

  • Every point is connected to every other by one line.

  • Every line has exactly two points.

  a. Draw a picture of this finite geometry.

  b. How many points are there?

  c. How many lines are there?

  d. Draw some triangles using this finite geometry.

Take It Further

A Square was not to be defeated by doubters, and so set out on another expedition, this time to the north, laying a trail of blue thread. When he returned two weeks later from the south, everyone was surprised. Most Flatlanders, of course, were surprised that he got back at all, and assumed that he got lucky again and veered off course. But A Square, too, was surprised. This second journey was much too short. Even stranger, he had never crossed the red thread that had been laid out during the first journey.

5. If A Square's world is a *plane,* as the Flatlanders imagined, then the first journey with the red thread must have somehow looped around to allow A Square to arrive from the west after having started out east. Experiment by drawing closed-loop paths on the plane—a red one starting east and returning from the west, and a blue one starting north and returning from the south.

  a. Can you explain the shorter time of the blue path?

  b. Can you find a way for the blue path not to cross the red path before returning home?

6. If A Square's world is what we would call a *sphere,* as A square theorized, then it is possible for the red journey to go *straight* around the globe, but it is still possible for it to have been some other kind of loop.

  a. How can you explain the shorter time of the blue path on a sphere?

  b. On a sphere, can you find a way for the blue path not to cross the red path before returning home?

7. **a.** Assuming the red thread was never broken or removed, what could account for A Square's results?

  b. Think about some other shape worlds. On what shape world might it be possible for A Square to make the second trip without passing the red thread?

8. **Write and Reflect** You explored the angles of a triangle on a sphere. You probably found that the smaller the triangle, the smaller the angle sum (though it never goes below 180°). Investigate this idea. Can you find a more precise relationship between the area of a triangle and its angle sum? Write about how you investigated the idea and what you found.

Perspective

Finite Geometries

William Kramer, a high-school mathematics teacher, decided to investigate a finite geometry with one of his classes. He and his class began an investigation of a 25-point geometry that he has continued with every geometry class for 28 years.

Mr. Kramer's class began with just six definitions:

Points There are exactly 25 points in the geometry, and they are given the letters *A, B, C, ..., Y.*

Lines There are exactly 30 lines in the geometry. Each row and each column from the blocks below is a line.

A	B	C	D	E		A	I	L	T	W		A	X	Q	O	H
F	G	H	I	J		S	V	E	H	K		R	K	I	B	Y
K	L	M	N	O		G	O	R	U	D		J	C	U	S	L
P	Q	R	S	T		Y	C	F	N	Q		V	T	M	F	D
U	V	W	X	Y		M	P	X	B	J		N	G	E	W	P

Segment A segment consists of two points and all of the points between them on a single line.

Length To specify the length of a segment, one must give two pieces of information: the number of steps it would take to go from one end to the other (the shortest route), and whether the segment is part of a row or column line. So, (2, *col*) and (2, *row*) are different lengths.

Perpendicular Two lines are perpendicular if one is a row and the other is a column *in the same block.*

Parallel Two lines are parallel if they are both rows or both columns *in the same block.*

From these definitions, students found midpoints of segments, perpendicular bisectors, polygons, and even circles. They found that the geometry has no rays. Mr. Kramer's students were accustomed to defining angles in terms of rays, and so they concluded that there were no angles, but they felt that they could still reason about *perpendicular* and *parallel* lines by thinking about distance and intersection

1. How many different segment lengths are there in this 25-point geometry?

2. **a.** How might a triangle be defined in this geometry?

   b. Are squares and rectangles possible?

3. **Write and Reflect** The wording of a definition may sometimes seem to be just a matter of style, but it can make all the difference in the world. Here are two reasonable ways one might define a midpoint of a segment.

> "Of all the points equidistant from the two endpoints of the segment, the midpoint is the one with the *shortest* equal distance from both ends."

> "The midpoint is the point on a segment equidistant from the two ends."

In the plane, these two definitions are equivalent. But on the sphere, they are not. Each definition identifies a unique point, but the point selected by one definition may not be the same as the point selected by the other. In 25-point geometry, only one of those definitions guarantees that every segment even has a midpoint.

Show why these are equivalent definitions on the plane but not on the sphere or in 25-point geometry. Explain the difference carefully. In the two geometries in which the choice of definition matters, which definition do you find more useful?

Unit 1 Review

1. For each description below, construct the shape that is described. Then name that shape.

   a. A quadrilateral with four 90° angles.

   b. A closed figure with three sides, each 5 centimeters long.

   c. A quadrilateral with two pairs of congruent sides.

   d. The set of points equidistant from a given point.

   e. The set of points equidistant from two given points.

   f. The set of points equidistant from a given line.

2. For each shape below, describe it by its features. Your descriptions should allow for no other shape to be confused with the one shown.

   a. an ellipse

   b. an isosceles trapezoid

   c. an isosceles triangle

   d. perpendicular lines

   e. parallel lines

   f. a rhombus

   g. a regular octagon

3. Follow the directions below to construct a shape.

   Step 1   Construct a square with sides 8 centimeters. Label the vertices of the square *ABCD*.

   Step 2   Find the midpoint of side $\overline{AB}$ of the square. Label the midpoint *M*.

   Step 3   Draw $\overline{CM}$ and $\overline{DM}$.

   Step 4   Draw the diagonals of the square. Label the point where they intersect *O*.

   Step 5   Construct a circle with center *O* and radius $\overline{OM}$.

   Step 6   Erase the diagonals of the square.

4. Figure out how to draw the shape below. Then write complete directions so that someone else could follow them. Without looking at the original, produce the same picture.

Hint: The triangle is equilateral, and the curves are arcs of circles.

5. State the Triangle Inequality.

6. Given the three lengths, decide if they will make a triangle. If they will not, explain why not. (Do not construct the triangle.)

  a. 3 cm, 3 cm, 3 cm

  b. 3 cm, 3 cm, 6 cm

  c. 3 cm, 4 cm, 5 cm

  d. 1 cm, 1 cm, 9 cm

  e. 9 cm, 9 cm, 1 cm

7. In your own words, describe what the term *invariant* means.

8. Give three examples of geometric invariants you have learned in this unit. Think about lengths, angle measures, areas (and ratios of those), collinearity, concurrence, shape, and so on. Describe the invariant carefully; include pictures with your descriptions.

9. What is the sum of the interior angles in a triangle? Prove what you say.

10. What is the sum of the interior angles in a polygon with *n* sides?

11. Give three examples of concurrence in triangles or other shapes. Describe what concurrence means, draw a picture of the situation, and explain which lines are concurrent.

12. a. Give two examples of situations that change in a continuous way.

  b. Give two examples of situations that change in discrete steps.

13. a. If something changes continuously, what information do you get from that?

  b. Why is it helpful to know about continuous change?

14. Choose one example of continuous change and one example of discrete change from your examples in Problem **12.** Draw graphs that might represent each situation.

UNIT 2

A Perfect Match
Investigations in Congruence and Proof

The Congruence Relationship

In this lesson, you will learn about a relationship called congruence and some notation to describe congruent figures.

To understand science or social relationships, mathematics or the arts, history or psychology, one must look at how things differ, and also at how they are the same. Mathematics looks at quantities, relationships in space, ways of classifying things, and processes by which things are done. The focus for this lesson is on shape: *What does it mean for two shapes to be the same?*

You will learn about another kind of sameness, called similarity, *later in your studies of geometry.*

Explore and Discuss

Mathematical language tries to be clear and precise, specifying exactly one meaning for each word. Sometimes this means inventing new words, or specializing the meanings of familiar words. Either way, the process involves people getting together and agreeing on what they will mean by the words they use.

a As a class, decide on what you will mean by "these two figures are the same."

To make a *wise* decision about what the statement should mean, you might consider some specific cases. For example, look at the four figures below and decide which ones you would call "the same."

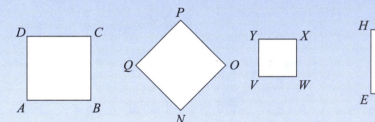

b In some ways, all of the figures above are the same. How?

c In some ways, they are all different. How?

So, mathematical language has several words to distinguish "how much the same" two figures are. When two figures are the same shape *and* size, regardless of location or orientation, they are **congruent.**

Congruence is such an important mathematical relationship, that it has its own symbol, $\cong$. The symbol is used this way:

$$\square ABCD \cong \square EFGH$$

d You can think of the symbol as composed of two parts: = and ~. What aspects of *congruence* might these two parts stand for?

Length, Measure, and Congruence

You can have congruent line segments and triangles, but you can also use the term *congruence* for figures that have more than just one or two dimensions. You can refer to congruent spheres and congruent tetrahedra, or congruent shapes in even higher dimensions.

1. **Write and Reflect** If you could use any tool or method you like, describe how you would decide if two line segments are congruent. What method would you use for angles? Triangles? Two rectangular solids (boxes)? Cones? Cylinders?

The word congruence *comes from the Latin word* congruens, *which means "to meet together."*

One class decided on this test for congruence: Given two shapes drawn on paper, if you can cut one out and fit it *exactly* on top of the other shape (edge to edge), then the two shapes are congruent. Their teacher supplied a name for the test: *superimposability.* Another way to say this is:

Two figures are congruent if they differ only in position or orientation.

2. Will the cut-and-move test work for line segments?

3. How could you adapt the test to decide whether 3-dimensional objects, such as spheres or rectangular boxes, are congruent?

A single geometric object—even a line segment—can have many different numbers attached to it, like length, slope, and location. So, when you are talking about a geometric figure, you must make it clear whether you are talking about a geometric object itself, like a point or a segment, or about one of the many numerical values that help describe it.

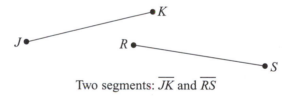

Two segments: $\overline{JK}$ and $\overline{RS}$

Symbols have been designed to help keep this distinction clear. For example, $\overline{JK}$ stands for a geometric object—the line segment joining points J and K. The symbol JK, without the overbar, stands for the *length* of that segment, a *number*.

You can compare two shapes to see if they are congruent, or two numbers to see if they are equal. But you cannot compare an object to a number. For example, look at the figures below. It looks like $\square ABCD \cong \square EFGH$.

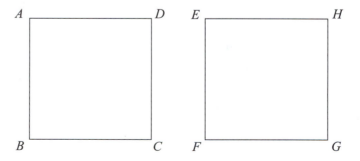

But what could "$\square ABCD = 4$" mean? The perimeter is 4? The area is 4? The length of a side is 4? Certainly, the *shape* does not equal 4.

4. Below are four statements about line segments. For each one, state whether numbers, points, segments, or other objects are explicitly mentioned in the description.

  a. $JK = RS$

  b. $\overline{JK} \cong \overline{RS}$

  c. $\overline{JK}$ and $\overline{RS}$ have the same length.

  d. The distance from J to K is the same as the distance from R to S.

5. Below are two segments and some statements about them. All are grammatical sentences—they have the right overall form—but only some are correct. Others may be false, or they may mix up different ideas in a way that leaves them with no clear meaning at all. For each statement, say whether or not it is correct. If it is *not* correct, explain what is wrong.

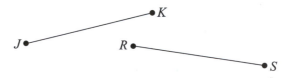

  a. $JK \cong RS$

  b. $\overline{JK} \cong \overline{RS}$

  c. $JK = RS$

  d. $\overline{JK} = \overline{RS}$

  e. $JK = 1"$

  f. $\overline{JK} = 1$

  g. $\overline{JK} \cong 1"$

6. If two segments have the same length, are they congruent? Why or why not?

7. If two segments are congruent, are they the same length? Why or why not?

The symbols are different for angles, but the distinction is the same: an angle is a geometric object; its measure is a number.

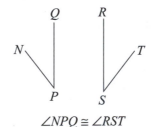

$$\angle NPQ \cong \angle RST$$

When you mean the *angle* of a geometric object, write $\angle NPQ$ or $\angle P$. When you mean the *size* or *measure* of an angle, write $m\angle NPQ$.

8. a. Why is the sentence "$\angle NPQ = 56.6°$" confusing?

  b. What is the correct symbolic way to write "angle NPQ has a measure of 56.6 degrees"?

9. If $m\angle NPQ = m\angle RST$, are the two angles congruent? Why or why not?

10. If $\angle NPQ \cong \angle RST$, do the two angles have the same measure? Why or why not?

11. Write and Reflect Write descriptions of the methods you can use to tell if

    a. two segments are congruent.

    b. two angles are congruent.

    c. two triangles are congruent.

    d. two cubes are congruent.

12. Is the following statement *true* or *false*? Justify your answer:

> *If two triangles are each congruent to the same triangle,*
> *then they are congruent to each other.*

The symbol ⊥ means "is perpendicular to."

13. ⬛ **Checkpoint** In $\triangle ABD$, $AD = BD$, $\overline{DC}$ is an altitude (that is, $\overline{DC} \perp \overline{AB}$), and F and E are midpoints. Judge each of the following statements as *true*, *false*, or *nonsensical*. Justify your answers. You may use measuring tools if you want.

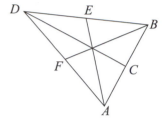

    a. $FD = DE$

    b. $\overline{FD} = \overline{DE}$

    c. $\overline{FD} = 1.5$ cm

    d. $\angle ACD = 90°$

    e. $\triangle DFB = \angle DEA$

    f. $\angle ACD$ is a right angle

    g. $\overline{FA} \cong \overline{BE}$

    h. $\overline{FA} \cong \overline{BD}$

    i. $\angle ADC = \angle BDC$

    j. $m\angle ADC = m\angle BDC$

    k. $m\angle DFB \cong m\angle DEA$

    l. $\angle DFB \cong \angle DEA$

    m. $\triangle DCA \cong \angle DCB$

    n. $\triangle DCA \cong \angle EAD$

ACTIVITY 2   Corresponding Parts

When drawing by hand, without any tools, it is hard to draw identical shapes. Sometimes you do not have tools, or do not want to take the time to measure others' drawings. So, tick marks are used when drawing identical shapes.

14. a. The picture below shows nine segments. Four pairs of segments are joined to make four angles. The segments and angles are specially marked. Explain what meaning you think the marks may have.

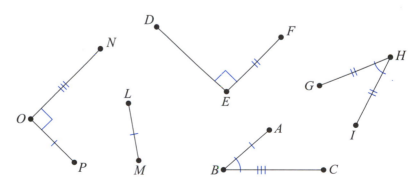

For example, if you think that $\overline{AB}$ is congruent to $\overline{LM}$, write $\overline{AB} \cong \overline{LM}$.

If segments have different *markings*, does that mean that they are not *congruent*?

b. Test the segments and angles for congruence. Then write congruence statements using the proper symbols.

c. Do your statements confirm your theory about the tick marks?

Symbols and pictures help clarify mathematical communication. For example, tick marks like the ones above indicate that line segments and angles are intended to be congruent. Even if someone's drawing isn't exact, identical tick marks on segments or angles will signify congruence.

15. The two triangles below are congruent, yet only *one* of the congruence statements is considered correct. Come up with your own reason for choosing one as *better* than all the others. Explain your reason.

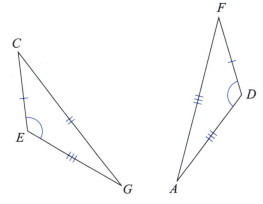

a. $\triangle DFA \cong \triangle GCE$

b. $\triangle DFA \cong \triangle EGC$

c. $\triangle DFA \cong \triangle CEG$

d. $\triangle DFA \cong \triangle ECG$

e. $\triangle DFA \cong \triangle GEC$

f. $\triangle DFA \cong \triangle CGE$

16. Even though only one of the *above* statements is correct, there are other correct congruence statements for these two triangles. How many? Write them.

17. Imagine two congruent irregular pentagons. How many pairs of corresponding parts do they have? Sketch your pentagons. Write a congruence statement.

18. Write and Reflect Explain in your own words—with pictures, too, if you like—what "corresponding parts of congruent figures are congruent" means.

19. Checkpoint Draw, label, and mark your own example of two differently oriented congruent triangles. Write the appropriate congruence statements for corresponding parts.

On Your Own

1. Define the following words or symbols.

  a. congruent

  b. $\cong$

  c. $\perp$

2. Are all equilateral triangles congruent? Explain.

3. When you read the sentence $\triangle CAT \cong \triangle DOG$, you know that the two triangles are congruent. Name all the corresponding parts.

Polyominoes are shapes that are built of squares, following one rule: When two squares are joined, their edges must fully coincide. So, the first three shapes below are polyominoes; the second three shapes are not, because the squares don't meet edge-to-edge.

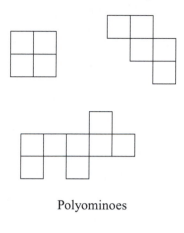

Polyominoes

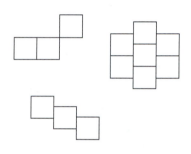

Not polyominoes

Polyominoes that are congruent are not different, even if they are drawn in a different orientation.

4. *Dominoes* How many different figures can be made from just two squares combined according to the rules?

5. *Triominoes* How many different figures can be made by connecting exactly *three* identical squares edge-to-edge and vertex-to-vertex?

6. *Tetrominoes* How many different figures can you create this way using four squares?

7. Make an eight-square figure by combining the T with some other tetromino. How many ways can you make your eight-square figure by combining the T with some *other* tetromino?

This is what the authors have called the "T tetromino":

Two eight-square shapes are "the same" if they are congruent.

8. There are at least seven triangles in the figure below. It is easy to find some pairs of triangles that are *not* congruent. Name any pairs of triangles that you think are congruent.

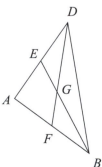

9. Look at the shapes below.

a. Decide whether the two shapes in each example "seem congruent" or are "surely not congruent."

b. List any measurements you could take, tests you could perform, or other information about the figures that would tell you if the figures are congruent.

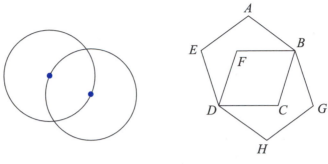

Two circles

Two pentagons:
ABCDE and *BGHDF*

Two cats

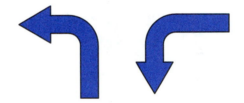

Two bent arrows

Two stars

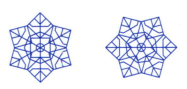

Two snowflakes

There are many other shape comparisons: Similarity *is* one. There are also many other size comparisons. One is perimeter.

10. There are many ways of comparing figures. For polygons, you have been looking at a *shape* and *size* comparison: congruence. Later, you will look at another size comparison: area. Use what you already know about area to answer the following questions.

    a. If two polygons are congruent, must they have the same area? Why or why not?

    b. If two polygons have the same area, must they be congruent? Why or why not?

11. The figure below contains three congruent triangles.

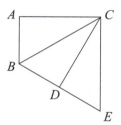

    a. Name the three congruent triangles.

    b. Copy or trace the figure. Mark congruent corresponding parts.

What is the mathematical definition of a kite?

    c. Shape *ABDC* in the figure is called a *kite*. In that kite, name the triangle that is congruent to △*ABC*.

    d. In △*BCE*, name the triangle congruent to △*ECD*.

12. The figure below is not drawn to scale. All parts marked as congruent are *intended* to be congruent. Segments that look straight, like $\overline{DH}$, are meant to be.

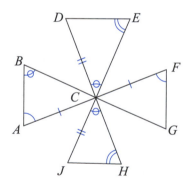

    a. If $m\angle F = 80°$, $m\angle H = 50°$, and $m\angle B = 40°$, then what are the measures of $\angle D$, $\angle E$, and $\angle A$?

    b. Redraw the figure. Draw all angles with the correct degree measure. Make all congruent segments actually congruent.

LESSON

2

Triangle Congruence

In this lesson, you will devise several tests for triangle congruence.

Why is so much attention given to triangles and their properties? One reason is that any polygon can be cut up into triangles, so properties of other shapes can often be derived by thinking about the triangles from which they are made.

This lesson will focus on deciding whether two triangles are congruent. Up to now, the methods you have used have been measurement or superimposition—seeing if one shape could fit exactly on top of the other. But these methods cannot always be used, and don't help you confirm strong statements.

It is often abbreviated as CPCTC.

How do you form the converse of a statement?

Explore and Discuss

A more powerful tool to use for congruent triangles is "corresponding parts of congruent triangles are congruent." It says that if you know that two triangles are congruent, then you also know that six pairs of parts are congruent—three pairs of sides and three pairs of angles. Maybe the *converse* holds as well. That is, maybe if the six pairs of parts are congruent, side-for-side and angle-for-angle, then the *triangles* will be congruent. In fact, this is true.

a Checking all corresponding parts is work. Could you check *fewer* than six pairs and still be sure that two triangles are congruent? For example, if you know that all three angles in one triangle are congruent to all three angles in another, can you be sure the triangles are congruent? What if three pairs of sides are congruent?

b "Information that is enough to specify one triangle exactly is also enough to ensure that two triangles are congruent." Do you believe this statement is true? Give reasons for your answer.

ACTIVITY 1

The Envelope Game

In other words, as long as the information you have could make more than one triangle, keep taking notes out of the envelope.

Your teacher will provide you an envelope containing six pieces of information about a triangle—three sidelengths and three angle measures—on separate notes. Take out one note at a time from your envelope until you have *just* enough information to build *only one* triangle with exactly those measurements. You can use any construction tools or techniques you want: compass, ruler, protractor, geometry software, paper folding, and so on.

When you are done, the triangle you have made should agree with all the information you pulled out of the envelope and all the information that's still in the envelope. So, check the remaining notes to be sure that you made the correct triangle.

1. How many pieces of information did you need? Draw your triangle. List the information you used.

Depending on the luck of the draw, you may need different numbers of pieces before you can "nail down" the triangle.

For example, would you ever need all six?

2. What is the largest number of notes you would ever have to take out of the envelope before you could construct the triangle? Give reasons for your answer.

3. a. What is the smallest number of notes you could take and still construct exactly one triangle that uses all the information?

Is one note ever enough?

b. Will *any* collection of that many notes *always* be enough?

4. **Checkpoint** One student playing the envelope game pulled out two notes: $m\angle A = 60°$ and $m\angle C = 90°$.

a. Construct two different triangles that fit this information.

b. Name one piece of information the student could pull out next so that only one triangle is possible.

ACTIVITY 2

Classifying Your Information

The six pieces of information about $\triangle ABC$ are the lengths of $\overline{AB}$, $\overline{BC}$, and $\overline{AC}$, and the measures of $\angle A$, $\angle B$, and $\angle C$.

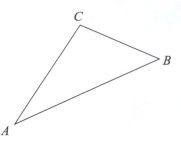

5. There are many *combinations of three* pieces of information about $\triangle ABC$. List all the possible combinations. The list is started for you.

- $m\angle A$, $m\angle B$, $m\angle C$

- AB, $m\angle A$, $m\angle B$

Your combinations could be classified in many ways. Here's a scheme that people have found useful, and a set of abbreviations—a kind of code—to help you remember the scheme.

Three angles: There's one combination of three angles.

AAA: The code uses "A" to stand for "angle."

Two angles and a side: Here you have two arrangements.

ASA: This symbol means that the side you know is between the two angles you know. For example, $m\angle A$ and $m\angle B$ and AB.

AAS: This means you know two angles and the side that is not between them. For example, $m\angle A$, $m\angle B$, and BC.

Two sides and an angle: Two arrangements again:

SAS: You know two sides and the angle between them. For example, AB, BC, and $m\angle B$.

SSA: You know two sides and the angle not between them. For example, AB, BC, and $m\angle C$. This abbreviation is used no matter which of the non-included angles you're talking about.

Three sides: Only one way:

SSS: You know all three sides.

A new envelope game: You will again get an envelope, but this time it will contain only three measurements. Different envelopes will contain different combinations of three pieces of information, so your class can compare them.

6. As in Problem **1**, the goal is to draw a triangle from the clues in your envelope. Call your set of clues "good" if it lets you make *exactly* one triangle. Call the set "bad" if it doesn't fit any triangle or if it fits more than one.

   a. List your clues. Classify them by one of the three-letter codes, SAS, AAA, and so on.

   b. Decide whether the set of clues is "good" or "bad." If it's good, show the triangle. If it's "bad," show why.

For Discussion

- Compare your clues to your classmates'. Which three-letter codes are "good" clue sets?
- List the "good" sets of measurements—the ones that uniquely determine a triangle. Call these the *Triangle Congruence Postulates*. Beside each postulate, write what it says and illustrate it with a picture.

"ASA" should be read as "angle-side-angle."

Order is important; ASA is not the same as AAS.

Again, the order will turn out to be important. SAS is not the same as SSA.

Information that is enough to specify one triangle exactly is also enough to ensure that two triangles are congruent.

The triangle congruence postulates say that if *two* triangles share certain pieces of information, for example, SSS, then they are congruent. They also say that it is possible to build *only one* triangle from these three pieces of information. Well, isn't there a contradiction? Are there *two* triangles or *only one*?

When people say "Given three sides, only *one* triangle can be built," they really mean that any *two* triangles built from these sides must be congruent.

7. Are triangles *ABC* and *ADC* congruent? If they are, which congruence postulate helped you decide?

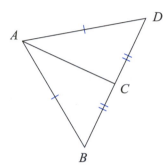

8. Are the two triangles pictured below definitely congruent or not necessarily congruent? If they are congruent, which congruence postulate helped you decide?

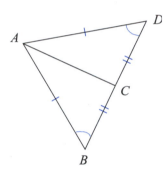

9. **Checkpoint** In two ways, show that a diagonal divides a square into two congruent triangles. Use two different congruence postulates to show that the triangles are, indeed, congruent.

ACTIVITY 3 # Counterexamples

Does an altitude to the third side of an isoceles triangle always do this? Can you write a proof for that fact right now?

Later on, you will take an in-depth look at how to use the triangle congruence postulates to prove strong statements like "in *any* triangle with two congruent sides, an altitude to the third side divides the triangle into congruent parts."

In disproving a statement, however, all you need is *one* counterexample: one case for which the statement doesn't hold. The conjecture may still ring true for a few, or even many other examples; but it is not a true statement until it is formulated in such a way that it is true in every possible case.

In the following problems, you are asked to show that the given information is not enough to define only one triangle. You need to show one counterexample, or one picture, for which the statement isn't true.

Your friend hasn't studied geometry yet. So, it isn't safe to assume your friend means "SAS" when you discover the message isn't stated clearly.

10. You and a friend are making triangular pennants for a school project. You want all the pennants the same size and shape. But you're not there when your friend calls you with the design. The message doesn't give you enough information. It states that the design has a 30° angle at the tip of the pennant, a 14-inch side, and an 8-inch side. Show why this is not enough information.

11. The picture below is a "proof without words" that two congruent corresponding sides and a congruent non-included angle (SSA) is not enough to guarantee two triangles to be congruent. Add the words, and labels if you want, that explain the proof.

Why is it convenient to draw a circle?

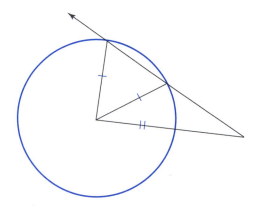

12. **Checkpoint** Find a counterexample to this statement: "AAA (all three angles have equal measure) is sufficient information to prove congruence in two triangles." Explain why it disproves the conjecture. Include a picture in your explanation.

On Your Own

1. Below are three pieces of information about $\triangle ABC$.

   • $m\angle B = 72°$

   • $m\angle C = 72°$

   • $m\angle A = 36°$

   Construct two triangles that fit the information.

2. Below are three pieces of information about $\triangle ABC$.

   • $m\angle A = 60°$

   • $AB = 8$ cm

   • $BC = 7$ cm

   Is there only one triangle you can construct? If yes, construct it. If no, construct two different triangles that both fit the information. Explain what additional piece of information would help you make just one triangle.

3. Suppose you played the envelope game and pulled out the following notes of information:

- $m\angle A = 27°$

- $m\angle B = 53°$

What are the possibilities for $m\angle C$? How do you know?

4. Write and Reflect In classifying the types of information about a triangle, this lesson listed AAS and ASA as two different possibilities. Some people say they are not really different. Explain that point of view. In other words, in what way is AAS the same as ASA?

For Problems **5–10,** do the following.

    a. Tell if the given information is enough to conclude that the triangles are congruent.

    b. If the given information is enough, describe a way to line up the vertices to make the triangles congruent. Then state the triangle congruence postulate that can be applied.

5.

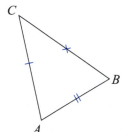
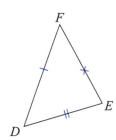

6.

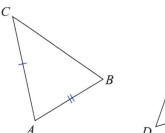
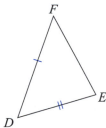

7.

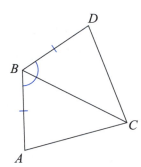

8.

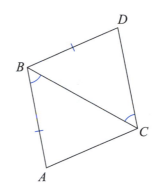

9.

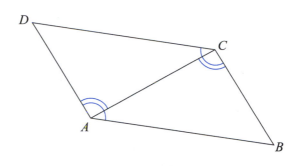

10.

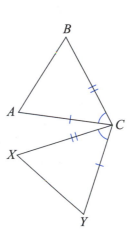

What does perpendicular bisector *mean?*

11. In the figure below, $\overline{BD}$ is the perpendicular bisector of $\overline{AC}$. With this information, two triangles can be proved congruent. Name them. Prove that they are congruent.

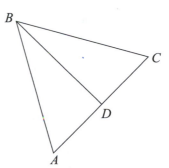

12. Does a diagonal of a rectangle divide the rectangle into two congruent triangles? What about the diagonals of parallelograms? Trapezoids? Kites? Explain your answers.

13. In a regular polygon, if all the diagonals from *one* vertex are drawn, will they divide the polygon into triangles that are all congruent to each other? Explain.

14. In a regular polygon, if *all* the diagonals are drawn, will every polygon formed have at least one other matching congruent polygon?

15. Write and Reflect If you want to prove something is true about all triangles, all circles, all numbers, and so on, thousands of examples won't prove your conjecture. Yet only one counterexample will disprove it. Explain.

Take It Further

16. In the figure at the right, $\overline{AD}$ is the perpendicular bisector of $\overline{BC}$. With this information, one pair of triangles can be proved congruent. (You did this in Problem **11.**)

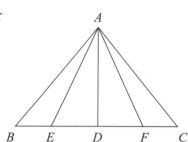

Adding just one more piece of information might make it possible to prove more triangles congruent. For each example below, decide whether or not the additional information makes that possible. If you think it does, name the new congruent triangles. State how you have proved them congruent.

a. What if $AB = AC$?

b. What if $\overline{AD}$ is the perpendicular bisector of $\overline{EF}$?

c. What if $\angle EAD$ is congruent to $\angle FAD$?

17. Imagine two parallel planes intersecting a cube. Is it possible for the cross sections to be congruent? Explain.

18. Draw a triangle on the coordinate plane. Call it $\triangle DEF$. Write the coordinates of points D, E, and F.

a. If you make a new triangle by taking points D, E, and F and adding 2 to every coordinate, will the new $\triangle DEF$ be congruent to the original $\triangle DEF$?

b. If you make another new triangle by taking the original points D, E, and F and multiplying every coordinate by 2, will the new $\triangle DEF$ be congruent to the original $\triangle DEF$?

Example: If D is the point (1, 4), adding 2 to the x- and the y-coordinates would give a new D at (3, 6).

19. At the beginning of this lesson, you read that "any polygon can be cut up into triangles." But it is not so easy to say how to do it. For example, in some polygons, one could "pick any vertex and draw all the diagonals from that vertex." But that rule would not work for the crazy polygon *ABCDEFGHJ* shown at the right. Can you find a method that will *always* work? Or is the statement, perhaps, not really true of *all* polygons?

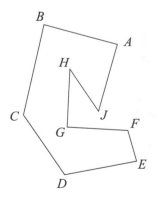

Which method makes it
easier to check for
congruence?

Cutting up, Congruently Each of the following problems asks you to cut a particular shape into some number of congruent pieces. Your teacher will provide you the shapes. First, trace the figure onto another piece of paper. Then either cut your copy into congruent pieces, or draw lines to show how such cuts could be made.

20. Use an $8\frac{1}{2}" \times 11"$ piece of paper as your rectangle. (You will need four sheets of paper.)

    a. Divide your rectangle into two congruent pieces.

    b. Divide another into three congruent pieces.

    c. Divide the third into four congruent pieces.

    d. Divide the fourth into six congruent pieces.

21. Make five copies of the hexagon.

    a. Divide one copy into two congruent pieces.

    b. Divide another into three congruent pieces.

    c. Divide the third into four congruent pieces.

    d. Divide the fourth into six congruent pieces.

    e. Divide the fifth into eight congruent pieces.

22. Make four copies of this shape:

    a. Divide one copy into two congruent pieces.

    b. Divide another into three congruent pieces.

    c. Divide the third into four congruent pieces.

    d. Divide the fourth into six congruent pieces.

23. Make two copies of the triangle.

    a. Divide one copy into two congruent pieces.

    b. Divide another into four congruent pieces.

Just to refresh your
memory, "justify your
answer" means provide
an explanation if your
answer is yes, or provide
a counterexample if your
answer is no.

    c. Could you divide *any* triangle into two congruent pieces? Four congruent pieces? Try to justify your answer.

24. **Project** Draw your own triangle. Then divide it into some number of congruent pieces. For example, you can divide the triangle into 2, 3, 4, or 5 pieces. Which numbers of congruent pieces can you *always* construct, no matter what kind of triangle you use? Which numbers require special triangles? Are there any that are impossible? Write about why you think some numbers require special triangles, and why some, if any, are impossible.

25. **Project** Draw a square. Divide it into two congruent triangles. In fact, try to divide it into 3, 4, 5, and 6 congruent triangles. For what numbers *n* is it easy to divide your square into *n* congruent triangles? For what numbers *n* is it difficult? For what numbers *n* is it impossible? Explain.

3

Warm-Ups for Proof

In this lesson, you will create convincing mathematical and nonmathematical arguments.

A **proof** in mathematics is a logical argument designed to explain a new observation in terms of facts one already understands. In a proof, you start with statements that everyone agrees with. From them, using only logic that everyone agrees with, you convince yourself, or others, that other statements *have* to follow. It must be clear that no alternate interpretations or counterexamples are possible.

What does a proof look like? That is trickier. In mathematics, not all proofs look the same because there are many different ways to make a convincing argument. A proof could include

What does hypothesis *mean in mathematics?*

- a picture or set of pictures,
- a set of equations,
- an essay,
- a series of statements accompanied by justifications,
- a construction using compass and straightedge or computer drawing tools,
- a computer program.

Explore and Discuss

You probably have found yourself in a situation where you wanted to convince other people of some fact. The fact may be a point of view that you hold, like "swimming is more fun than soccer", or it may be something more objective, like "smoking cigarettes causes health problems." In the latter case, you may use arguments that are close to mathematical *proof*.

In the second week of school, a P.E. teacher sends you a note saying that you've been absent from her third period P.E. class. Your schedule, that you have been following, says you have geometry third period and P.E. fifth period with another teacher. How would you convince the teacher that her schedule is the one that is wrong?

Mathematical Arguments

How is the word "argument" used differently in mathematics than in everyday language?

The following problems will help you get started writing mathematical arguments.

1. Divide a square into four pieces so that each piece has the same area. Write an argument that will convince a friend that the four pieces really do have the same area. Try it out and see if the argument is convincing.

2. Divide a square into five pieces so that each piece has the same area. Write an argument. Check to be sure that your argument is convincing.

3. What kind of number, odd or even, do you get if you add *any* odd number to *any* even number? Write two convincing arguments for this fact.

   a. Write one argument that would convince someone in the fourth grade.

   b. Write another argument that would be understood by someone who knows algebra.

4. Checkpoint The sum of the measures of the angles in a triangle is 180°. Use this fact to show that the sum of the measures in a quadrilateral is 360°.

1. Ruth claims that, in the picture below, $\triangle ABC \cong \triangle DBC$.

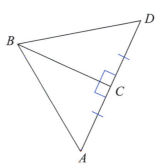

   Her argument says:

   - The triangles share side $\overline{BC}$.

   - $\angle ACB \cong \angle DCB$ because they are both right angles.

   - The picture says $\overline{AC} \cong \overline{DC}$.

   Is Ruth's argument correct? Which triangle congruence postulate should she use to conclude that $\triangle ABC \cong \triangle DBC$?

2. Ruth was given the picture below. She was asked to make a conjecture and prove it. The given information: $\angle ABC \cong \angle ACB$ and $\overline{BE} \cong \overline{CD}$.

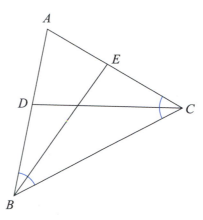

   She decided that $\triangle BDC \cong \triangle CEB$ by the following argument:
   - The triangles share side $\overline{BC}$.
   - This, with the given information, shows that $\triangle BDC \cong \triangle CEB$ by SAS.
   Critique Ruth's argument. What mistake did she make?

3. Write an argument that the measure of an external angle in a triangle is equal to the sum of the two opposite internal angles. In other words, prove that $m\angle DAB = m\angle B + m\angle C$.

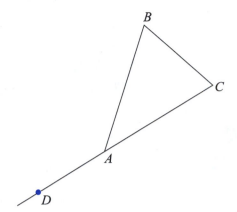

4. Prove each of the following using algebra.

a. The sum of two even numbers is even.

b. The sum of two odd numbers is even.

c. The product of an even number with any number is even.

d. **Challenge** The product of two odd numbers is odd.

LESSON 4

Writing Proofs

In this lesson, you will practice writing proofs using congruent triangles.

As you might imagine, the way a proof is expressed—the way you make your argument—can vary quite a bit. The *logic* of a proof is determined completely by mathematics and rules of reasoning, but the *look* of a proof is part of the customs and culture of the people who are its audience. For high school students, this depends to some extent on the country in which they study. Schools in China, Israel, France, or Russia sometimes teach ways to present proofs that are quite different from those taught in most American schools.

Explore and Discuss

Below you will find one simple proof and several different correct ways to write it. Read this section with two goals in mind:

- to compare the different methods of writing, and

- to pull from all of these proofs what they have in common—the essential elements of proving triangles congruent.

Given that $\overline{AB}$ and $\overline{CD}$ are parallel and that E is the midpoint of $\overline{AD}$, prove that $\triangle ABE$ is congruent to $\triangle DCE$.

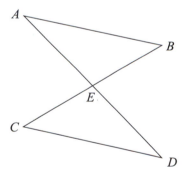

In this **paragraph style proof,** a series of statements fit together logically to establish that the two triangles are congruent. The statements are written in paragraph form.

Because $\overline{AB} \parallel \overline{CD}$, the alternate interior angles are congruent. So, $\angle ABE \cong \angle DCE$ and $\angle BAE \cong \angle CDE$. Also, because E is the midpoint of $\overline{AD}$, then $\overline{AE} \cong \overline{ED}$. Therefore, by AAS, $\triangle ABE$ and $\triangle DCE$ are congruent.

A **two-column statement-reason proof** is a set of statements written in the left column. Each statement has a reason why it must be true: that reason is written in the right column. The last statement is what is being proved. The reasons are axioms, theorems, definitions, or "givens"—statements that everyone assumes for this problem.

Statements	Reasons
1. $\overline{AB} \parallel \overline{CD}$	1. Given
2. $\angle ABE \cong \angle DCE$	2. Parallel lines form congruent alternate interior angles with a transversal.
3. $\angle BAE \cong \angle CDE$	3. Parallel lines form congruent alternate interior angles with a transversal.
4. E is the midpoint of $\overline{AD}$	4. Given
5. $\overline{AE} \cong \overline{ED}$	5. The midpoint is defined as the point that divides a segment into two congruent parts.
6. $\triangle ABE \cong \triangle DCE$	6. AAS

Some students who studied in China use an **outline style proof.** The symbol $\because$ means *because,* and is used for information that is given. The symbol $\therefore$ means *therefore.* Some important reasons are included in parentheses.

$\because$ $\overline{AB}$ is parallel to $\overline{CD}$

$\therefore$ $\angle BAE \cong \angle CDE$

$\therefore$ $\angle ABE \cong \angle DCE$ (alternate interior angles)

$\because$ E is midpoint of $\overline{AD}$

$\therefore$ $AE = ED$

$\therefore$ $\triangle ABE \cong \triangle DCE$ (AAS)

Another outline style was used by some students who studied geometry in Russia. A statement is written and justifications for it are recorded below in outline form.

Given that $\overline{AB}$ is parallel to $\overline{CD}$ and E is the midpoint of $\overline{AD}$, $\triangle ABE \cong \triangle DCE$ by the AAS postulate because

1. $\angle BAE \cong \angle CDE$ (alternate interior angles),

2. $\angle ABE \cong \angle DCE$ (alternate interior angles), and

3. $AE = ED$ (E is midpoint of $\overline{AD}$).

> **a** All four proofs show that triangle ABE is congruent to triangle DCE using the AAS triangle congruence postulate. What else do all these proofs have in common?
>
> **b** What are the advantages and disadvantages of the different styles?

ACTIVITY 1 ▶ Why Proof?

In *Explore and Discuss,* you saw four different ways to present a proof. But that might be getting ahead of the story. A more basic question is, "Why bother with proof at all?" A surprisingly complex answer comes from a mix of tradition, necessity, and culture.

Mathematicians have always been experimenters: they perform thought experiments, they build and experiment with models, and they gather data. From the data they make conjectures. But for conclusions they must rely on *deduction* and *proof*. New results come from reasoning about things that *must follow logically* from what is already known or assumed. The mixing of deduction and experiment is one of the distinguishing features of mathematical research.

Deduction and experiment are different ways to think about things. As an example of how they are different, consider the following conjecture:

If two lines intersect, the angles that aren't adjacent to each other, the *vertical angles*, have the same measure.

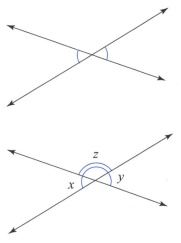

1. Set up an experiment using geometry software or paper folding to test the conjecture.

2. **a.** Write a deductive proof of the fact that vertical angles always have the same measure.

   b. What is the relation of x to z? Of y to z?

3. **Write and Reflect** Compare the methods used in Problems **1** and **2.** Which is more convincing to you? What are the advantages of each? Disadvantages?

4. Checkpoint The sum of the interior angles of an n-gon is $(n - 2) \cdot 180$.

   a. Describe an experiment to test this conjecture.

   b. Write a deductive proof.

ACTIVITY 2 ## A Beginner's Manual

Problems that ask you to prove that two triangles are congruent are often studied early in geometry not only because triangles are important, but because the *structure* of these proofs is relatively simple. In most of these problems, you can follow a straightforward plan.

- Identify the parts of the two triangles that are known to be congruent.

- Determine whether there is enough information to prove that the triangles are congruent, and which reason, SSS, SAS, ASA, or AAS, will be used.

- Organize the information and write the proof.

Many students feel like they are wandering around lost when they are first asked to write their own proofs. After some practice, though, finding a proof can have the feel of being a detective or working a good puzzle: one looks for clues, finds connections, and puts the pieces together.

Write a proof for each of the following statements. Use one of the styles for presenting proofs described earlier.

5. **Given:** $\overline{AB} \cong \overline{BC}$ and $\overline{BD} \cong \overline{BE}$

   Prove: $\triangle ABD \cong \triangle CBE$

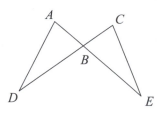

6. **Given:** $\overline{SV} \cong \overline{UT}$ and $\overline{ST} \cong \overline{UV}$

   Prove: $\triangle STV \cong \triangle UVT$

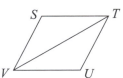

7. **Given:** *SEBW* is a square.

   Prove: $\triangle SWB \cong \triangle EBW$

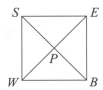

8. **Checkpoint** Below is a sketch of a proof. Study it. Then write the whole proof in one of the styles, two-column, paragraph, or outline.

   Given: $\overline{XE}$ is a median in $\triangle XMY$ and $\overline{XY} \cong \overline{XM}$.

   Prove: $\triangle XEM \cong \triangle XEY$.

   - Segment *XE* is a median, so *E* is the midpoint of $\overline{MY}$. That means $ME = YE$.

   - $XY = XM$ is given.

   - Segment *XE* is in both triangles.

   - The triangles have three pairs of congruent sides, so they are congruent.

On Your Own

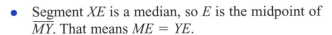

1. **Write and Reflect** One of the reasons that the deductive method became so important in mathematics is that there are many mathematical problems for which it is extremely difficult, or even impossible, to perform a complete experiment. Give an example of a mathematical problem for which deduction would be the only feasible approach.

2. Draw an isosceles triangle and the bisector of the vertex angle. Prove that the two small triangles formed are congruent.

In an isosceles triangle, the vertex angle is the angle whose sides are the congruent sides of the triangle.

3. Below is a proof that shows that any pair of lines is parallel. What's wrong with it?

Suppose $\overleftrightarrow{EF}$ and $\overleftrightarrow{GH}$ are lines. Draw transversal $\overleftrightarrow{LM}$.

∵ ∠ELM and ∠HML are alternate interior angles

∴ ∠ELM ≅ ∠HML

∴ $\overleftrightarrow{EF}$ ∥ $\overleftrightarrow{GH}$ (alternate interior angles)

4. a. What's wrong with the following proof?

Given: $\overline{AB}$ ∥ $\overline{ED}$ and $\overline{AB}$ ≅ $\overline{ED}$

Prove: △ABF ≅ △DEF

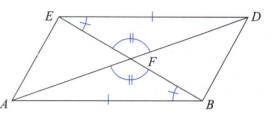

It is given that $\overline{AB}$ ∥ $\overline{ED}$, so ∠DEB ≅ ∠ABE because parallel lines make congruent alternate interior angles with a transversal. And ∠AFB ≅ ∠DFE because they are vertical angles, and vertical angles are congruent. It is also given that $\overline{AB}$ ≅ $\overline{ED}$, so △ABF ≅ △DEF by ASA.

b. If the proof is incorrect, does that mean that the statement cannot be proved? If the statement can be proved, prove it.

5. To be sure that a statement holds, mathematicians require reasoned proof. How could you convince someone that a statement like "all horses are the same color" *doesn't* hold?

What is a prime number?

6. Here is a famous conjecture about prime numbers. "Take any whole number, square it, add the original number, and add 41. The result is always prime." See the example at the right. Is the conjecture true? Justify your answer.

n	$n^2 + n + 41$
1	43
2	47
3	53
4	61
5	71

7. Use circle P below to prove that the three triangles shown are congruent.

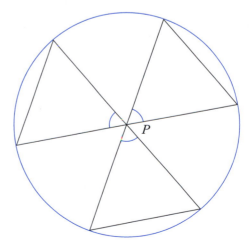

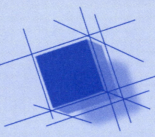

Perspective

Deduction and Experimentation

When mathematicians work on problems, they almost always use a mixture of deduction and experiment. But a tradition grew up in Greek mathematics around 300 B.C. that survives even today:

A conclusion is not accepted in mathematics unless someone can produce a deductive proof for it.

Why is such rock-solid assurance so important? Mathematics builds new ideas by fitting together older ideas in new ways. The new ideas will not stand up if their foundations—the older ideas on which they are based—are not solid.

Some people feel that one disadvantage of this tradition is that while mathematical results are presented as a beautiful flow of deduction, they are almost never *obtained* this way. The presentation hides the author's research process. You, the reader, see the results, but not how they were discovered. You don't get to see the false starts, the special cases, the reasons for taking certain steps, or the sudden insight that caused everything to fall into place. High school textbooks have also tended to follow this tradition, with the result that many show just the facts, procedures, and perhaps their proofs, but not how they were discovered or invented. Hiding the human processes that created our mathematics can make it mysterious and forbidding, and can give the impression that memory, rather than reasoning, is the key to success in mathematics. There is also a difficulty in maintaining the tradition of deductive proof. Mathematics is getting to be a huge enterprise. The number of new results published each year is quite large, and each new result is based on the validity of other research papers. No one can check all this, and there are many examples of incorrect results that are published, complete with alleged proofs that contain subtle errors.

Still, the deductive tradition for presenting mathematics has been responsible for much of the success and the appeal of the subject over the past 2000 years, and there's little evidence that this tradition will change any time soon.

Over the centuries, there have been several attempts to give deductive treatments for whole areas of mathematics. The idea is to state a few simple axioms and then to deduce from these axioms everything that is currently known about some field. Two very famous and influential works of this type are Euclid's *The Elements* (300 B.C.) and Bourbaki's *Elements of Mathematics*, a project that was begun in the middle of this century and continues today.

1. **Project** Research Ramanujan, Euclid, or Bourbaki. Write a report or make a presentation on the style of mathematics, both research and reporting, that your person or group used.

2. A mathematician who consulted with the authors of this book said that she hates TV ads because they insult people's intelligence by making claims and not offering a bit of evidence to support the claims. Write about some experience or situation in which *you* were especially bothered by people not justifying what they said.

> In mathematics, there are several important conjectures that are widely believed to be true on the basis of experimental evidence, but no one has yet been able to prove them.

> It is not certain whether the famous Euclid was a person or a pseudonym for a group of people.

> Bourbaki was a pseudonym for a group of French mathematicians that formed after World War II. Srinivasa Ramanujan (1887-1920) was an Indian mathematician who came up with results via a process that few people understand even today.

LESSON

5

Analysis and Proof, Part 1

In this lesson, you will learn some techniques for coming up with mathematical proofs of geometric facts.

How does a doctor figure out what's causing your cough? How does a mechanic figure out why your car won't start? How does a chess player decide on the next move? How does a mathematician find a proof for a conjecture? These are not easy questions to answer.

Explore and Discuss

On the humorous National Public Radio call-in program *Car Talk,* the following conversation took place between a caller and the hosts, "Click" and "Clack."

Caller: Yesterday the van did the same thing. As soon as we got above 45 miles per hour, it started shuddering again, only worse. Immediately and more dramatically.

Clack: I could've guessed that . . . You said you feel the vibration in the steering wheel?

Caller: Yes, very much in the steering wheel.

Click: Good, good, good. What color is this van? Is it one of the earth tones?

Think about a time when you had to diagnose a situation and find its underlying causes.

a Explain the situation.

b Write about what you did and how you looked for clues.

Source: Reprinted by permission of *Car Talk. Car Talk* is hosted by Tom and Ray Magliozzi, produced by WBUR Boston (90.9 FM), and distributed by National Public Radio. For further information, visit the Car Talk web site: www.cartalk.com

ACTIVITY 1 ▶ The Visual Scan

You have had some practice writing proofs. The *real* question is:

> *How do you* come up *with a proof in the first place?*

Creating a proof is sometimes called the "analysis of the proof." Analysis is always necessary before writing a proof, but sometimes it is very brief and goes on in your mind, almost without you noticing it. If you see the logic underlying the proof, then writing the proof is just a matter of expressing yourself clearly. But what if you *don't* yet see the connections? What if you see a lot of facts and clues but they don't seem to point to a solution? How, then, do you begin?

In this unit, you will learn three techniques for analyzing proofs:

- the visual scan,
- the flow chart, and
- the reverse list (you will see this in the next lesson).

You may find that one method makes the most sense to you and becomes your main tool for analysis. To be skilled at analysis, however, you will need to use all three techniques. In fact, if you are having trouble analyzing a proof using one method, it can be quite helpful to switch from one technique to another.

A *visual scan* is a careful examination of the model, usually a drawing, created for the problem. After a sketch is drawn, mark it to show all known congruent parts and any other known information. Mark additional parts that you conclude are congruent. In studying the picture, a strategy for the proof may become clear to you.

For example, given that E is the midpoint of $\overline{HF}$ in $\triangle FGH$, and that $\overline{EG}$ is perpendicular to $\overline{HF}$, prove that $\overline{HG} \cong \overline{FG}$.

This method is the simplest. To some people, it's like doing a calculation in your head; you just "see" what's supposed to happen.

1. First, sketch a figure showing the given information. Studying the marked figure at the right suggests a strategy for this proof.

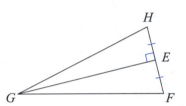

   - Show that $\triangle HEG \cong \triangle FEG$ by SAS.
   - Then conclude that $\overline{HG}$ and $\overline{FG}$ must be congruent because they are corresponding parts of congruent triangles.

   Write the full text of this proof as outlined above.

2. Prove that the base angles of an isosceles triangle are congruent. Below is an outline for your proof.
   - Sketch a figure.
   - Drop an altitude from the vertex to the base.
   - Show the two new triangles formed are congruent.
   - Conclude that the base angles of the original triangle are congruent.

In the problems above, the two segments or two angles were proved congruent by showing that they are corresponding parts of congruent triangles. The CPCTC strategy is used in many proofs. In order to show that two segments or angles are congruent:

- First, find two triangles that contain the segments. Then prove the triangles congruent.
- Conclude that the segments are congruent because they are corresponding parts of congruent triangles.

Keep this strategy in mind as you work through the rest of this lesson. See how often it appears as you continue your study of congruence.

For Discussion

The CPCTC statement serves a somewhat different purpose in proofs from statements like SSS or SAS. Discuss how each kind of statement is useful in proofs, and how their uses differ.

3. **Checkpoint** Below is a figure that someone marked for a *visual scan,* but an error was made.

  a. Find the error in the marking.

  b. What incorrect conclusion in a proof might result from this error in marking the figure?

  Given that $\overline{AC}$ and $\overline{BH}$ bisect each other and $\overline{AB} \cong \overline{CH}$, what type of quadrilateral is *ABCH* ?

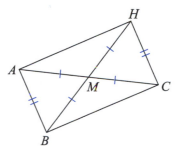

ACTIVITY 2 **The Flow Chart**

The *flow chart* is a "top-down" analysis technique. At the top of the chart, write what you know is true. Below each statement, write conclusions based on that statement. Keep working down until the desired conclusion is reached. For example:

> **Given:** Isosceles triangle *ABC* with
> $AB = BC$ and $AE = DC$
>
> **Prove:** $\angle BDE \cong \angle BED$

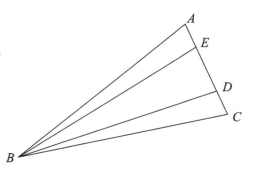

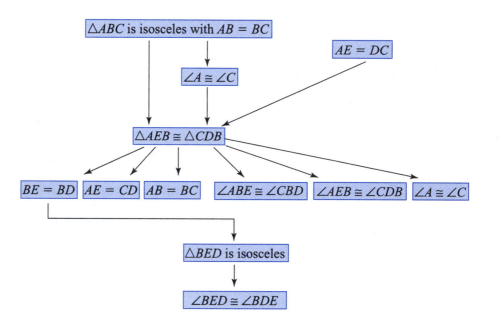

Leave out unnecessary steps and information. A readable proof will take the most direct route through the flow chart.

4. Write the proof that is described by the flow chart above.

A flow chart has several advantages. It gives you a way to start investigating and organizing what you know, even if the entire proof is still unclear to you. When it is done, it forms an outline for the steps in a written proof. Also, extra information that you might write down may be helpful in generating ideas for alternate ways to write the proof.

5. Make a flow chart for the proof below.

Given: $\angle 1 \cong \angle 2$, $\overline{XY}$ bisects $\angle MXT$

Prove: $\overline{MY} \cong \overline{TY}$

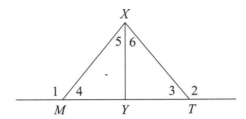

6. **Checkpoint** Below is a flow chart outlining a proof that the diagonals of a parallelogram bisect each other.

a. Copy and complete the flow chart.

b. Write the proof that the diagonals of a parallelogram bisect each other.

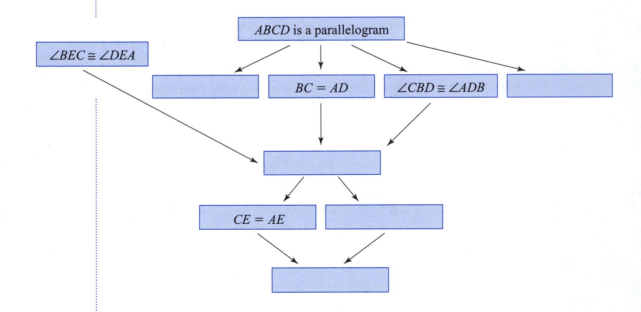

1. Use a visual scan to analyze the proof below. Copy the figure. Mark the given information on your copy. Then write an outline for the proof.

    Given: $\overline{HJ} \cong \overline{HL}$ and $\overline{JK} \cong \overline{LK}$

    Prove: $\triangle HJM \cong \triangle HLM$

    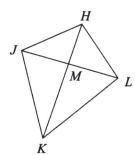

2. **Given:** $\overline{GF} \perp \overline{GH}$ and $\overline{GF'} \perp \overline{GH'}$

    Prove: $\angle F'GF \cong \angle H'GH$

    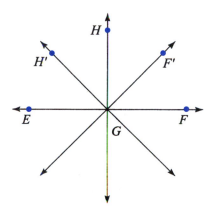

3. **Given:** Quadrilaterals *FACG* and *DABE* are squares.

    Prove: $\triangle FAB \cong \triangle CAD$

    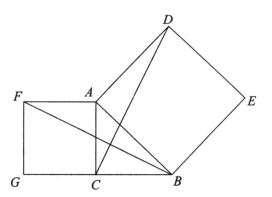

4. Given that $\triangle LJM$ and $\triangle LNK$ are equilateral, prove that $\overline{MK} \cong \overline{JN}$.

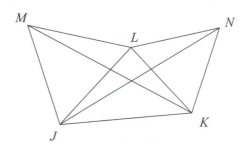

5. In isosceles triangle QRT, $QT = TR$. Segments VR and UQ are medians. Prove that they have the same length.

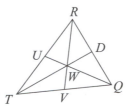

6. Point P is on the perpendicular bisector of $\overline{LM}$. Prove that $PL = PM$.

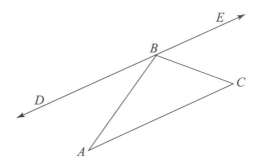

7. In the plane, it is possible to construct a parallel to any side of any triangle, as shown in the figure below, in which $\overleftrightarrow{DE} \parallel \overline{AC}$. Use this construction to write a proof that $m\angle A + m\angle C + m\angle ABC = 180°$.

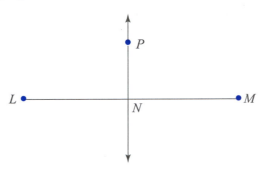

Take It Further

In Problem **7** you showed that all triangles have angle measures that sum to 180°. So you would expect that any triangle you drew, no matter how big, would have angle measures that sum to 180°.

8. Suppose that your school paved all of the northern hemisphere. On this enormous parking lot, two trucks at the North Pole start at right angles to each other and drive due south to the equator, painting lines as they go. A third line is painted along the equator to connect the two free ends, completing a triangle.

   a. What is the sum of the angles in the triangle?

   b. Explain what you found in part **a.** How can it contradict your *proof* that the sum of the angles in a triangle is 180°? What is going on?

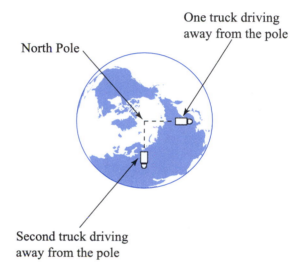

One truck driving away from the pole

North Pole

Second truck driving away from the pole

9. In the figure below, *QSTR* and *STVU* are parallelograms. Prove that $\triangle QSU \cong \triangle RTV$.

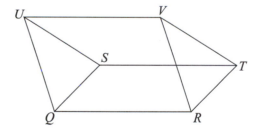

10. Quadrilateral *FGJH* is a rectangle and *L* and *K* are midpoints.
Prove that ∠*LMH* ≅ ∠*LFK*.

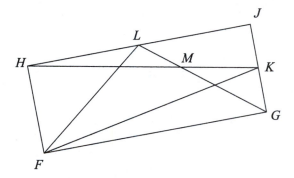

Source: This problem is from *Quantum,* September/October 1994, Volume 5, #1, p.17. Reprinted by permission of Springer-Verlag New York, Inc. Any further production is prohibited.

L E S S O N

6

Analysis and Proof, Part 2

In this lesson, you will learn more strategies for coming up with proofs, as well as explore the question, "What am I trying to prove?"

Usually statements don't come your way in the form

> **Given:** Isosceles triangle ABC with $AB = BC$
>
> **Prove:** $\angle A \cong \angle C$.

Instead, you are trying to be certain about things that you suspect are true, maybe from the results of an experiment. You say things like "vertical angles are congruent," "base angles of an isosceles triangle are congruent," or "If it rains this afternoon, then there won't be soccer practice."

When you see a statement written like these, how do you know what to prove? One way is to break the sentence into a *hypothesis* and a *conclusion*. The sentence "vertical angles are congruent" can be rewritten as: "If two angles are vertical angles, then they are congruent." The hypothesis is "two angles are vertical angles" and the conclusion is "the angles are congruent."

Here are two rules of thumb for recognizing each part of a statement:

- If the sentence is in "if-then" form, the "if" clause is the hypothesis and the "then" clause is the conclusion.

- If the sentence isn't in if-then form, the hypothesis is formed from the subject and the conclusion is formed from the predicate, generally at the conclusion of the sentence.

The hypothesis is what you are assuming. The conclusion is what you are trying to conclude from the hypothesis.

Often the word "then" is understood and not stated, as in "If it rains this afternoon, there won't be soccer practice."

Explore and Discuss

The fact that a sentence states a conclusion doesn't mean that the conclusion is *true*. Carefully, read the table below.

Sentence	Hypothesis	Conclusion
If two parallel lines are cut by a transversal, the alternate interior angles are congruent.	two parallel lines are cut by a transversal	the alternate interior angles are congruent
The base angles of an isosceles triangle are congruent.	base angles of an isosceles triangle	are congruent
Two triangles with the same area are congruent.	two triangles with the same area	are congruent
Congruent triangles have the same area.	congruent triangles	have the same area
People with large hands have large feet.	people with large hands	have large feet
Out of sight, out of mind.	something is out of sight	it is also out of mind

ACTIVITY 1

What Am I Trying to Prove?

Recall, two things are congruent if they differ only in position or orientation.

If you are given a statement to prove, then the hypothesis of the statement is what is "given," and the conclusion of the statement is what you have to prove. For example, suppose you want to prove the statement "the base angles of an isosceles triangle are congruent." The given would be that you *have* an isosceles triangle. The thing you would want to prove is that the base angles are congruent.

For each of the following problems below, draw a picture illustrating the hypothesis (the given). Then decide whether or not the statement is true. For those that are not true, give a counterexample. For those that are true, give a proof.

1. Two lines that are perpendicular to the same line are parallel to each other.

2. If a line bisects an angle in a triangle, then it bisects the opposite side.

3. Any two medians of an equilateral triangle are congruent.

4. Equilateral triangles are equiangular.

5. Equilateral quadrilaterals are equiangular.

6. If a triangle has two congruent angles, it is isosceles.

7. **Checkpoint** Consider these problems.

    a. Equiangular triangles are equilateral.

    b. Equiangular quadrilaterals are equilateral.

The Reverse List

In the visual scan and the flow chart, you start with what you know and work toward the final conclusion you are trying to prove. The *reverse list* works in the opposite direction. Start at the end and work backwards, repeatedly asking "What do I NEED?" and then "What can I USE to prove that?"

Before getting more precise, study this example:

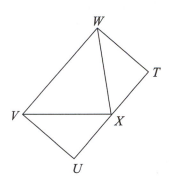

Given: Quadrilateral *TUVW* is a rectangle; *X* is the midpoint of $\overline{TU}$.

Prove: Triangle *XWV* is isosceles.

To prove that triangle *XWV* is isosceles:

- NEED: Triangle *XWV* is isosceles.
- USE: A triangle is isosceles if two sides are congruent.
- NEED: $\overline{XW} \cong \overline{XV}$
- USE: CPCTC

How do I choose which sides for this "need"? A careful sketch often helps.

How do I choose which triangles for this need?

  – NEED: congruent triangles; choose $\triangle WXT \cong \triangle VXU$
  – USE: SAS
    ∗ NEED first side: $\overline{TW} \cong \overline{UV}$
    ∗ USE: Opposite sides of rectangle are congruent.
      • NEED: *TUVW* is a rectangle.
      • USE: Given
    ∗ NEED angle: $\angle T \cong \angle U$
    ∗ USE: All the angles of a rectangle are right angles, so they are congruent.
      • NEED: *TUVW* is a rectangle.
      • USE: Given
    ∗ NEED second side: $\overline{TX} \cong \overline{UX}$
    ∗ USE: Midpoint divides segment into congruent parts.
      • NEED: *X* is the midpoint of $\overline{TU}$.
      • USE: Given

8. A complete analysis outlines the proof for you in reverse order. Write the proof for the example given above.

9. **a.** Create a reverse list for this problem:

   If P is any point on the perpendicular bisector of $\overline{AB}$, then $\triangle APB$ is isosceles.

   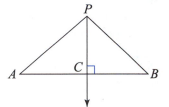

   b. Write the proof using the reverse list you wrote in part **a.**

How do you decide what statements to put in the USE spots? There is no foolproof method, but there *is* a straightforward way to narrow down the number of statements to try.

What you do is look in your notes, the unit, and your memory, and find all the previously-established results that have what you NEED as a conclusion. For example, one of the NEED statements in the above analysis is $\overline{TX} \cong \overline{UX}$. You should say to yourself, "I'm looking for a previously-established result that has congruent segments as a conclusion."

Here are a few:

- An isosceles triangle has two congruent sides.

- Corresponding parts (sides) of congruent triangles are congruent.

- The midpoint of a segment divides it into two congruent segments.

Now that you have the narrowed list, you take each statement, one by one, and decide if you really can use it. How? Take its *hypothesis* and turn it into a NEED. See if you can establish *that*. In this case, you'd go down the list and ask:

- Can I get $\overline{TX}$ and $\overline{UX}$ as sides of an isosceles triangle? No. They are not even sides of the same triangle.

- Can I get $\overline{TX}$ and $\overline{UX}$ as corresponding sides of congruent triangles? The only ones that would work are $\triangle XTW$ and $\triangle XUV$. But the reason I want $\overline{TX} \cong \overline{UX}$ in the first place is so that I can *prove* these triangles congruent.

- Can I get $\overline{TX}$ and $\overline{UX}$ as pieces of a segment that is divided by a midpoint? Yes, I'm given that X is the midpoint of $\overline{TU}$.

When you are doing this kind of analysis, here are some points to remember:

- "Given" information can be used anywhere you want as a USE.

- CPCTC can be a USE for both congruent segments and congruent angles.

- So far, you know of only four ways to get congruent triangles: SSS, SAS, ASA, and AAS.

- The *reverse list* analysis almost always works, but before you hit on the right path, you may go down some dead ends.

10. Below is a *reverse list* analysis that has led to a dead end. Study it. Then finish the analysis by choosing a different path that works better.

Given: Quadrilateral *TUZW* is a parallelogram.

Prove: $\triangle TXB \cong \triangle ZYB$

NEED: $\triangle TXB \cong \triangle ZYB$

USE: SSS

- NEED: side $\overline{TB} \cong \overline{ZB}$

- USE: Diagonals of a parallelogram bisect each other.

- NEED: side $\overline{TX} \cong \overline{ZY}$

- USE: ???

Analyzing proofs can be like driving without a roadmap. It is easy to take wrong turns, or find that the path you're on doesn't go anywhere.

Dead end.

11. Checkpoint In the figure at the right, $\overline{AD}$ is the perpendicular bisector of $\overline{BC}$. With this given information, one pair of triangles can be proven congruent.

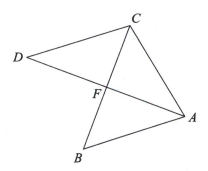

a. Name the pair.

b. Prove that they are congruent.

For Problems **1–7**, prove each of the following statements. Practice using reverse lists.

1. Opposite angles of a parallelogram are congruent.

2. Opposite sides of a parallelogram are congruent.

3. The diagonals of a parallelogram bisect each other.

4. If a triangle is isosceles, then the medians drawn to its legs are congruent.

5. If a triangle is isosceles, then the angle bisectors of the base angles are congruent.

6. If two altitudes of a triangle are congruent, then the triangle is isosceles.

7. If a triangle is isosceles, then the altitudes drawn to its legs are congruent.

8. **Write and Reflect** Which of the three analysis techniques, visual scan, flow chart, or reverse list, do you like the best? The least? Explain why.

9. **Write and Reflect** Flow charts are used in computer programming, debates, and many other fields as a way to organize information. Find an example of a flow chart. Write an explanation of its purpose and contents.

Use reverse lists to prove each of the following statements.

10. If two medians of a triangle are congruent, then the triangle is isosceles.

11. If two angle bisectors of a triangle are congruent, then the triangle is isosceles.

7

Investigations and Demonstrations

In this lesson, you will combine experimentation and proof to help you refine your understanding of congruent triangles.

Explore and Discuss

Below are some general guidelines to follow as you work on any of the investigations.

- *Explore* the problem. Use hand or computer drawings to help you under-stand what the problem statement means.

- *Explain* what you observe. If you can, justify what you say with a proof. If you can't do a complete proof, describe what you did figure out, and say what is missing from your proof.

- *Summarize* your work. Include drawings, conjectures that you made, a list of important vocabulary words, theorems, rules or ideas that came up in the investigation, and questions that require further exploration.

ACTIVITY 1 ▶ ## Perpendicular Bisectors

To get the converse, switch the "if" (hypothesis) and "then" (conclusion) parts of the theorem:
"If a point is equidistant . . . then the point is on the . . . ."

1. Verify and prove this theorem: "Any point on the perpendicular bisector of a line segment is equidistant from the endpoints of the line segment."

2. You have proved the theorem: "If a point is on the perpendicular bisector of a segment, then the point is equidistant from the segment's endpoints." Can you prove the theorem's converse?

3. **Checkpoint** In the figure below, $\overleftrightarrow{ST}$ is the perpendicular bisector of $\overline{RQ}$. Prove that $\angle SRT \cong \angle SQT$.

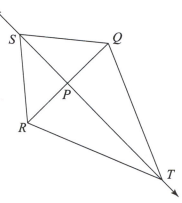

Angles and Sides in Triangles

4. The sizes of the angles in a triangle are related to the relative sizes of the sides. For example, if a triangle has two sides congruent, it must also have two congruent angles.

   a. Prove this fact.

If you have written this proof before, look up your old proof.

   b. Strengthen the conclusion. You can specify *which* angles must be congruent.

   c. What if a triangle is scalene? Could it have two angles the same size? Develop a conjecture about how to locate the largest, smallest, and "middle sized" angles in a triangle. How could you prove your conjecture?

5. **Checkpoint** The SAS postulate implies that if two sides and an included angle of one triangle are congruent to two sides and the included angle of the other triangle, then the third sides are also congruent.

Why is this result sometimes called the "hinge theorem?"

   Suppose, instead, that two sides of one triangle are congruent to two sides of another, but the included angle in the first triangle is larger than the included angle in the second. Then what can you conclude about the third sides?

Isosceles Triangle Proofs

Below are four statements:

1. Triangle *ABC* is isosceles (with base $\overline{AB}$).

2. Segment *CD* is a median.

3. Segment *CD* is an altitude.

4. Segment *CD* bisects ∠*ACB*.

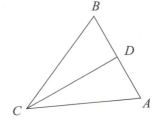

6. Show that if any two of the statements are given, the other two statements can be proved. For example:

   Given: statements 1 and 2

   Prove: statements 3 and 4

   Continue writing statements and proofs until you have verified that any two will prove the remaining two. How many theorems do you have? Write up your work for the investigation. Organize your sketches, notes, questions, ideas, and proofs.

7. **Checkpoint** The investigation lists four statements. Invent *one* statement, about △*ABC* or about △*ADC* and △*BDC*, that guarantees the four statements in the investigation are true.

1. Describe the set of triangles which have

 a. one side (but no other) whose perpendicular bisector passes through the opposite vertex.

 b. two (but not three) sides whose perpendicular bisectors pass through the opposite vertex.

 c. three sides whose perpendicular bisectors pass through the opposite vertex.

 It may help to construct examples of triangles in which the perpendicular bisector of at least one side does pass through the opposite vertex, and other triangles in which it does not. Explain what you find.

2. Take a cup, glass, or can and trace out a circle on paper. Explain how to find the center of the circle.

3. Give an algorithm for constructing a circle that will pass through the vertices of a given triangle.

What does algorithm mean?

4. Draw several triangles. In each, construct the perpendicular bisectors of all three sides. Notice that all three perpendicular bisectors meet at a point. Does this always happen? Provide a proof or a counterexample.

If you use geometry software, you can make one construction and just drag it around.

5. A circular saw blade fell and broke. All you can find of it is a piece that looks like the one at the right. Explain how to find the diameter of the blade so you can buy a new one.

6. Show that if two *right* triangles have the hypotenuse and a leg of one congruent to the hypotenuse and a leg of the other, the triangles are congruent.

This test is sometimes called Hypotenuse-Leg and is abbreviated HL.

7. An isosceles trapezoid has congruent base angles. Prove that the legs of an isosceles trapezoid are congruent. (Hint: The lines in the picture may help.)

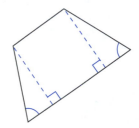

8. Prove that the diagonals of an isosceles trapezoid are congruent.

9. Carefully locate several points that are equidistant from the *sides* of an angle. What would you get if you were able to locate *all* the points equidistant from the sides of an angle? Prove what you say.

10. Draw a triangle. Figure out how to *inscribe* a circle in it. Write an explanation of how you did it.

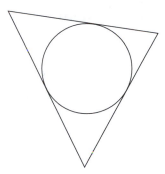

11. Below are six statements about quadrilateral *WPDK*.

- $WP = KD$
- $WK = PD$
- $ED = EW$
- $KE = EP$
- $\overline{WP} \parallel \overline{KD}$
- $\overline{WK} \parallel \overline{PD}$

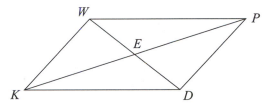

a. Find two statements that give sufficient information about *WPDK* to allow you to prove the rest.

b. Are *any* two statements sufficient to prove the rest?

c. Invent a seventh statement about *WPDK* that would make the other six statements true.

For Problems **12–14,** do the following.

a. Make a careful drawing of the figure that is described.

b. Use your drawing and your own words to explain what the statement says about the figure.

c. Write what you know about the figure. Make conjectures about things you think might be true, and explain why.

d. Try to prove the statement using any style of proof you like.

12. From □*ABCD*, extend $\overline{AB}$ through *B* to point *E*, extend $\overline{BC}$ through *C* to *F*, extend $\overline{CD}$ through *D* to *G*, and extend $\overline{DA}$ through *A* to point *H*, so that $BE = CF = DG = AH$. Quadrilateral *EFGH* is a square.

13. If the diagonals of a quadrilateral are congruent and also one pair of opposite sides are congruent, then at least one of the triangles into which the quadrilateral is divided by the diagonals is isosceles.

14. On a given isosceles triangle, pick any point along the base. Draw lines parallel to the congruent sides, forming a parallelogram. The perimeter of the parallelogram is fixed, regardless of which point is picked along the base.

This paper-folding construction starts with a rectangular sheet of paper. Fold *A* onto *B* and crease along the dotted line.

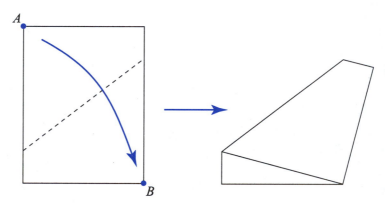

Then fold *C* onto *D* and crease along the dotted line.

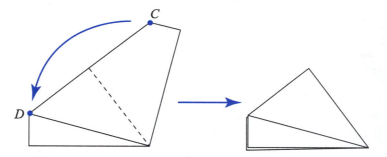

Unfold the paper. The creases should look like this:

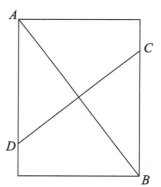

15. Prove that any point on $\overline{CD}$ is the same distance from point *A* as from point *B*.

There is no "SSA" triangle postulate because two sides and a non-included angle of one triangle can be congruent to two sides and a non-included angle of the other *without* the two triangles being congruent.

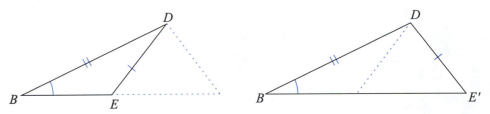

If this is so, this would be a generalization of HL. Why?

16. Could there be an "SS**A**" postulate? What if you knew not just the non-included angle's measure but also that it was the *largest* angle in each triangle? That is, what if two sides and the largest non-included angle of one triangle were congruent to the corresponding two sides and largest angle of the other? Then, can you conclude that the triangles are congruent?

17. Is there an "SS**a**" postulate? That is, what if you also knew that the non-included angle was the *smallest* angle in the two triangles? Could you then conclude that the triangles are congruent?

Congruence in Quadrilaterals and Beyond

In this lesson, you will explore possible congruence tests for quadrilaterals and other figures.

Chances are you know a great deal about quadrilaterals. From the earliest elementary grades, you have constructed, measured, cut, and folded squares and rectangles. Use this practical experience now as you begin a more formal study of quadrilaterals. While the focus in this lesson is on congruence, all of the properties of these shapes are important in understanding what makes each one unique and what factors determine their congruence.

Explore and Discuss

Even if triangles are the simplest polygon, quadrilaterals may be the easiest to find in the real world. Quadrilaterals are everywhere: in street maps, window panes, books, packaging, and so on.

a What kinds of quadrilaterals appear regularly in the world around you?

b Make a list of several "real-life" examples of each type of quadrilateral you listed in part **a.**

c Find an example of each in your school.

ACTIVITY 1 ## Families of Quadrilaterals

For Problems **1–10,** construct an accurate drawing for each figure described below. As you draw each figure, do the following.

a. Review the definition of that type of quadrilateral.

b. Study the figure. List additional properties that appear to be true.

c. Compare your drawing to those made by other students. Decide whether that set of directions describes a unique figure or many possible correct figures.

1. a **parallelogram** with one pair of sides measuring 2" and one pair of sides measuring 3"

2. a parallelogram with one pair of opposite sides measuring 3" and one angle measuring 45°

3. a **rectangle** with a 3-inch diagonal and a 1-inch side

4. a rectangle with diagonals meeting at a 120° angle and a side 2 inches long

5. a **rhombus** with a 4-inch side and a 2-inch diagonal

6. a rhombus with diagonals measuring 3" and 2"

7. a **kite** with a 2-inch and a 3-inch side, and the angle between them 120°

8. a **trapezoid** with two right angles, legs measuring 1" and 1.5", and one base of 2"

9. a parallelogram with a side 2", the length of a diagonal 5", and the altitude to the given side 3"

10. a **square** with a diagonal measuring 2 inches

11. For each type of quadrilateral, list the special properties it *appears* to have. For each property, find a logical explanation or proof to show why it *must* be true for *all* quadrilaterals of that type. For example, if you noted that opposite sides are the same length in the parallelogram, prove that this will be true for all parallelograms.

12. **Write and Reflect** Make a map, flow chart, outline, family tree, or other system to classify all the quadrilaterals. Show how they are related and what features they share. If you make a map, think about which quadrilaterals belong in each region and how the regions are connected. Use both logic and imagination in designing your diagram.

13. Checkpoint Consider these problems.

    a. Construct a rhombus with each side measuring 5 centimeters and one 90° angle.

    b. Write another description, using a different name for the quadrilateral, for the figure you just constructed.

    c. Construct a rectangle with diagonals that are perpendicular and measure 4 centimeters each.

    d. Write another description, using a different name for the quadrilateral, for the figure you just constructed.

ACTIVITY 2 ▶ # To Be Sure or Not To Be Sure . . .

A carpenter accurately cuts four boards to frame a door: two sides of 80" (on the edge that fits against the door), and a top and bottom of 30" (again at the edge by the door). Using a carpenter's square, one side piece is set at a right angle to the floor piece. This guarantees that the frame has one right angle and opposite sides of equal length. Is that enough to ensure that the *other* three corners are right angles, and, therefore, that it will fit a rectangular door?

Judging the shape of a quadrilateral from incomplete information is not always easy, but your intuition is often correct. In each problem below, decide whether or not the information given is sufficient to ensure the shape of the figure. Write a convincing argument. You may answer from intuition, but then hold yourself to more rigorous standards of logical argument in your explanations.

14. Complete the problem of the carpenter and the door. Is the information given sufficient to determine that the frame is rectangular?

15. Is a quadrilateral with congruent opposite sides guaranteed to be a parallelogram?

16. If a quadrilateral's diagonals bisect each other, what shape must it be?

17. If a quadrilateral has one pair of opposite sides parallel and the same length, must the other pair of sides also be parallel?

18. Checkpoint A quadrilateral with four congruent sides is a rhombus, by definition. Find some other set of information that would be enough to prove that a quadrilateral is a rhombus.

ACTIVITY 3 **Applications of Analysis and Proof**

Quadrilaterals are more complicated than triangles. They have diagonals, more sides, and more angles. Greater complexity means greater difficulty in proofs and problems. As you work, think of conjecture, analysis, and proof as three aspects of the same problem-solving process. But remember, they won't always occur in the same order. Sometimes the conjecture will appear first, and you will try to prove it. Sometimes an analysis for the proof of one conjecture will suggest a new conjecture.

Actually, for some rectangles, the angle bisectors don't create a quadrilateral! Can you discover which ones?

19. In a triangle, the angle bisectors all intersect at one point. In a rectangle, the angle bisectors intersect to form a quadrilateral. What type of quadrilateral? Support your conjecture.

20. What type of figure is formed by the angle bisectors of other types of quadrilaterals? Write up your conjectures, including sketches and justifications.

21. What type of quadrilateral is formed by connecting the midpoints of a rectangle's sides? Prove your conjecture.

22. In rhombus *CDEF*, $\overline{CH}$ and $\overline{JE}$ are altitudes.

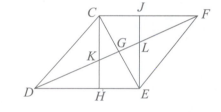

   a. What type of quadrilateral is *CJEH*? Prove it.

   b. Prove that $\triangle GLE \cong \triangle GKC$.

   c. What other triangles can be proven congruent?

23. The sides of parallelogram *MNPQ* have been trisected and four of those points of trisection have been connected to form quadrilateral *STWX*.

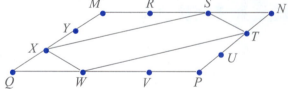

   a. List some things that can be proven about the figure described.

   b. What kind of quadrilateral is *STWX*? Prove your conjecture.

   c. Which of your results could you still prove if *MNPQ* were not known to be a parallelogram?

24. Checkpoint Two opposite vertices of a square, *B* and *D*, are connected to the midpoints of the opposite sides.

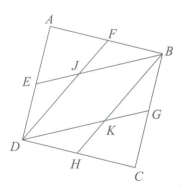

   a. Name all the quadrilaterals that are formed.

   b. Conjecture about their shapes. Prove what you say.

Proving Congruence for Quadrilaterals

Each of the recent problems has focussed on the properties of *one* quadrilateral. In working the problems, you probably needed to use congruent triangles more than once as a justification or explanation. How does one prove *two* quadrilaterals are congruent to each other? What determines congruence for two quadrilaterals?

The basic definition of congruence for polygons still applies. Two quadrilaterals are congruent if all corresponding sides and angles are congruent.

Quadrilaterals' "parts" include diagonals as well as sides and angles.

For Discussion

For triangles, the congruence postulates are SSS, SAS, AAS and ASA. Information about three parts can totally determine the shape of a triangle and is therefore the basis of the congruence postulates. How many parts must be known to determine fully the shape of a quadrilateral?

25. For polygons with more than three sides, it's more difficult to be sure that the sides and angles are *corresponding*. The figure at the right shows one order in which segments of lengths 5, 6, 7, and 8 can be arranged to form a quadrilateral. How many other orders are there?

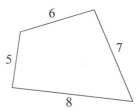

26. Is there an SSSS test for quadrilaterals? In other words, is there a quadrilateral congruence test equivalent to SSS? That is, if two quadrilaterals have congruent corresponding sides, are they congruent?

27. Is SAS a congruence test for quadrilaterals? That is, will knowledge about two sides and the included angle fully determine the shape and size of a quadrilateral? Develop a quadrilateral version of SAS: A congruence test for quadrilaterals that uses sides and included angles.

28. Is ASA a congruence test for quadrilaterals? Develop a quadrilateral version of ASA: A congruence test that uses angles and included sides.

29. The congruence tests you have been working on are for all quadrilaterals. Develop a congruence test that will work specifically for rectangles. The goal is to find the minimum information that will completely determine the shape of a rectangle, or will show that two rectangles are exactly the same size and shape.

30. Each congruence rule below is incomplete. Add as little information as you can to make each a true test for congruence for that type of quadrilateral.

   a. Two parallelograms are congruent if they have congruent diagonals and . . .

   b. Two rhombuses are congruent if they have one pair of corresponding sides congruent and . . .

   c. Two trapezoids are congruent if they have the same base angles and height (height is measured between the parallel sides) and . . .

How many angles do you need to know in a trapezoid to know all the angles?

31. **Checkpoint** Below are two quadrilaterals. Decide whether or not each set of given information is sufficient to prove that the two quadrilaterals are congruent. Justify your decision. (Note: The figures are NOT drawn to scale.)

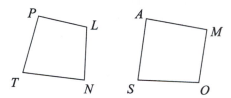

a. Is it correct to conclude that $PLNT \cong AMOS$ if

- $PT = AS$,
- $LN = MO$,
- $\overline{PT} \parallel \overline{LN}$, and
- $\overline{AS} \parallel \overline{MO}$?

b. Is it correct to conclude that $PLNT \cong AMOS$ if

- $\angle P$ and $\angle A$ have the same measure,
- $\angle N$ and $\angle O$ have the same measure, and
- $TN = SO$?

c. Is it correct to conclude that $PLNT \cong AMOS$ if

- $PLNT$ is equilateral,
- $AMOS$ is equilateral, and
- $PT = AS$?

ACTIVITY 5 ▶ **Congruence in Three Dimensions**

Extend your ideas about congruence from two to three dimensions. Try to answer each question just by visualizing and creating images in your mind. After you have made a mental picture, make a model or sketch to illustrate your thinking.

32. Write and Reflect What does it mean for two 3-dimensional figures to be congruent?

33. Are two spheres that have the same volume congruent?

a. What measurements are needed to guarantee congruence in two spheres?

b. Do congruent spheres have the same surface area?

34. Imagine two rectangular solids that are 12 inches tall and 5 inches wide. Are they congruent?

35. Imagine two noncongruent cylinders that have the same height. Name three measurements that could be taken for the cylinders that would be different. Are there any measurements, besides height, that could be the same, even if the cylinders are not congruent?

36. A tetrahedron has four triangular faces. Imagine a tetrahedron whose faces are all equilateral.

   a. How many edges does it have?

   b. How many different edge *lengths* are there?

   c. Picture all of the angles that could be measured for this solid. How many different angle sizes are there?

37. Define a box as a 3-dimensional object with six faces, all of them quadrilaterals but not necessarily rectangles or even parallelograms. Design a way to test two boxes for congruence without measuring any of the angles.

38. Following are four sets of unfolded solid objects. Recall that these are called *nets*. For each set, imagine how they could be refolded. Decide if the objects in each set would fold to form congruent solids.

   a. Set A

   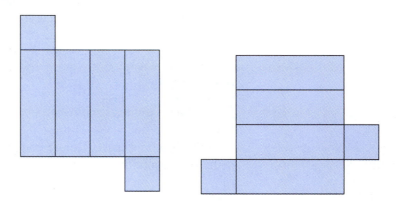

   b. Set B

   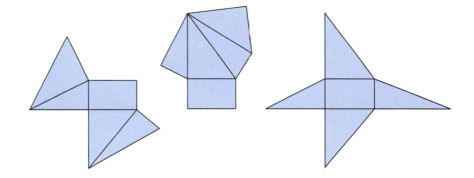

   c. Set C

   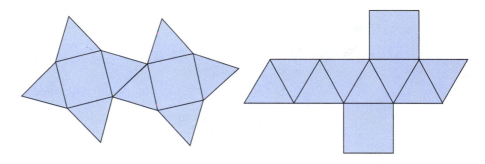

d. Set D

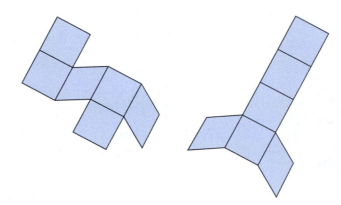

39. Draw two different nets that will fold to make congruent triangular prisms.

40. It is possible to create two noncongruent polygons from the same set of (more than three) edges. Is it possible to create two noncongruent solids from the same set of faces? Try to picture these in your mind.

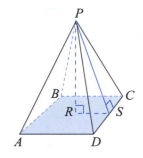

A square right pyramid:
***ABCD** is the square
base, PR is the height,
PS is the slant height.*

41. **Checkpoint** Some of these statements contain too much information. Rewrite each statement leaving out anything that is extra.

a. Two cubes are congruent if they have the same surface area and height.

b. Two rectangular solids are congruent if they have the same volume, length, width, and height.

c. Two square right pyramids are congruent if they have the same slant height, height, and base area.

d. Two cylinders are congruent if they have the same height, circumference, and surface area.

e. Two regular tetrahedra are congruent if they have the same height and volume.

1. Answer each question. Write a detailed explanation with your answer.

a. Is every square a rectangle?   **d.** Is every rhombus a square?

b. Is every square a rhombus?   **e.** Are any kites square?

c. Is every rectangle a square?

2. **a.** Decide which of the following quadrilaterals will make each of the Venn diagrams complete: kite, trapezoid, isosceles trapezoid, parallelogram, rectangle, rhombus, square.

i.   **ii.**   **iii.**

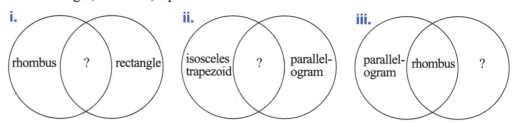

b. Copy or trace the family tree shown below. Place the names of the seven quadrilaterals in the appropriate places.

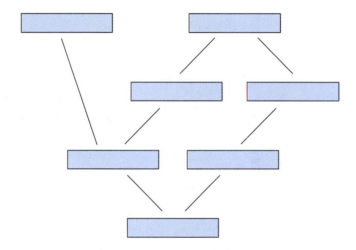

3. The figure below shows parallelograms *MNPQ* and *MPRQ*.

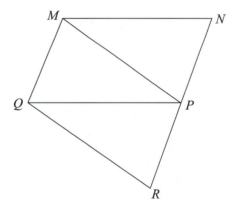

a. Are the two parallelograms congruent? Explain.

b. What congruent triangles are there in the figure? Justify each choice.

c. As drawn, $\overline{NR}$ looks straight. From the information you have, *MNPQ* and *MPRQ* are both parallelograms. Is it possible to prove that *N*, *P*, and *R* are collinear? Explain why or why not.

d. If ∠*N* measures 70°, what other angle measures can be determined?

e. If *MN* = 4.0 cm and *PR* = 2.5 cm, what is the perimeter of quadrilateral *MNRQ*?

4. In quadrilateral *ABCD*, *AF = FE = EG = GC* and *JE = EH*. Also, $\overline{AC}$ bisects ∠*DAB* and ∠*BCD*. What conjectures can you make about the characteristics of *ABCD* and *FHGJ*? Write proofs to support your conjectures.

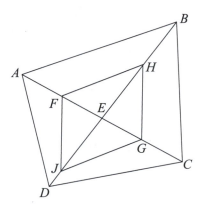

5. Imagine two empty boxes. Both are 10 inches tall and both contain exactly the same volume of air. Are the two boxes congruent?

6. Are all cubes congruent? What is the minimum number of measurements required to insure congruence for two cubes?

7. Design nets for two noncongruent cubes.

8. In the figure at the right, a plane intersects a cube at opposite vertices.

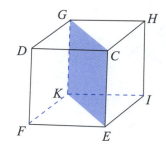

   a. If the plane cuts the cube into two solids in this way, will the two solids be congruent? Justify your answer.

   b. There is more than one way for a plane to intersect a cube only at midpoints of its edges. Sketch the different ways. For which cuts, if any, are the two parts congruent solids?

   c. If a plane divides a cube into two congruent solids, will these solids have equal volumes? Equal surface areas?

Some solids, like the cube, are made from a set of plane surfaces called *faces*. Other solids, like cylinders and spheres have curved surfaces. Of the solids that are built of curved surfaces, some can be built from plane figures, suitably bent. The plane figures for such a solid or surface are called its *development*. Below, for example, is the development for a cylinder.

9. Draw a net for a rectangular pyramid. Use tick marks to indicate all of the segments that *must* be congruent for the net to fold up into the pyramid.

10. In the development of a cylinder, indicate what measurements must be equal for the development to "work."

11. Draw two different developments that will make congruent cylinders.

12. Some solids built of curved surfaces *cannot* be developed from plane figures. For each of the following, either show the plane figures from which it can be built, or explain why you believe it cannot be built from plane figures.

   a. a sphere

   b. a cone

   c. the bell (horn part) of a trumpet

   d. the body of an acoustic nonelectric guitar

   e. a football

   f. a soup bowl

   g. a playground slide

Take It Further _____

13. If you know the lengths of the sides of a particular polygon, how many diagonals' lengths do you need to know to be able to construct a congruent copy of the polygon? Develop a congruence rule for quadrilaterals and other polygons that uses sides and diagonals, but no angles. Be specific about *how many* and *which* diagonals must be known.

Unit 2 Review

1. Define congruence.

2. Describe one way to decide if two figures are congruent if you don't know anything about them.

3. **a.** Name the triangle congruence postulates.

   b. Explain what they mean.

   c. Draw a picture to illustrate each postulate.

4. For the following pairs of triangles, decide whether or not they are congruent. If they are, prove it using the triangle congruence postulates.

   a.

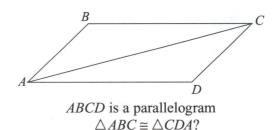

   ABCD is a parallelogram
   $\triangle ABC \cong \triangle CDA$?

   b.

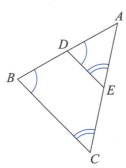

   $\triangle ABC \cong \triangle ADE$?

   c.

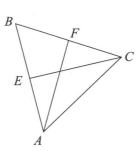

   $\overline{AB} \cong \overline{BC}$
   E and *F* are midpoints
   $\triangle EAC \cong \triangle FCA$?

d.

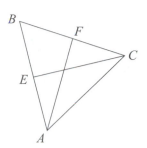

$$\overline{AB} \cong \overline{AC}$$
E and F are midpoints
$$\triangle EAC \cong \triangle FCA?$$

e.

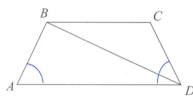

$$\triangle ABD \cong \triangle CBD?$$

5. Recall the definition of an equilateral triangle.

    a. Experiment to decide if the three altitudes of an equilateral triangle are congruent. Describe your experiment. Include pictures with your description.

    b. Prove that the three altitudes of an equilateral triangle are congruent.

    c. Experiment to decide whether or not the three altitudes of a nonequilateral triangle are congruent. If they are not, make another conjecture about the heights.

    d. Can you prove your conjecture?

6. Find the values for x, y, and z.

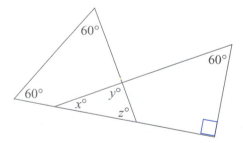

7. Describe at least one way to test that two quadrilaterals are congruent.

UNIT

3

The Cutting Edge
Investigations in Dissections and Area

Cut and Rearrange

In this lesson, you will find ways to dissect a parallelogram into a rectangle, a triangle into a rectangle, and a trapezoid into a rectangle.

You will learn to change one shape into another by cutting the first shape into parts and rearranging the parts to form the second shape. Cutting and rearranging a given shape will not convert it into just any other imaginable shape. For example, you would not expect it to work if the resulting shape had more or less total "stuff" in it! Chih-Han Sah, a mathematician who specializes in dissections of this sort, calls two figures that can be cut into each other *scissors-congruent*.

Explore and Discuss

Geometric language helps you give clear descriptions of how you move pieces to change one shape to another. Imagine you have two isosceles right triangles to build shapes. You can use three basic moves to change one shape to another.

Slide or **Translation** Below, a parallelogram becomes a square as one of the small triangles slides parallel to the base of the parallelogram.

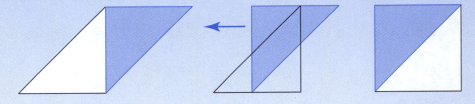

Flip or **Reflection** Below, the parallelogram becomes a triangle when one of its halves, a triangle, is flipped.

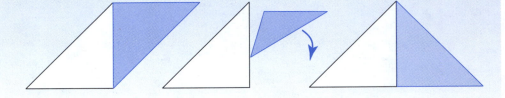

Rotation A piece can be rotated around a point. Below, a square turns into a triangle when one of the small triangles is rotated around a vertex.

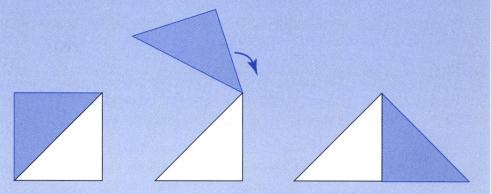

For each of the pairs of figures below, do the following.

a Figure out how to change the first shape into the second shape by reflecting, rotating, or translating one or two of the pieces.

b Write a description, telling which piece or pieces you moved and how you moved them.

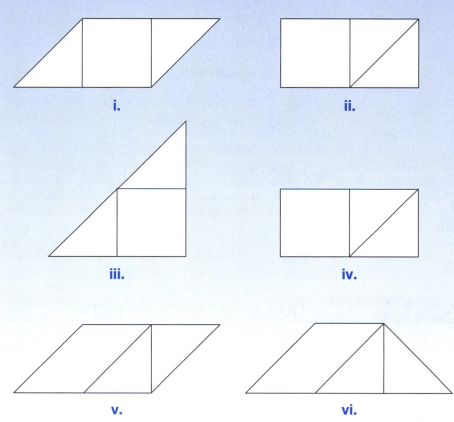

i.

ii.

iii.

iv.

v.

vi.

Some Cutting Problems

Each of the following problems can be solved in more than one way. Work with others to find a few solutions. Then pick one that you like best. *Save your written work. You will need it later.*

For Problems **1–5**, start with the shape described. Find a way to cut that shape into pieces that can be rearranged to form a *scissors-congruent* second shape described in the problem. Write a description of the cuts you made, and how you moved the pieces.

1. Start with a parallelogram. Find a way to cut your parallelogram into pieces that can be rearranged to form a rectangle.

2. Start with a right triangle. Dissect the triangle so that the pieces can be rearranged to form a rectangle.

A scalene triangle has no two sides the same length.

3. Start with a scalene triangle (not right). Cut it into pieces that will form a parallelogram.

4. Start with a scalene triangle (not right). Cut it into pieces that will form a rectangle.

5. Start with a trapezoid. Dissect the trapezoid into pieces that will form a rectangle.

6. **Checkpoint** Start with two congruent isosceles right triangles and a square with sidelength equal to the legs of the triangles like the ones below. Put all three pieces together in different ways to form

   a. a trapezoid.

   b. a right triangle.

   c. a parallelogram that is not a rectangle.

   d. a rectangle.

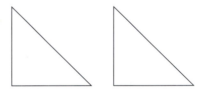

Two small isosceles right triangles

Do the Cuts Really Work?

Have you ever expected a dissection to work, but then discovered that the pieces didn't quite fit? Or perhaps the pieces *looked* like they fit, but it was difficult to be sure? When you can explain *why* a cut works, you can know for sure that it does.

Properties of Parallelograms To understand why a dissection works, you must know the *properties* of the shapes you are cutting. Start with parallelograms.

When mathematicians talk about dissecting, they often don't stop at cutting up a shape—they want to rearrange the pieces into something new!

7. **Write and Reflect** List all of the properties of parallelograms you can think of. The list is started below. See how many properties you can add to it.

• Parallelograms have exactly four sides.

• Opposite sides are parallel.

Share your ideas with others. You will need as complete a list as possible to help you answer the next few questions.

Parallelograms to Rectangles Below is one student's method for dissecting a parallelogram into a rectangle.

First, cut out the parallelogram. Then fold over vertex A, so you fold at vertex D and A lines up on $\overline{AB}$.

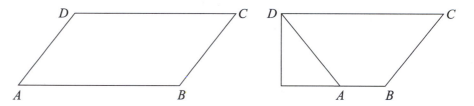

Then unfold and cut along the crease. Slide the triangular piece along $\overline{AB}$ so that it matches up with the other side of the parallelogram. Segment AD matches up with segment BC. Then you have a rectangle.

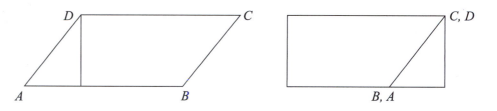

The student's cut created two pieces: a triangle and a trapezoid. The student then described how to rearrange these pieces. But what guarantees that the rearrangement has four sides? Below are two ways that the dissection might fail.

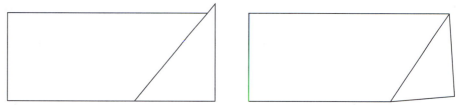

The newly glued edges might not match.

The newly created bottom might be bent.

8. Explain how the properties of parallelograms assure you that when you slide the triangle to the opposite side of the trapezoid,

   a. the two pieces will fit together exactly.

   b. the new bottom edge will be straight.

Triangles to Parallelograms Below is a student's method for dissecting a triangle into a parallelogram.

First, cut out the triangle. Find the midpoints of $\overline{AC}$ and $\overline{BC}$. Let M and N be the midpoints of $\overline{AC}$ and $\overline{BC}$, respectively. Cut along $\overline{MN}$.

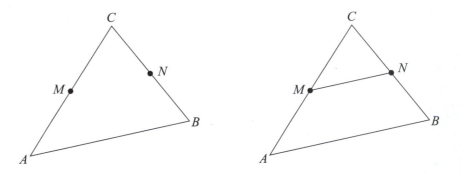

Rotate $\triangle MCN$ around M, until $\overline{MC}$ matches up with $\overline{MA}$. Then you have a parallelogram.

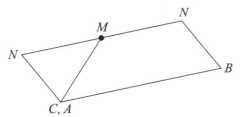

Use the pictures and description, and your list of properties of parallelograms to answer Problems **9–11.**

9. If the final figure is to have just four sides, then $\overline{MC}$ and $\overline{MA}$ must be congruent and not look like the picture below. What part of the student's procedure guarantees that these segments match?

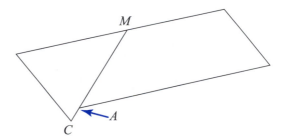

10. If the final figure is really a parallelogram, then the top must be straight, and not look like the picture below. How does the student's method *guarantee* that it will be straight?

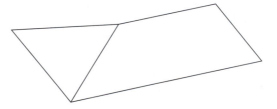

11. If the final figure is really a parallelogram, then in the picture below, it must be true that $\overline{AY} \cong \overline{BX}$. How does the student's method *guarantee* that $\overline{AY} \cong \overline{BX}$?

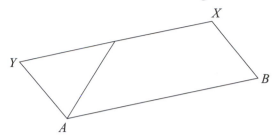

Properties of Rectangles You probably know even more about the properties of rectangles than you do about the properties of parallelograms. Rectangles are even more special than parallelograms—they are a special type of parallelogram—and so they will have more properties on their list.

Listing all the properties—even obvious or redundant ones—can help you notice things you might otherwise overlook.

12. Write and Reflect Write a list of all the properties of rectangles that you can think of. The list is started for you below. See how many properties you can add.

- Rectangles are parallelograms.

- Rectangles have exactly four sides.

- All angles measure 90°.

Rectangles to Triangles Below is one student's method for changing a rectangle into an isosceles triangle.

Start with a rectangle. Cut along a diagonal.

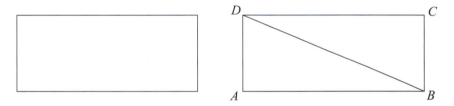

Slide △*ABD* to the right, along the bottom, so $\overline{AD}$ lines up with $\overline{BC}$. Then flip △*ABD* so you have a triangle instead of a parallelogram.

The student's diagonal cut creates two triangles. What guarantees that the final rearrangement of these pieces has three sides?

Below are two ways that the dissection might fail.

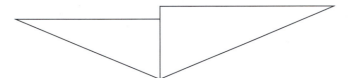

The newly glued edges might not match.

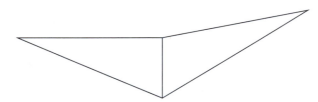

The newly created top might be bent.

13. Explain how the properties of rectangles assure you that when you slide △ABD to the right and flip it according to the student's instructions,

   a. $\overline{AD}$ will fit $\overline{BC}$ exactly.

   b. the new top edge will be straight.

14. If you cut along the other diagonal, will you get a different triangle? Explain your answer. Be sure to use what you know about rectangles.

15. [Checkpoint] Name each of the following figures using only the markings given. Be as specific as possible. Justify each decision.

   a.    **b.**   **c.**   **d.**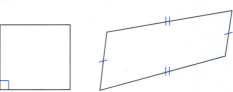

On Your Own

1. Make a second copy of the trapezoid you used for Problem **5**. Dissect it so that the pieces form a triangle.

2. Can you reverse a dissection process? Choose three shapes from the list below. For each of the shapes you chose, start with a rectangle and dissect it into pieces you can rearrange to form that shape.

   a. an isosceles triangle   **d.** a scalene triangle

   b. a right triangle   **e.** a trapezoid

   c. a nonrectangular parallelogram

3. Show that a parallelogram can be dissected and rearranged to form a rectangle having one base congruent to a diagonal of the parallelogram.

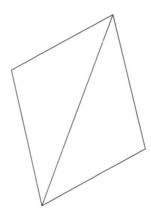

4. Make a cutting argument to demonstrate that △ABC is scissors-congruent to some right triangle.

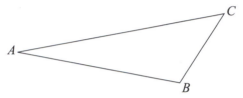

5. Cut a square into a rectangle that has a base congruent to one of the square's diagonals.

6. Start with a scalene triangle. Find a way to dissect it into *two pieces* that you can rearrange to form a new triangle, of equal area, with no angles congruent to any of the angles of the original triangle. Outline a presentation that will persuade the class that your dissection works.

   To be persuasive, you should show three things:

   • the final figure is a triangle (it has exactly three sides),

   • the two joined edges match, and

   • all three angles differ from those of the original triangle.

7. Again, start with a scalene triangle. Use what you know about the area formula for triangles to think of a way to draw, by hand or with geometry software, another triangle of equal area that has no angles congruent to the angles of △ABC.

8. Start with another scalene triangle. Find a way of dissecting it into a triangle, of equal area, with no sides congruent to the sides of the original triangle. Justify your method.

9. Again, start with a scalene triangle. Use a method other than cutting and rearranging to create a triangle of equal area with no sides congruent to the sides of △DEF.

10. Start with a rectangle. Make a same-area rectangle with no sides congruent to the sides of the original figure.

11. Below is a copy of the answer that one group of students wrote for Problem **5** on page 164. From this explanation, figure out what the students must have done. Then rewrite the explanation to clarify it. Add mathematical language where it is useful.

First, we started out with a trapezoid.

Next, we folded $\overline{AB}$ down to $\overline{CD}$.

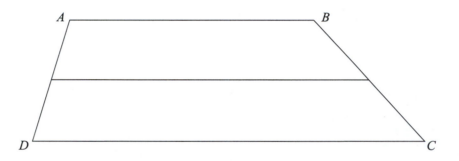

Then we cut on the line that we folded. Next, we folded the left-end corner and cut it to make the other side a straight end so it would look like a rectangle.

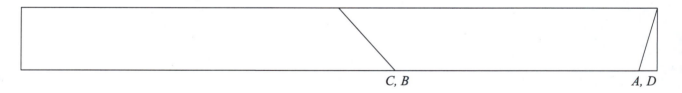

12. You have been showing that two different shapes can have equal area by finding a way to cut up one and rearrange its pieces to form the other. If you suspected that two particular shapes had *unequal* area, but you did not know area formulas, how could you show that one had a greater area than the other?

13. Can an arbitrary quadrilateral be dissected into a rectangle? Approach the problem experimentally. Trace the two figures below. Then try to cut them into rectangles.

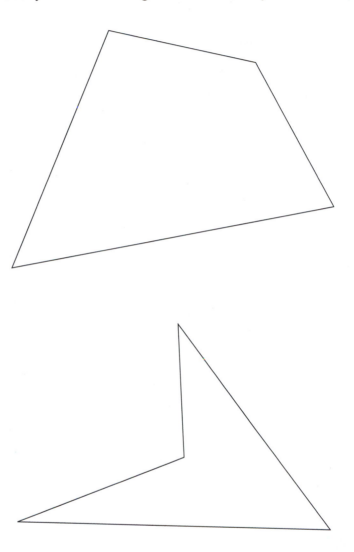

If your experiment succeeded, you have shown that you can cut *these two* particular quadrilaterals into rectangles. If the experiment failed, might there not be *some* method—perhaps a very complicated one—that might succeed?

14. Try to come up with a *reasoned argument* to say why any quadrilateral can or cannot be dissected into rectangles.

LESSON 2

Cutting Algorithms

In this lesson, you will create reliable algorithms for dissecting any parallelogram into a rectangle, any triangle into a rectangle, and any trapezoid into a rectangle.

An **algorithm** is a process—a set of steps—that is completely predetermined. There is no unpredictability, no dice roll, no doubt that it will do precisely what it did the previous time. There are also no hard judgment calls—no more or less about it—just clear, explicit instructions. Many computer programs are written as algorithms, with every step precisely spelled out.

Historical Perspective The word *algorithm* is a distorted transliteration of a great Islamic mathematician's name. abu-Ja'far Muhammad ibn Mūsā was known as al-Khwārizmī, which means "the one from Khwarazm." The Latin attempt at spelling al-Khwārizmī was Algorismi, which later became algorism and then algorithm.

Explore and Discuss

A process can be quite reliable without doing what you want done. Being reliable and predictable is not the same as being successful. A successful algorithm can be used to solve a problem, to make something, or to accomplish a task. For example, you probably know an algorithm for adding two fractions with different denominators. There is often more than one algorithm for a particular purpose.

Write an algorithm for doing a familiar task, like making a peanut butter and jelly sandwich. Your algorithm should contain everything you need to know, including when to open jars, how much peanut butter to use, and so on.

ACTIVITY 1 — Describing the Steps

In Problems **1–5** from Lesson 1, you were asked to "Write a description of the cuts you made, and how you moved the pieces." You may have written your descriptions as algorithms.

Now you will develop *general* algorithms for successfully dissecting any triangle, trapezoid, or parallelogram into a rectangle, and you will prove that your algorithms will always work. Remember that, for each problem, there may be very many algorithms that work, but there are many more that don't, or that work only for special cases.

Later, you will use successful algorithms to develop formulas for the area of triangles, parallelograms, and trapezoids.

1. The pictures below show a triangle being dissected into a rectangle. Write complete and precise instructions for the cuts and rearrangements that are illustrated. Try to make each step clear enough for someone else to follow.

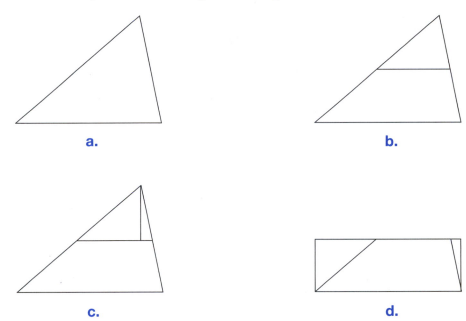

a.   b.

c.   d.

You have done a lot of thinking since Problems **1, 4,** and **5,** in Lesson 1, where you first wrote about how you dissected a parallelogram, triangle, and trapezoid into rectangles. The next problems ask you to edit these old descriptions or write new ones to be as clear and precise about the steps as you now can be.

2. For each dissection—parallelogram, triangle, and trapezoid to rectangle—use two pages to rewrite your algorithm.

    a. On one page, draw pictures that illustrate the steps of your method.

    b. On the other page, describe the same steps precisely using *words only.*

3. Find a partner. Exchange the words you have written (no pictures) describing how to dissect a parallelogram into a scissors-congruent rectangle.

    a. Follow the directions you get *exactly as they are written* to try out your partner's algorithm. Does the algorithm work? Are the directions clear? Were any directions confusing?

    b. Give your partner feedback on the algorithm you tried, and listen to your partner's feedback on yours. How can the instructions be refined?

4. Work with a partner in the same way to refine your algorithms for dissecting a trapezoid into a rectangle.

5. **Checkpoint** Based on your experiences giving and getting feedback on two of the algorithms, rewrite at least one of your dissection algorithms to be as clear as possible. Write a final version of your algorithms, with pictures, for dissecting

    a. a triangle into a rectangle.

    b. a parallelogram into a rectangle.

    c. a trapezoid into a rectangle.

Checking an Algorithm

Sometimes the best way to know whether or not an algorithm works for all possible cases is to try it on all possible cases. That won't work for the three algorithms you just wrote. Why not? In cases like these, you need another way to check. Sometimes the best approach is to check the justification for each step. When that seems too difficult, you may begin by looking for counterexamples—that is, special cases in which the algorithm will fail. But how do you know where to look for special cases when there is an infinite number of cases from which to choose?

6. Study the pictures and justifications for each step of the following algorithm.

    a. Show why this method won't work for all triangles. For example, you can give an example of a triangle for which it won't work, and explain how it will fail.

    b. Explain what special feature or features a triangle must have to allow this method to work.

    Step 1 Cut out the triangle. Draw an altitude from the top vertex to the base. This makes two right triangles because the altitude forms 90° angles.

    Step 2 Slide one of the triangles along the base.

    Step 3 Flip one of the triangles so that two sides of the original triangle match up.

The final shape is a rectangle because

• the two vertical sides are congruent since they were made by the same cut, and

• two opposite angles are right angles since they were made by an altitude cut.

7. Below is another algorithm, this time for cutting a parallelogram into a rectangle. This also has a problem, but it is quite subtle. It seems to work perfectly, but there are parallelograms for which it fails. Fix it.

Step 1 Draw the perpendicular bisector of one side of the parallelogram.

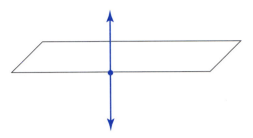

Step 2 Cut along the perpendicular bisector. This perpendicular cut guarantees right angles.

Step 3 Slide one piece parallel to the bisected side until the uncut ends match.

The sides *will* match—the properties of parallelograms guarantee that opposite sides are congruent and that the sum of measures of consecutive angles is 180°.

One very natural pitfall in writing an algorithm is basing it on a drawing or a situation that is too *standard*.

For example, most people tend to picture fairly symmetric figures.

But a less symmetric figure is less special.

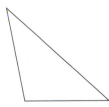

Most people also tend to picture the figures with one side horizontal, so that they look like they won't fall over. That also is too standard.

A rule that works well for a stable figure may not work so well for one that looks unbalanced.

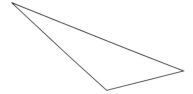

This is why it is so important to have good strategies for checking and *debugging* a process.

For Discussion

One way of finding important special cases with which to test an algorithm is to stretch a definition to its *extreme*.

- Discuss the difference between the *definition* of a trapezoid and your usual *mental picture* of a trapezoid.

- With other students, find three or four trapezoids that are very *nonstandard* and also quite different from each other.

8. **Write and Reflect** Study your algorithms for dissecting parallelograms, triangles, and trapezoids into rectangles. Check some nonstandard cases for each algorithm. If there *are* cases for which your algorithm does not work, explain what *feature* or *features* of them foiled your algorithm. And, of course, the mathematical thing to do is revise the method and test it again.

9. Checkpoint The following argument was given by a student when asked to dissect a parallelogram and rearrange the pieces into a rectangle.

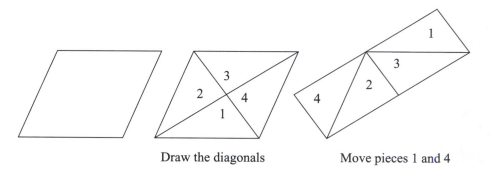

Draw the diagonals   Move pieces 1 and 4

   a. What is wrong with this argument?

   b. For which type(s) of parallelograms will this dissection work?

Justifying the Cuts

You now have algorithms to cut and rearrange triangles, parallelograms, and trapezoids into scissors-congruent rectangles; and you are reasonably convinced that these algorithms work in all cases. It's time to justify your work geometrically.

In Lesson 1, you listed some properties of rectangles. To be sure that a given shape is a rectangle, you could check every property you know, but that's probably not necessary.

For Problems **10–12**, you may want to make hand drawings or use geometry software. In constructing figures, see if the given information forces you to draw a rectangle or not. If you can make a figure that meets the criteria but is not a rectangle, then the given information is not sufficient.

10. What does it take to make a parallelogram? Are any of the listed items below enough information to determine a parallelogram?

- a quadrilateral with one pair of sides congruent
- a quadrilateral with opposite sides congruent
- a quadrilateral with opposite sides parallel

Which of the above are easy to check for with cutting and rearranging?

11. What does it take to make a rectangle? Are any of the listed items below enough information to determine a rectangle?

- a quadrilateral with all angles measuring 90°
- a quadrilateral with opposite sides congruent
- a quadrilateral with congruent diagonals

Which of the above are easy to check for with cutting and rearranging?

12. Review your cutting algorithms once more. Justify each step to show why your sequence of steps reliably produces the desired result.

   a. Why do pieces match up?

   b. Why are line segments straight?

   c. Why are angles right angles?

   d. Once more, pay close attention to see if the justification you use is true in general, or only in certain cases.

13. **Checkpoint** Below is a way to turn a scalene triangle into a right triangle. Justify each step.

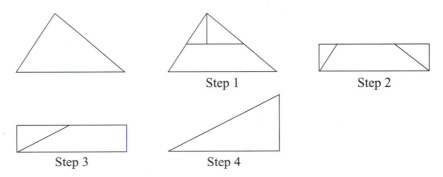

   a. Why do pieces match up?   **c.** Why is there a right angle?

   b. Are segments straight?   **d.** Is this a general or a specific case?

On Your Own

1. Define each of the following.

   algorithm
   trapezoid
   dissection

2. Write an algorithm for tying your shoelace. Be precise. Do not leave out any steps.

3. Write an algorithm for multiplying a three-digit number by a two-digit number.

Assume you know how to multiply two one-digit numbers.

4. Write an algorithm for dissecting a parallelogram into a triangle.

5. Justify each step you used in the dissection of the parallelogram in Problem **4** above.

6. Write an algorithm for dissecting a triangle that is isosceles (but not equilateral) and rearranging it into a scalene triangle.

7. Justify each step you used in the dissection of the isosceles triangle in Problem **6.**

8. Suppose someone followed your algorithm for cutting a triangle into a rectangle and then handed you the rectangle. How would you dissect the rectangle back into the original triangle?

9. Suppose that Jane has an algorithm for dissecting a trapezoid into a rectangle. Explain how you could use Jane's steps to dissect a rectangle into a trapezoid.

10. Use your algorithms for dissecting a trapezoid into a rectangle and dissecting a rectangle into a parallelogram to create an algorithm for dissecting a trapezoid into a parallelogram.

11. Use your algorithms for dissecting a trapezoid into a rectangle and dissecting a triangle into a rectangle to create an algorithm for dissecting a trapezoid into a triangle.

Perspective

Grace Murray Hopper

Have you ever wondered why computer programmers talk about "debugging" a program. Here's the answer!

The word *debug* is in common use today. It means to fix the problems in an algorithm, usually a computer program. The term is derived from the phrase "a bug in the program" or "a bug in the system," which is widely attributed to Grace Hopper, a pioneer in the computer field. She was once described as "the third programmer on the first computer in the United States." She was also the inventor of COBOL, one of the first computer programming languages.

Grace Hopper worked on huge computers. Some were the size of a room and they were filled with moving parts. One of her programs was not running correctly, and she couldn't figure out why. Finally, she went into the computer and found a large moth in the machinery. When she removed this bug from the system her program ran well, and computer scientists everywhere had a new phrase to use.

Area Formulas

In this lesson, you will use dissection algorithms to find the area formulas for parallelograms, triangles, trapezoids, and circles.

Explore and Discuss

You probably know the area formula for a rectangle. It no longer matters whether you remember the area formulas for triangles, parallelograms, and trapezoids. By formalizing the cutting and rearranging ideas you have developed, you are ready to discover these formulas on your own, and you can prove that they're correct.

a What is the area formula for a rectangle?

b Can you make a mathematical argument that it is correct?

ACTIVITY 1

Parallelograms

The first problems in Lesson 1 asked you to dissect a parallelogram into a rectangle. By the end of Lesson 2, you had refined your dissection into a *general* solution that reliably rearranges any parallelogram, no matter what its dimensions, into a scissors-congruent rectangle. In the next few problems, you will use your dissection to discover an area formula for parallelograms.

1. Make a parallelogram. Dissect it into a rectangle using an algorithm that has been proven to work with any parallelogram.

   a. Measure and record the length and width of the rectangle.

   b. As you carefully rearrange your pieces to reconstruct your original parallelogram, notice what measurements on the parallelogram correspond to the rectangle's length and width.

   c. Discuss your results with others. What conjectures do you make about measurements on a parallelogram compared with the dimensions of a scissors-congruent rectangle?

Is there a difference between the image you have of the "height of a parallelogram" and this formal definition?

Definition

The **height** of a parallelogram is the perpendicular distance (shown as dashed lines in the following figures) between any pair of parallel sides chosen as the bases (the heavy lines).

The height is not the same as the length of the remaining sides, unless the parallelogram happens to be rectangular. Depending on the choice of the bases, the height must sometimes be measured outside or partly outside the figure.

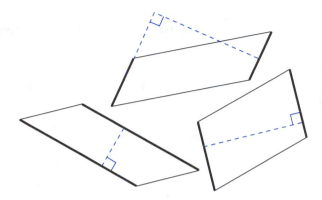

2. Draw a new parallelogram. Label a base b and the corresponding height h.

  a. Dissect the parallelogram and rearrange the pieces into a rectangle.

  b. Explain how your dissection ensures that the base and height of the parallelogram are the same as the base and height of the rectangle. Do not use any numbers in your explanation.

3. Based on your work and what you know about the area of rectangles, find a formula for the area of a parallelogram.

4. **Checkpoint** Find the area of the following parallelograms.

a.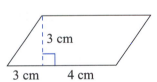
3 cm
3 cm   4 cm

b.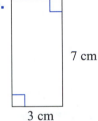
7 cm
3 cm

c.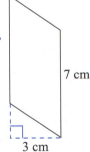
7 cm
3 cm

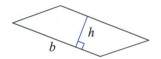

In your dissection, does it matter which side of the parallelogram you call the base?

ACTIVITY 2 ▶ **Triangles**

In Lesson 2, you worked to develop an algorithm for dissecting any triangle into a scissors-congruent rectangle. From any correct algorithm, you can discover an area formula for triangles.

5. Draw a triangle. Dissect it into a rectangle using an algorithm that has been proven to work with any triangle.

  a. Record the length and width of the rectangle.

  b. Carefully rearrange your pieces to reconstruct the triangle, noticing how the rectangle's length and width relate to the triangle.

Definition

Designate any side of a triangle as its base. The **altitude** of a triangle is the perpendicular distance (shown as dashed lines in the following figures) from that base (the heavy line) to the opposite vertex.

Depending on the choice of the base, the height sometimes must be measured outside the figure.

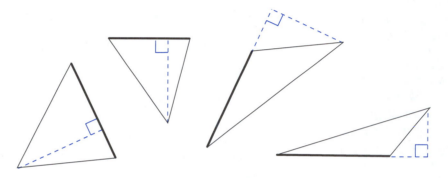

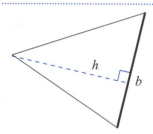

Does it matter which side of the triangle you choose as the base?

6. Draw a new triangle. Choose a base so that you can sketch the altitude *inside* the triangle. Label this base and the corresponding height.

   a. Dissect your triangle into a rectangle.

   b. Explain what happens to the base and height at each stage of the dissection. What are the dimensions of the rectangle in terms of b and h?

7. Based on your work, find a formula for the area of a triangle.

8. **Checkpoint** Use the picture below to find the area of the following triangles.

   a. $\triangle LAU$

   b. $\triangle LUE$

   c. $\triangle LAE$

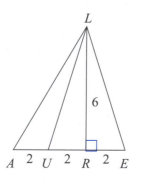

Trapezoids

In Lesson 1, you dissected a trapezoid into a rectangle. In designing an area formula for trapezoids, you may want to use that algorithm and what you know about rectangles. Or, you might choose to dissect a trapezoid into a parallelogram or triangle and use what you now know about them.

For Discussion

Decide on a method for changing a trapezoid into some other figure for which you know an area formula. Consider the ease of dissection and of comparing the dimensions.

9. Draw a trapezoid (you might try one with strange dimensions). Trace at least two copies of it to dissect.

   a. Dissect the trapezoid into another figure whose area you know how to compute.

   b. Measure and record the dimensions of the new figure, from which you could calculate its area.

   c. Carefully rearrange your pieces to reconstruct the trapezoid, noticing how the dimensions of the new figure compare to the dimensions of the trapezoid.

   d. How do the measurements on the trapezoid compare to the dimensions of the new figure you made?

Definition

The two parallel sides (shown as heavy lines) of a trapezoid are called **bases.** The **height** of the trapezoid is the perpendicular distance (shown by the dashed line) between the bases.

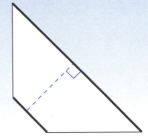  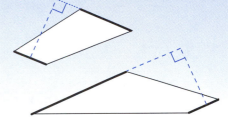

How do you normally picture the base of a trapezoid? How is the formal definition the same as your picture? How is it different?

10. Using methods similar to those you used for parallelograms and triangles, find a formula for the area of a trapezoid.

11. `Checkpoint` Find the area of the following trapezoids.

a.

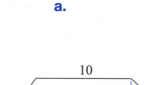

b.

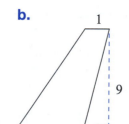

c.

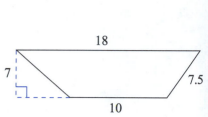

ACTIVITY 4 ▶ Circles

For Discussion

Can a circle be dissected, with a finite number of cuts, into a rectangle?

The problems at the very end of this section will lead you toward that thinking. If you like thinking on your own, don't look too far ahead.

You can get useful *ideas* about the area of a circle by thinking about how it might be dissected into a rectangle, even if it would take more cuts than you could ever actually perform.

Below is a set of dissections that suggests a way to think about the area of a circle. Read it carefully and go through the process yourself, performing each dissection. As you work, try to figure out for yourself where this process is going, and what line of thinking might lead you to develop an area formula.

You will need at least three copies of a circle to dissect in this experiment. You can copy this one, or make your own with a compass or geometry software.

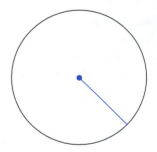

Step 1 Cut one circle into four congruent wedges. Rearrange them as shown below. Tape the pieces together. Label this new shape *Shape 1,* and set it aside.

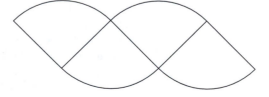

Step 2 Start with a second copy of the original circle. Divide it into eight congruent wedges. Rearrange the parts as shown below. Tape the pieces together. Label this new shape *Shape 2,* and set it aside.

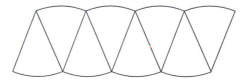

Step 3 Repeat the process with a new circle, dividing it into 16 congruent wedges. Rearrange the parts in a similar way, and tape them together. Label this new shape *Shape 3,* and add it to your collection.

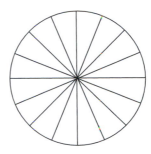

The process you started can be thought of as the first three steps in a potentially endless process. Each step produces a shape that has two scalloped sides (the top and bottom in our pictures) and two straight sides.

12. Write and Reflect Here is your last chance to see where this is heading before the hints come. Look at your three shapes. It is not practical to continue the process by hand, but your *mind* can continue it. Think about what *Shape 10* might look like. Sketch the picture you have made in your mind. Using *words* (no formula for now), explain how this diagram might lead to a way of finding the area of a circle.

13. Call the **radius** of your original circle r and the **circumference** C. Using those variables, describe

> *What do* radius *and* circumference *mean?*

   a. the length of the scalloped sides of *Shape 3.*

   b. the length of the straight sides of *Shape 3.*

14. Think about changes and what does not change from step to step in the process.

> *Think about formal things like lengths, and also about informal things like "rough appearance."*

   Scalloped-side invariants List a few things that *do not* change about the scalloped sides from step to step in this process.

   Scalloped-side changes List things that *do* change about the scalloped sides.

   Straight-side invariants List things that *do not* change about the straight sides.

   Straight-side changes List things that *do* change about the straight sides.

   Area How does the area of each shape in your series compare to the area of the original circle?

Where necessary, explain the items in your lists to make them clear.

15. a. Mathematicians describe the process as producing a "sequence" of shapes that "approaches a rectangle." Explain what "approaches a rectangle" might mean for this sequence.

b. In what way do the shapes more and more resemble a parallelogram?

c. Why do the shapes resemble a *rectangular* parallelogram?

16. a. If you carried the process far enough so that the new shape is "almost a rectangle," what would be the approximate dimensions (base and height) of that rectangle in terms of r and C?

b. Using the approximate dimensions you just found and the area formula for a rectangle, what is the approximate area of *this* "almost rectangle"?

See Unit 4, A Matter of Scale, *for an in-depth study of similarity and more information about π.*

All circles are *similar* (they have the same shape). Therefore, the ratio of the circumference to the diameter in one circle is the same as the ratio of the circumference to the diameter in any other circle. This ratio is invariant: it doesn't depend on the size of the circle. The ratio $\frac{C}{d}$ named by the Greek letter π (pi). That is, π is defined to be the ratio of the circumference to the diameter of a circle.

The value π is approximately $\frac{22}{7}$ or 3.14.

$$\pi = \frac{C}{d} = \frac{C}{2r}$$

17. Use the results of Problem **16** and the formula $C = 2\pi r$ to write a formula for the area of the "almost rectangle" in terms of r. (That is, eliminate the C in the formula.)

For Discussion

Mathematical thinking *must* be precise. So what sense does it make for us to be talking about "almost rectangles" as if "almost" was just as good as the real thing?

While your results actually should be correct, the thinking outlined above makes some questionable assumptions and does leave out some steps. Discuss the strengths and weaknesses of this argument as you see them. What holes in the argument would have to be filled in order to make it completely convincing?

18. You can measure a line segment with a ruler. How would you measure the circumference of a circle?

19. Write and Reflect Suppose you had *only* a ruler. Write an algorithm that would allow you to take measurements and to get as close as you wanted, within the limits of the ruler's accuracy, to the circumference of a circle.

20. Checkpoint Assuming your formula is correct, find the area of the following circles.

a. a circle with a radius of 6 inches

b. a circle with a diameter of 15 centimeters

c. a circle with a circumference of 14π feet

On Your Own

How can you show that the two triangles are equal in area?

1. Start with a scalene triangle. Divide it into *two* triangles of equal area by making only one cut. Explain your solution.

2. Start with a scalene triangle. Divide it into *four* triangles of equal area. Prepare an explanation of your solution.

3. Problem **2** above has at least ten different solutions. By yourself or with a small group, find as many different solutions as you can.

4. Divide $\triangle ABC$ into *three* triangles that are equal in area. Prepare an explanation of your solution.

5. Using half sheets of standard $8\frac{1}{2}'' \times 11''$ paper as your starting rectangles, cut one rectangle into

a. two rectangles that are equal in area.

b. four rectangles that are equal in area.

c. five rectangles that are equal in area.

6. When cutting a rectangle into equal pieces, is there any number of pieces that is *not* possible? Explain your answer.

7. For each statement below, decide whether or not it is true for *all* cases. If you decide it is *not* generally true, clarify whether it is *never* true, or whether it *can* be true, but only in special cases. Justify your answer with an explanation and examples.

a. If a triangle is cut along a median, then it is divided into two triangles of equal area.

b. If a triangle is cut along an altitude, then it is divided into two triangles of equal area.

c. If a triangle is cut along an angle bisector, then it is divided into two triangles of equal area.

d. If two triangles have the same angles and the same area, they are congruent.

e. If two triangles have the same sidelengths, they have the same area.

f. If two triangles have the same area, then they have the same sidelengths.

g. If all three sides of one triangle are different in length from the three sides of another triangle, then the triangles have different areas.

h. If two triangles have congruent angles, then their areas are equal.

8. Refer to the seven shapes on page 188 to answer the following problems. Be sure to provide reasons for your answers.

a. Find two shapes that have the same area. Explain how you decided their areas were the same.

b. Group the shapes by area.

c. Is the area of shape **i** greater than, less than, or equal to the area of shape **iv**?

d. Compare the areas of shapes **i** and **iii** as you did for shapes **i** and **iv**.

e. Compare the areas of shapes **ii** and **iii**.

f. Compare the areas of shapes **ii** and **v**.

g. Compare the areas of shapes **vi** and **vii**.

h. Compare the areas of shapes **ii** and **vii**.

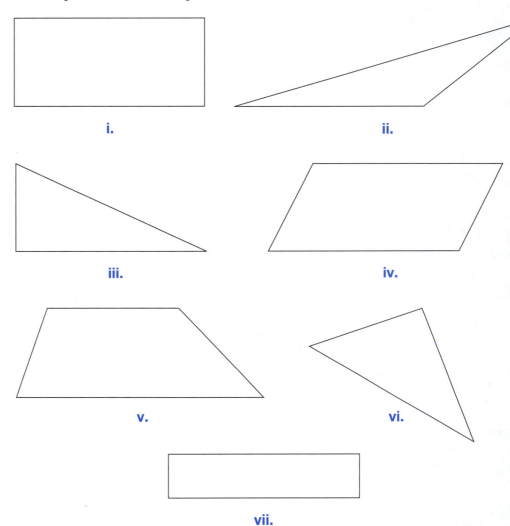

i.

ii.

iii.

iv.

v.

vi.

vii.

9. In the rectangle below, is the sum of the areas of $\triangle ACX$ and $\triangle BCY$ greater than, less than, or equal to the area of $\triangle ABC$? Justify your answer.

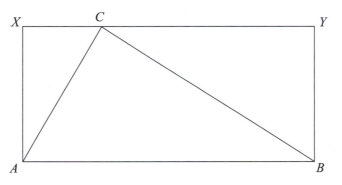

10. Below is a picture of a right triangle.

    a. What is the height from vertex A to base $\overline{BC}$?

    b. What is the area of the triangle?

    c. What is the height from vertex C to base $\overline{AB}$?

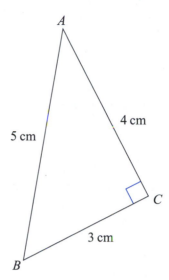

11. In the figure below, $m\angle ABC = 90°$ and h is the altitude to base $\overline{AC}$ of $\triangle ABC$. Compare the quantities $AC \cdot h$ and $AB \cdot BC$. Explain.

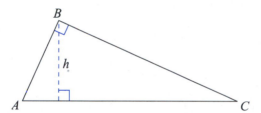

Take It Further

12. Segment CM is a median. Point P is on $\overline{CM}$. Show that $\triangle APC$ has the same area as $\triangle PBC$.

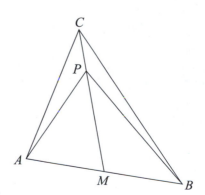

13. Quadrilateral *ABCD* is a rectangle; *ABEF* is a parallelogram; *C*, *D*, *E*, and *F* are collinear; $\overline{AD}$ and $\overline{BE}$ intersect at *X*. Show *BXDC* has the same area as *AXEF*.

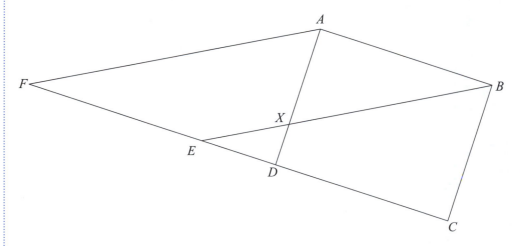

14. Points *S* and *T* are on $\overline{PQ}$ and $\overline{PR}$ of △*PQR*. Segment *ST* is parallel to segment *QR*. Segments *SR* and *QT* intersect at *U*. Show that △*SQU* has the same area as △*TRU*.

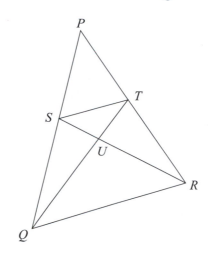

Two points are equidistant from a line if the perpendiculars drawn from the points to the line are equal in length.

15. Quadrilateral *ABCD* is a parallelogram with diagonal $\overline{BD}$. Points *P*, *S*, and *R* are on $\overline{BD}$, $\overline{CD}$, and $\overline{DA}$, respectively, such that *S* and *R* are equidistant from $\overline{BD}$. Show that *BARP* has the same area as *BCSP*.

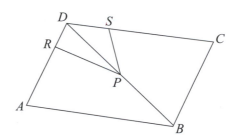

Problems **16–18** ask you to think about how the area formulas for triangles, trapezoids, and parallelograms are related. Below is a series of pictures, showing one way a trapezoid can be dissected into a parallelogram.

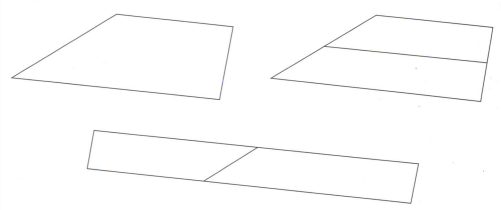

16. a. Describe the dimensions of the parallelogram in terms of the dimensions of the original trapezoid.

b. Describe the area of the parallelogram in terms of the area of the original trapezoid.

c. Derive an area formula for trapezoids from this relationship.

17. a. In your head, imagine redoing Problem **16** with a trapezoid in which one base is a normal size, but the other is so small that it can hardly be seen. What does such a situation tell you about the area formula for triangles?

b. What happens to the area formula for trapezoids if one of the bases is extremely small—practically a single point—compared to the other?

18. What happens to the area formula for trapezoids if both of the bases are the same size? What kind of figure is a trapezoid with congruent bases?

19. This problem asks you to think *intuitively* about the area of triangles, without using formulas.

The fact The area of the triangle may change when one of its vertices is moved or dragged around. However, there are some directions to move the vertex so that the area of the triangle does not change.

The challenge Pick a vertex. Imagine moving it about, or actually move it about using geometry software. Do the following *without* measuring the area of the triangle.

a. Find directions of movement that you are certain will increase the area. Explain why you are certain they will. Do the same for directions that will decrease the area.

b. If there are directions in which to increase the area and directions in which to decrease it, there might be directions in which the area is invariant. Find a direction that seems likely to keep the area invariant. Explain why you picked it.

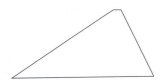

Picture a trapezoid like this, only with an even smaller base on top: "infinitesimally small," a mathematician might say.

Geometry software might help you to get a picture of this dragging—varying one part of a figure while keeping other parts the same.

c. Check your direction in part **b.** Correct it if necessary. Relate what you have seen to the area formula for a triangle.

In previous problems you have dissected parallelograms into rectangles without restrictions. What if, however, the rectangle must have a specific baselength? Problems **20** and **21** address this problem.

20. Show that the parallelogram can be dissected and superimposed onto a rectangle with the same base and height as shown below. Trace the figures. Cut them out to work with, or use geometry software.

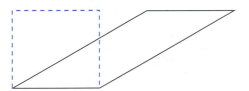

Many students find it helpful to draw a rectangle with the appropriate base and height. Then try to fill it in with pieces of the parallelogram.

21. The parallelogram below is an extreme example of the one in Problem **20.** Trace the parallelogram. Then figure out how to cut it up and rearrange the pieces into a rectangle with one side equal to the *smaller* side of the parallelogram.

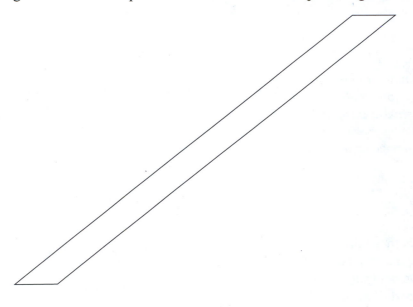

The Midline Theorem

In this lesson, you will create a dissection-based proof of the Midline Theorem: "In a triangle, a segment connecting two midpoints is parallel to the third side and half as long."

Explore and Discuss

One student used the following method to dissect a triangle into a parallelogram.

Start with a triangle. Cut between the midpoints of two sides.

Rotate the top triangle around one of the midpoints. The two segments will match because there is a cut at a midpoint.

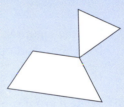

Quadrilateral *ABCD* is a parallelogram because the opposite sides are congruent; $\overline{AD} \cong \overline{BC}$ because they were made by cutting at a midpoint; and $\overline{AB} \cong \overline{CD}$ because a midline cut makes a segment half as long as the base.

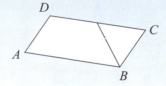

This student made use of an idea that you may or may not have encountered: a midline cut. The student claims that cutting along the midline of a triangle produces a segment that is half as long as the base. Does that seem reasonable?

Proving the Midline Theorem

If the student's statement about the midline cut seems reasonable, the next step should be to try and prove it. You have all the tools necessary for such a proof: ideas about creating and writing a proof, information about triangle congruence, and knowledge of the properties of parallelograms. The next few problems will help you construct a proof. Your job at the end of the problem set will be to write up a formal proof of the midline cut conjecture, which you can then call the Midline Theorem, using one of the methods of proof presentation that you learned about earlier.

1. **Write and Reflect** Explore the midline cut. You can do this by cutting triangles and comparing lengths, by constructing triangles and taking measurements, or any other way that you like. The idea is to do enough so that you either definitely believe or definitely don't believe the student's statement about the midline cut. When you are done, write a few paragraphs to explain how you investigated the problem and what you decided.

2. To prove the midline cut works, you need a slightly different construction. In $\triangle ABC$ below, D is the midpoint of $\overline{AC}$; E is the midpoint of $\overline{BC}$; D, E, and F are collinear; and $\overline{DE} \cong \overline{EF}$. Prove that $\triangle DEC \cong \triangle FEB$.

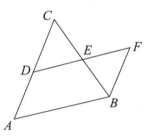

3. Use the figure in Problem **2** to show that, if $\triangle DEC \cong \triangle FEB$, $ABFD$ is a parallelogram. Don't use any information about the length of $\overline{DE}$ or $\overline{DF}$.

You can answer this question, even if you haven't solved Problem 3 yet. Just pretend you have solved it.

4. If $ABFD$ is a parallelogram, list everything you can conclude about $\overline{DE}$ and $\overline{AB}$.

5. Write a proof of the statement below, using a style of proof that you have learned.

Theorem 3.1 The Midline Theorem

In a triangle, a midline is parallel to the third side and half as long.

6. **Checkpoint** Find lengths x and y and angle z in the picture below.

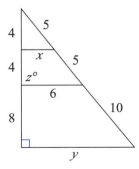

For Problems **1–3,** use quadrilateral *ABCD.* The midpoints of each side, points *E, F, G,* and *H,* have been connected in order.

1. Use the picture below to explain why $\overline{EF}$ is congruent to $\overline{GH}$.

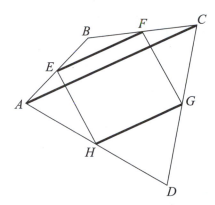

2. Use the picture below to explain why $\overline{EH}$ and $\overline{FG}$ are congruent.

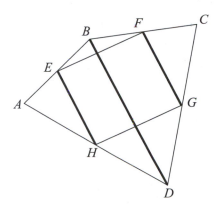

3. What kind of shape is *EFGH?* Prove it.

4. a. One side of a triangle has length 12. How long is the segment that joins the midpoints of the other two sides?

b. What if the side had length 10? 18? 19?

A kite is a quadrilateral with two pairs of congruent adjacent sides.

5. Suppose a kite has diagonals with lengths 5 and 8. A quadrilateral is formed by joining the midpoints of the kite's sides. Give the perimeter and describe the angles of that inner quadrilateral.

6. The diagonals of a quadrilateral measure 12 and 8. What is the perimeter of the inner quadrilateral you get by joining up the midpoints of the first one?

7. What kind of a parallelogram do you get when you connect the midpoints of a kite? Feel free to experiment.

8. Draw a few quadrilaterals that have the property that when you connect the midpoints of the sides, you get a rhombus. Is there some way to tell whether or not a particular quadrilateral will generate a rhombus in the middle without actually testing it? Explain.

Take It Further

The idea of a **midline**—a segment joining two midpoints in a triangle—can be generalized to quadrilaterals. Below are a few pictures of possible midlines for quadrilaterals.

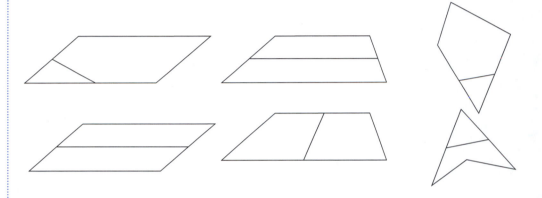

9. Experiment with the various possible meanings for *midline* of a quadrilateral. Can you find any relationship between a midline and the sides of a quadrilateral? Between a midline and a diagonal? Answer the questions below.

a. How did you define *midline* for a quadrilateral? Joining any two midpoints? Two opposite midpoints? Two adjacent midpoints?

b. With what kinds of quadrilaterals did you experiment?

c. What did you find? Are there any special properties of a midline of a quadrilateral? Can you make any conjectures?

The Pythagorean Theorem

In this lesson, you will explore dissection-based proofs of the Pythagorean Theorem.

Reasoning about dissection and rearrangement has been useful throughout mathematics. Many proofs of the Pythagorean Theorem, a famous and valuable fact about right triangles, use dissection. In Euclid's classic texts, *The Elements,* the theorem is stated this way:

> ### Theorem 3.2 The Pythagorean Theorem
>
> *In a right triangle, the square built on the hypotenuse is equal to the squares built on the other two sides.*

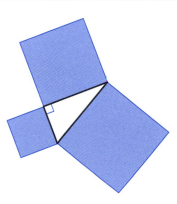

Euclid viewed the Pythagorean Theorem differently from the way we view it today. The "Perspective: The Pythagoreans," will show you some of the ways conceptions of area and length have changed over the years.

As it is worded, the theorem is about the relationship among three squares, and that is how Euclid meant it.

Today, most people think of the theorem as stating a relationship among three numbers, a, b, and c, which represent the lengths of the sides of a right triangle.

If c is the length of the hypotenuse, then c^2 is the area of the square on that hypotenuse. The theorem states that the area of *that* square is the same as the combined area of the other two squares: $c^2 = a^2 + b^2$.

Many people have written proofs of the Pythagorean Theorem. Below is a famous one-word proof written by Bhāskara Acharya (1114–1185), a Hindu mathematician. He published a book called *Lilavati,* named after his daughter. Bhāskara's work was primarily in algebra—for example, solving quadratic equations and looking for Pythagorean triples.

Study the proof below and explain each step.

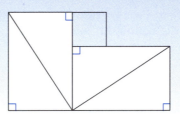
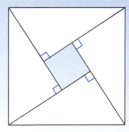

Behold!

A Dissection Proof

The proof outlined below is probably from China, about 200 B.C. Most likely, the author of the proof developed the theorem independently, rather than learning of it from the Pythagoreans.

1. For the proof outline, follow the directions at each step, and answer the questions as you work. When you are finished, you will have constructed a proof of the theorem.

    Step 1 Construct an arbitrary right triangle that is not isosceles. Label the short leg a, the long leg b, and the hypotenuse c.

    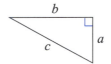

    (Note: This proof will still work if the triangle is isosceles; then $a = b$. In this case, you cannot talk about a "short leg" and a "long leg.")

    Step 2 Construct two squares whose sides have length $a + b$.

    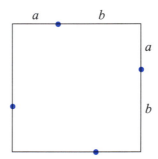
    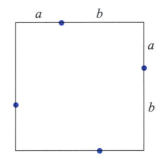

Step 3 Dissect one of the squares as shown below.

- a square with sidelength a in one corner
- a square with sidelength b in the opposite corner
- two rectangles cut at their diagonals into four triangles

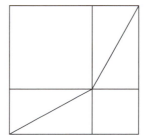

2. Show that each of the four triangles you have just created is congruent to the original right triangle.

Step 4 Dissect the other square into five pieces as shown in the picture below.

- four triangles congruent to the original right triangle
- remaining piece in the center

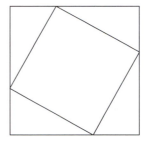

3. Show that the piece in the center is a square with sidelength c.

   a. Explain why all of its sides have length c.

   b. Explain why all of its angles are right angles.

Step 5 The two original squares have the same area.

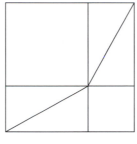 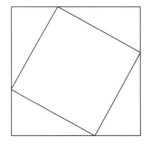

Area of this square   equals   area of this square.

The eight triangles are congruent. So, the four from the first square are equal in area to the four from the second square.

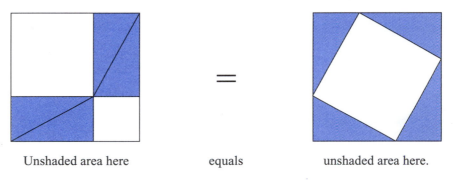

Shaded area here   equals   shaded area here.

Step 6 Remove the four triangles from each square. What remains in the first square will have the same area as what remains in the second square.

Unshaded area here   equals   unshaded area here.

The *geometric* equality—the Pythagorean Theorem as Euclid knew it—has been shown.

Our more modern algebraic interpretation

$$a^2 + b^2 = c^2$$

follows from the algebraic formulas for the areas of squares. The areas of the two squares on the left are a^2 and b^2. The area of the square on the right is c^2. Geometric reasoning tells you that the areas on the left ($a^2 + b^2$) and right (c^2) are equal: $a^2 + b^2 = c^2$.

4. **Write and Reflect** Test your own understanding. Put away these pages and see if you can write this entire proof from memory. Don't worry about the wording as long as your writing is clear and the reasoning is complete.

The Pythagorean Theorem could be stated this way: "If $\triangle ABC$ is a right triangle, then the sum of the squares on the legs equals the square on the hypotenuse." The *converse* of this theorem would be this: "If for $\triangle ABC$ the sum of the squares on two sides equals the square on the third side, then $\triangle ABC$ is a right triangle."

This converse of the Pythagorean Theorem is actually what the Egyptians were using when they marked off right angles with (3, 4, 5) triangles. You have the tools you need to prove the converse of the Pythagorean Theorem, using what you know about triangle congruence.

5. An outline for a proof of the converse of the Pythagorean Theorem follows. You need to fill in the gaps to complete the proof.

To form the converse of a statement, switch the **if** *and* **then** *parts. The converse of a true statement is not always true.*

Given: $\triangle ABC$, where the lengths of the sides satisfy $a^2 + b^2 = c^2$

Prove: $\triangle ABC$ is a right triangle.

   a. Construct a new right triangle with legs of lengths a and b.

   b. What is the length of the hypotenuse of your new triangle? Why?

   c. Your new triangle and $\triangle ABC$ must be congruent. Why?

   d. Triangle ABC must be a right triangle. Why?

6. Use the outline and your answers for Problem **5** to write a formal proof for the following: If a triangle has sides whose lengths satisfy the equation $a^2 + b^2 = c^2$, then the triangle is a right triangle.

7. [Checkpoint] Verify the Pythagorean Theorem numerically by testing a specific case.

   a. Construct a right triangle with one leg 3 inches long and the other leg 4 inches long. What is the area of your triangle?

   b. Construct a square with sides of length 7 ($a + b = 3 + 4$) inches. What is the area of your square?

   c. Dissect a square into five pieces as shown at the left.

     • four right triangles congruent to the original one

     • one square in the middle

   Find the area of the square in the middle by subtracting the areas of the four right triangles.

   d. Calculate $a^2 + b^2 = 3^2 + 4^2$. Is the sum equal to the area of the middle square?

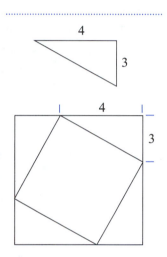

ACTIVITY 2 ▶ # Pick–a–Proof

Choose one of the following famous proofs of the Pythagorean Theorem to study and explain.

The first dissection proof is based on Euclid's proof of the Pythagorean Theorem, presented as Proposition 47 in Book 1 of *The Elements,* Euclid's classic geometry texts. *The Elements* (13 books in all) are the best-selling mathematics books of all time.

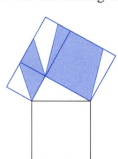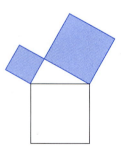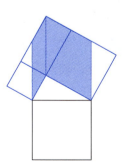

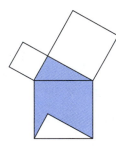 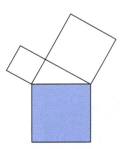

The second proof is by Henry Perigal. Henry Perigal was a stockbroker who lived in London in the 19th century. He discovered this dissection proof of the Pythagorean Theorem around 1830. He liked it so much that he had the diagram printed on his business cards. Perigal was also an amateur astronomer, and his obituary in the 1899 notices of the Royal Astronomical Society of London described one of this pet peeves: He wanted to convince people that it was terribly incorrect to say the moon rotates rather than revolves around the earth.

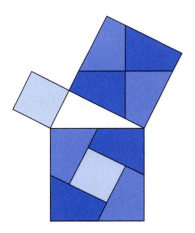

James Garfield (1831–1881) was the 20th President of the United States. Five years before becoming President, he discovered the third proof of the Pythagorean Theorem.

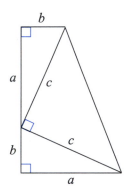

$$\tfrac{1}{2}(a + b)(a + b) = \tfrac{1}{2}(ab) + \tfrac{1}{2}(ab) + \tfrac{1}{2}c^2$$

$$a^2 + 2ab + b^2 = 2ab + c^2$$

$$a^2 + b^2 = c^2$$

8. **Checkpoint** Your earlier proofs were all to show that things were what they seemed to be. Here is a place where proof is essential, but in an upside-down kind of way. You *don't* want to believe the result, so you want to show that something is *not* what it seems to be.

Carefully copy the square below. You may want to use graph paper. It is eight units on a side. *Carefully* cut your copy into four pieces as indicated by the sketch.

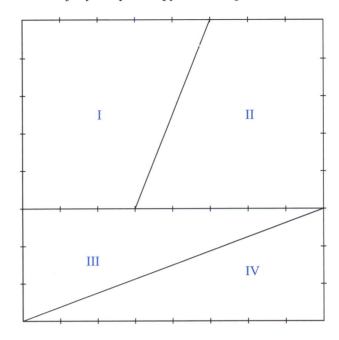

a. What is the area of an 8 × 8 square?

Now rearrange the pieces into the following figure.

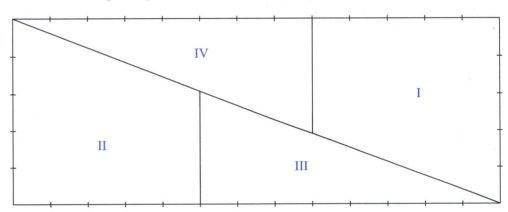

b. What is the area of a 5 × 13 rectangle?

c. The same four pieces seem to make either a square or a rectangle. Yet, if you compute the two areas by formula, the areas are not the same! Figure out what has gone wrong. Explain how it happened.

1. Try Problem **7** on page 201 again, but start with a triangle whose legs are 5 centimeters and 12 centimeters long.

If the area of a square is 4, how long are its sides?

2. How long is one side of the middle square in Problem **7c?**

3. Draw a right triangle with legs 5 centimeters and 12 centimeters. Draw a square whose side is the hypotenuse of this triangle.

How long is one side?

    a. Use the Pythagorean Theorem to find this square's area.

    b. What is the square's perimeter?

4. The picture below shows squares on the sides of a right triangle. It gives the area of two of the squares.

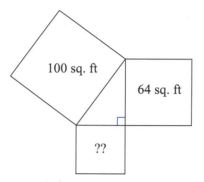

    a. Find the missing area.

    b. Find the lengths of the three sides of the triangle.

This triangle is not easy to construct. You will have to be a little creative.

5. Construct a right triangle with a 17-centimeter hypotenuse and a leg of length 8 centimeters. Draw a square on the *other* leg of the triangle.

    a. What is the area of the square you have drawn?

    b. What is the length of the other leg of the triangle?

By reasoning about squares on the sides of a right triangle, and by using the area formula for a square, you can find the length of any side of a right triangle if you know the other two sides. But how often does one need to know the sides of a right triangle? Remarkably, the Pythagorean Theorem is one of the most widely-applied facts in mathematics and throughout our daily lives. Below are a few examples of its use.

6. You are standing at one corner of a rectangular parking lot measuring 100 feet by 300 feet.

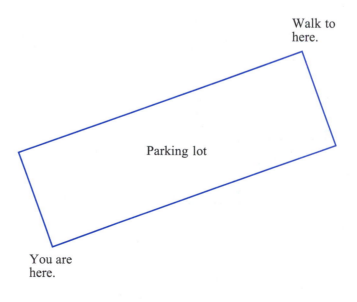

a. If you walked along the sides of the parking lot to the opposite corner, how far would you walk?

b. If you walked diagonally across the parking lot to the opposite corner, how far would *that* be? How much shorter or longer is it than walking along the sides?

c. Of course, there might actually be cars in the parking lot, blocking you from walking directly on the diagonal. In this case, how might the two paths—walking along the sides of the whole parking lot, or zig-zagging through the lot—differ in length? Explain.

7. Find the length of the diagonal of a square whose sides have length

    a. one foot.   **d.** ten feet.

    b. two feet.   **e.** 100 feet.

    c. four feet.

8. Find a pattern in the lengths you calculated in Problem **7.** Write a simple rule relating the diagonal of a square to its sides.

9. A baseball diamond is really a square 90 feet on a side. How far is second base from home plate?

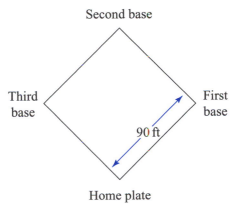

10. In the familiar configuration at at the left, the large square *ABCD* has two smaller squares drawn inside it, □*CGEF* and □*AHEJ*. If *CF* = 1 and *BF* = 3, find the following lengths and areas.

    a. *AB*   **e.** *CE* (not drawn)

    b. *BE*   **f.** area of *AHEJ*

    c. *BJ* (not drawn)   **g.** area of *CGEF*

    d. *HJ* (not drawn)   **h.** area of △*BEF*

11. You have seen the picture at the left before, too. If *AE* = *BF* = *CG* = *DH* = 3, and *EB* = *FC* = *GD* = *HA* = 1, find the following.

    a. *EF*

    b. perimeter of *ABCD*

    c. area of *ABCD*

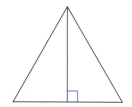

12. Find the height of an equilateral triangle with sides of

    a. one centimeter.   **d.** ten centimeters

    b. two centimeters.   **e.** 100 inches.

    c. three centimeters.

13. Find a pattern in the lengths you found in Problem **12.** Write a rule relating the altitude of an equilateral triangle to the sides of the triangle.

14. Find the lengths of all the segments that are not labeled in the picture below. Describe a pattern in the lengths.

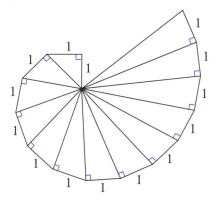

15. At the right is a picture of a cube with one of its diagonals. If the edges of the cube are 10 inches long, how long is the diagonal?

16. An airplane left Los Angeles. After flying 100 miles north, it turned due east and flew 600 miles. Then it turned north again and flew 350 miles. About how far was the airplane from its starting point? (Why does the Pythagorean Theorem *not* give the precise distance here?)

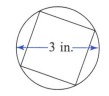

17. A square is inscribed in a circle with a 3-inch diameter. What is the area of the square?

18. Triangle *ABC* below is *not* a right triangle. Find its area.

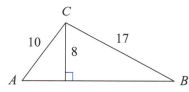

19. What is the area of the plot of land shown in the diagram? Explain any assumptions you make.

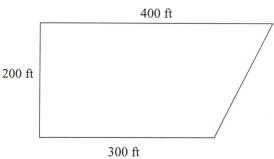

20. The picture below began with a big rectangle whose sides measured 24 and 32. The midpoints of the sides were connected over and over again, to make the design.

   a. List as many facts as you can about this figure.

   b. What is the perimeter of the smallest quadrilateral in the figure? How did you arrive at that answer? Did you, for example, make use of things you knew about the perimeters of the intermediate quadrilaterals?

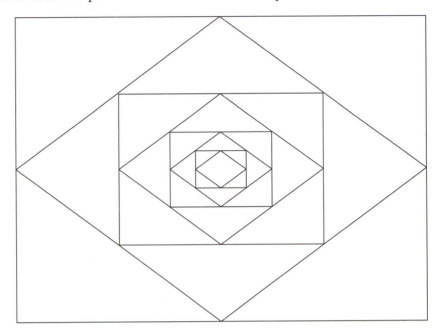

A *Pythagorean triple* is a set of three positive integers, such as (3, 4, 5), which satisfy the equation $a^2 + b^2 = c^2$. If you have a Pythagorean triple, you can build a right triangle with sides of lengths a, b, and c.

21. Look back through your work in this lesson and find two more Pythagorean triples.

22. The following triples are members of a *family* of Pythagorean triples. There are other Pythagorean triples that do not belong to this family. Check that each triple listed below *is* a Pythagorean triple. In what way are these triples enough alike one another to warrant calling them a family?

   a. (3, 4, 5)   **d.** (45, 60, 75)

   b. (6, 8, 10)   **e.** (300, 400, 500)

   c. (30, 40, 50)

23. Draw triangles with the following sidelengths. What do the triangles have in common?

   a. 6 cm, 8 cm, 10 cm

   b. 3 in., 4 in., 5 in.

   c. 15 cm, 20 cm, 25 cm

24. Find a Pythagorean triple that is *not* part of the (3, 4, 5) family. List four of its family members.

25. Below is a picture proof. The statement of the theorem it claims to prove is this:

> ### Theorem 3.3
>
> *The angle bisector of the right angle of a right triangle divides the square constructed on its hypotenuse into two equal-area quadrilaterals.*

Write an explanation, using facts you know about dissection, about why this picture is a convincing argument for the theorem. In other words, complete the proof by adding a written argument.

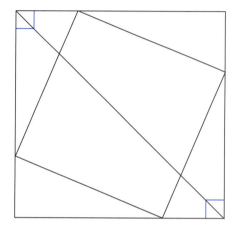

26. Below is another, more difficult picture-proof of the Pythagorean Theorem. This was probably devised by George Biddel Airy (1801–1892), an astronomer. The poem that accompanies the picture appears with it. See if you can reason through this proof and poem.

Here I am as you may see,
$a^2 + b^2 - ab$.

When two triangles on me stand,
Square of hypotenuse is planned.
But if I stand on them instead,
The squares of both the sides are read.

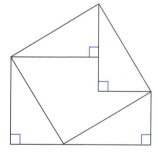

27. The Pythagorean Theorem makes a statement about the areas of squares built on the sides of a right triangle, namely that the two smaller areas sum to the third. What if the shapes built aren't squares? Suppose, instead, you constructed semi-circles on the sides of a right triangle? An equilateral triangle? A rectangle? Construct various shapes on the sides of a right triangle. Explore the cases when it is possible to relate the three areas in some way.

28. A water lily is growing in a pond, rooted to the pond floor. A blossom is 10 inches above the water when it's standing. When it's pulled 20 inches to the side, the blossom just touches the surface of the water. How deep is the pond?

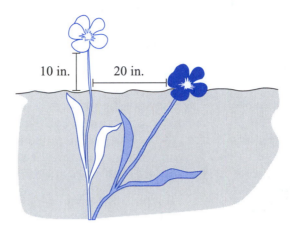

10 in.   20 in.

29. Project Find out more about Pythagoras and the secret Society of Pythagoreans. Besides the theorem that bears his name, Pythagoras and his students made important discoveries in the areas listed below. Choose one of these topics to research. Find out as much as you can about it, and write a short paper to explain your findings to the class.

- irrational numbers, and the proof that $\sqrt{2}$ is irrational

- the golden ratio

- the five regular (Platonic) solids

- intervals on the musical scale

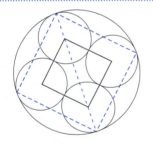

Tangent circles touch at only one point.

30. Challenge Four wires run through a cable. In the cross section of the cable shown at the left, the wires are tangent congruent circles centered on the vertices of a square. The outside shield of the cable is the large circle, which *circumscribes* the other four. The dashed lines suggest some relationships that you need to know, but proofs of these relationships might not be obvious. For this problem, assume these relationships without proof.

a. If the small circles have a diameter of 1 mm, what is the diameter of the large circle?

b. Why is this *not* the same question as the following: "If the wires are 1 mm wide, how wide is the cable?"

31. The ratio of width to height for a television screen is required to be 4:3. The length of a diagonal of the screen is used to measure the advertised size. If you buy a 50-inch TV, the screen has a 50-inch diagonal. What are the width and height of this screen? What is its viewing area?

32. Project Find a way to generate Pythagorean triples.

Perspective

The Pythagoreans

Do you know who the Pythagorean Theorem is named for? This essay will introduce you to Pythagoras and the secret Pythagorean Society of ancient Greece.

The Pythagorean Theorem is named for one of the most famous of the ancient mathematicians, the Greek, Pythagoras. He lived about 569–500 B.C., around the same time as Lao-Tse, Buddha, and Confucius. Pythagoras spent much of his early life traveling. During that time he studied with Thales, the so-called "Father of Geometry." Thales was known for insisting on proofs rather than accepting intuition about geometric ideas. His student Pythagoras also took proofs very seriously.

After his travels, Pythagoras settled in Italy (then a part of the Greek Empire). He started a secret society known today as the Pythagoreans. The Pythagoreans met to study mathematics, music, and astronomy. There were religious and political aspects to their Society as well. For example, the Pythagoreans believed that when a person died, the soul moved into another body. Like many religions, they also had special rules about foods. For them, eating beans and drinking wine were prohibited.

Each member of the Society was sworn to secrecy, and all mathematical discoveries were attributed to Pythagoras. It is possible that Pythagoras discovered his famous theorem on his own, or perhaps he learned of it during his early travels and merely discovered a new proof. Or it could be that one of his students in the Society discovered the theorem. Unlike most academic groups throughout history, the Pythagoreans valued women and their thinking. Theano, the wife of Pythagoras, was a mathematician in her own right.

In any case, the theorem was known to others before the Pythagoreans discovered it. Babylonians knew the theorem around 1500 B.C., nearly 900 years before the birth of Pythagoras. Mathematicians in other cultures had probably discovered it as well. Ancient Egyptians used the related fact—that a triangle with sides of 3, 4, and 5 units is a right triangle—to mark off right angles for property boundaries. Mathematicians in ancient India discovered several of what are now called *Pythagorean triples*, sets of three integers that satisfy the relationship $a^2 + b^2 = c^2$.

The vow of secrecy and the Pythagorean Theorem may have caused the death of at least one member of the Society of Pythagoreans. Pythagoras and his students believed that everything in the universe depended on whole numbers. They believed that the number 1 was divine because it was the building block for all other numbers. Rational numbers were acceptable to the Society because they represent a *ratio* of two whole numbers. Irrational numbers, however, went against their religion in a fundamental way. The Pythagoreans learned that $\sqrt{2}$ is not a rational number; it cannot be expressed as a fraction $\frac{a}{b}$. The discovery that $\sqrt{2}$ is not rational remained a religious secret until Hippasus, a member of the Society, disclosed the secret to outsiders. It is said that members of the Society drowned him in the sea as punishment.

The Pythagoreans' political activity—at one time they held power in several cities in southern Italy—worried the government, and eventually the Society was forced to disband. Pythagoras died in exile.

LESSON 6

Analyzing Dissections

In this lesson, you will critique cutting arguments, explore area and perimeter through dissection, and solve challenging cutting problems.

Explore and Discuss

Jeremy was presented with the following problem.

Construct a triangle *DEF* that is equal in area to △*ABC* below, but that has no angle congruent to any angle of △*ABC*.

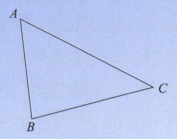

In his response, Jeremy said:

> I measured the sides of the triangle. They were 5 cm, 4.2 cm, and 3.3 cm, which added up to 12.5 cm. So I made a new triangle that had sides 1.5 cm, 5.25 cm, and 5.75 cm. All the angles are different, but the sides add up the same.

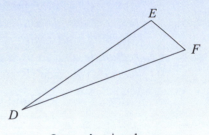

Jeremy's triangle

Do you believe that Jeremy's construction solves the problem? Write either a defense or a critique of his approach.

ACTIVITY 1 ▶ A Parallelogram Cut to a Rectangle

Scene 1: Students in Ms. Engle's geometry class have been working on ways to dissect a parallelogram and rearrange the pieces to form a rectangle of the same area. Ann explains her solution.

Ann I took the parallelogram and cut a triangle off one end, like this:

Now I have a triangle, and another shape that looks like a trapezoid.

Next I cut the triangle in two, like this:

To make a rectangle, I put one of the triangles onto one end of the trapezoid, and the other triangle onto the other end of the trapezoid.

Ms. Engle How do you know that the rectangle has the same area as the parallelogram?

Ann It must, since I used up all the pieces, and didn't overlap any of them.

Ann's explanation was very clear and the final figure certainly looked like a rectangle. But Ms. Engle had appointed a committee—Rafael, Michelle, and Emily—to look for real proof that the final figure is a rectangle, and to ask questions after the presentation. Answer the problems as you read the dialogue.

Rafael How did you decide where to make the first cut? It looks like you were cutting off an isosceles triangle. Is that what you were trying to do?

Ann I did try to make it isosceles. I thought it might not work if it were uneven. So I measured the end of the parallelogram with my ruler and then found a place to cut that had that same length.

1. Did Ann need to make the triangle isosceles? Explain.

Michelle When you cut the triangle in two, did you cut at any special place?

Ann I knew that I wanted to make the rectangle corners out of the pieces of the triangle, so I made my cut along an altitude.

2. a. Define an *altitude* of a triangle.

   b. Explain why Ann chose to cut along an altitude.

   c. How might Ann have constructed the desired altitude accurately if she didn't have a protractor?

> *Emily*   I can see why the top angles are right angles, but how do you know the bottom ones are?
>
> *Ann*   Oops. I didn't think of that. I'm pretty sure they are, but I don't know why.

3. Are the lower angles, ∠*RSB* and ∠*YBS*, in the figure below right angles? Justify your answer.

Scene 2: Peter had worked on the same problem as Ann, but used a different method. Peter drew a sketch on the blackboard and then explained his method.

> *Peter*   I just made one cut, right through the parallelogram. That is the dashed line on my picture. Then I moved the left half over to the right side, and it made a rectangle.

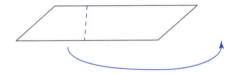

> *Aaron*   I don't get it. Where is the rectangle? Could you just show us with the paper. The arrow doesn't make sense.
>
> *Peter*   OK. Here is before . . . and here is after.

> *Aaron*   I see it now. But I still don't think it is right. That doesn't look like a rectangle. It doesn't even have right angles.
>
> *Peter*   Well, maybe I cut it a little crooked. But they really are supposed to be right angles. I know because the way I got the dashed line was by folding my parallelogram over on top of itself like this:

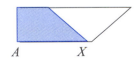

> The edges line up along $\overline{AX}$, so it has to make right angles.
>
> *Aaron*   That makes sense. I guess they must be right angles then.

Once the class understood what Peter had done, the debate centered on whether or not the final figure was actually a rectangle. Here are some of the issues in that debate. All of these questions must be resolved if you are to be certain the last figure is a rectangle. Read each one carefully, and answer it in writing. After you have written your answers, trade with another student, and critique your partner's answers thoughtfully.

4. Peter folded the bottom edges carefully to make a right angle at *A*. How do you know that this also makes right angles along the top edge at *B*?

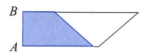

5. Once Peter establishes that his figure has four right angles, has he proved that it is a rectangle?

6. Peter joined the two pieces at $\overline{CD}$. What proof do you have that the two edges joined at $\overline{CD}$ are the same length?

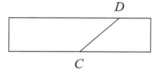

7. What might Peter's rectangle look like if the edges he joined were different lengths? Make a sketch.

8. When Peter put the two pieces together at $\overline{CD}$, he assumed they made a straight line along the top and the bottom of the rectangle. Prove that joining angles 1 and 2 makes a straight line, and that joining angles 3 and 4 makes a straight line.

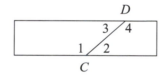

9. In a rectangle, opposite sides have the same length. Explain how you know that this is true in Peter's rectangle.

AB = EF because:

AE = BF because:

Here are two figures which each have four right angles.

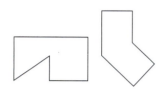

10. **Checkpoint** Below is Noriko's method for cutting a trapezoid to a triangle. Answer the questions to help her justify her method.

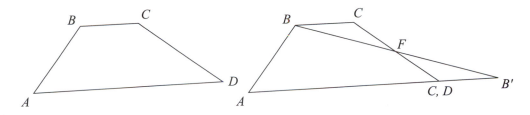

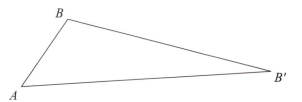

a. What point on $\overline{CD}$ does she need to find so that $\overline{CF} \cong \overline{DF}$?

b. Will $\overline{BB'}$ be a straight line? How do you know?

c. Will $\overline{AB'}$ be a straight line? How do you know?

ACTIVITY 2 ## Area and Perimeter

When you cut and rearrange a triangle, a parallelogram, or a trapezoid into a rectangle or vice versa, the area is unchanged. What other properties of the figure are invariant through cutting and rearranging? You've already verified that angle measure and sidelength are not always preserved. What about the **perimeter,** the *sum* of all side-lengths? You probably already have some intuition about that. Now you will check that intuition.

11. a. Construct a parallelogram. Measure the lengths of its sides. Calculate its perimeter.

    b. Dissect your copy into a rectangle.

    c. Using measurements, calculate the perimeter of the rectangle. Compare it with that of the original parallelogram.

12. Compare the perimeters of the starting and ending figures when you use your algorithms for

    a. dissecting a triangle into a rectangle.

    b. dissecting a trapezoid into a rectangle.

13. Write and Reflect Summarize the results of the experiments you have just performed. Each experiment is a *specific* case, but try also to generalize from what you have seen. For each algorithm, try to answer the following questions.

    a. Does *that* algorithm

        i. reliably preserve the perimeter,

        ii. sometimes preserve the perimeter and sometimes change it, or

        iii. reliably change the perimeter?

b. If you are sure that the algorithm reliably changes the perimeter, does it

    i. reliably increase the perimeter,

    ii. sometimes increase the perimeter and sometimes decrease it, or

    iii. reliably decrease the perimeter?

With dissections, if you know the area of the original figure, it is no effort to find the area of the new figure. But the perimeter is not so easily determined. You now have decided, in a *qualitative* way, how perimeter is affected by each algorithm, but you have no *quantitative* way of computing a new perimeter from the old one.

Below is a dissection that changes perimeter in a regular way that allows you to compute the new perimeter from the old. In general, however, computing change in perimeter is not so straightforward.

Step 1 Construct a fairly large paper square. You will find 8 inches on a side particularly convenient.

Step 2 Record both the area and perimeter of your square. *Save your data. You will need it later.*

Step 3 Construct the midpoint of the top side of your square. Using the left half of that top side as the hypotenuse, construct an isosceles right triangle.

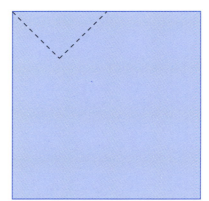

Step 4 Cut out the triangle. Rotate it around the midpoint. Tape its hypotenuse onto the remaining half of that side of the square.

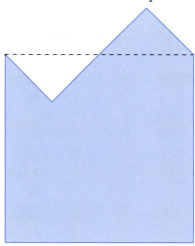

Step 5 Move clockwise to the next side of the original square. Repeat the process: cut out the isosceles right triangle, rotate it, and tape it to the remaining half of that side.

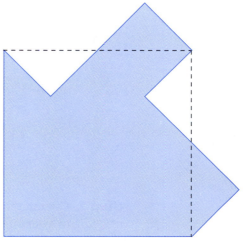

Step 6 Do the same to the remaining two sides of the original square. Your figure should now look like this:

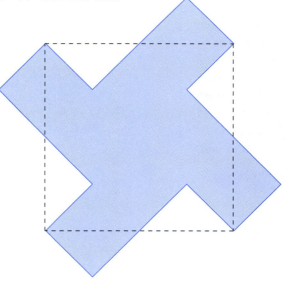

14. a. What is the area of this new shape?

b. How can you find it without further computations?

15. a. Is the perimeter of the new shape greater than, less than, or the same as the perimeter of the square?

b. How can you compute the perimeter of the new shape without measuring anything?

Repeat the process on the new figure.

This is a difficult problem. Spend some time thinking about it. Study the construction and the picture of the final figure.

Step 1 Pick a side on the new shape. Construct its midpoint. Then construct an isosceles right triangle inside the figure with its hypotenuse along the left half of the side you just bisected as viewed from inside the figure.

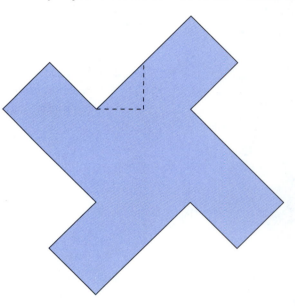

Step 2 Cut out the triangle, rotate it, and tape it onto the other half of the side of your figure.

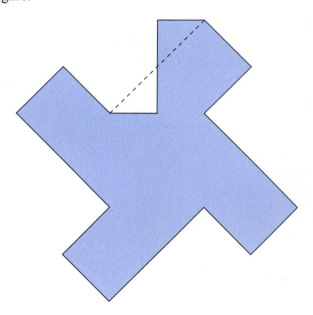

Make sure to move systematically (clockwise is probably easiest) around the figure, cutting the triangle out of the left-hand half of each side as viewed from inside the figure, and taping it to the right-hand half.

Step 3 Repeat this for each side of your figure.

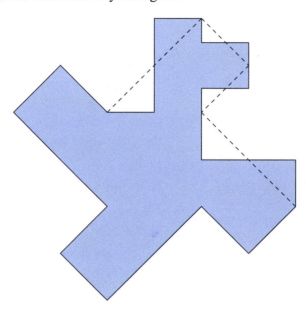

The dashed lines show sides that are done. See if you can complete the process.

16. **a.** What is the area of the new shape that you created?

   b. How does it compare with the area of the square you started with?

17. **a.** What is the perimeter of your new shape?

   b. How does it compare with the perimeter of the square you started with?

   c. Explain how you found the perimeter. Did you measure it, use some particular process of computation, or find another way?

18. Checkpoint Write a paragraph to explain what this dissection shows you about area and perimeter.

For Problems **1–3**, use the picture and the information given below.

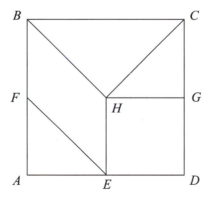

Quadrilateral *ABCD* is a square, *EHGD* is a square, and *E*, *F*, and *G* are midpoints.

1. Justify each step in the argument below, showing *DEHG* has the same area as *EFBH*.

   a. Area of trapezoid *ABHE* = Area of trapezoid *DEHC*. Why?

   b. Area of △*AFE* = Area of △*GHC*. Why?

   c. So, area of *DEHG* = area of *EFBH*. Why?

2. Make a similar argument, justifying each step, to show that the area of △*BCH* is the same as the area of *EHGD*.

3. Make a similar argument, justifying each step, to show that the area of △*BCH* is the same as the area of *EFBH*.

4. For each statement below, decide whether it is true for all cases or false. Justify your answer with an explanation and examples.

   a. If every side of triangle *B* is congruent to or longer than every side on triangle *A*, then triangle *B* has the greater area.

   b. If every side on rectangle *B* is congruent to or longer than every side on rectangle *A*, then *B* has the greater area.

   c. It is possible to decrease the area of a triangle while increasing the length of every side.

   d. It is possible to increase the area of a rectangle while decreasing the length of every side.

   e. It is possible for one circle to have a greater circumference but less area than another circle.

   f. It is possible for one shape to have greater perimeter but less area than another shape.

For Problems **5–8,** do the following.

   a. Construct the given figures if possible.

   b. If you cannot construct the figure, explain why not.

5. two triangles have the same area and different perimeters

6. two rectangles have the same area and different perimeters

7. two squares have the same area and different perimeters

8. two circles have the same area and different circumferences

9. Farmer Nancy and Farmer Michelle are in a land dispute. Michelle bought half of Nancy's land but the plots were divided into different shapes. In dividing up the land, Nancy used the same length of fence around each farm to mark off two plots. Michelle says it isn't fair because she has less land.

 a. Do the farms have the same area? If so, draw a situation where you have two farms with different shapes, the same amount of fencing around each, and the same area.

 b. Could Michelle be right that the farms have the same amount of fencing around them but they have different areas? If so, draw a situation that represents Michelle's point of view.

Take It Further

10. Imagine continuing the process described on pages 216 and 217 performing it on the figure you just created, and then performing it again on the results of that dissection. How would the perimeter grow? Explain how you figured out this answer.

Unit 3 Review

1. **a.** Dissect the figures below into rectangles.

   b. Draw pictures to illustrate the algorithm you used to dissect each figure.

   c. Justify each step of the algorithm you used to dissect each figure.

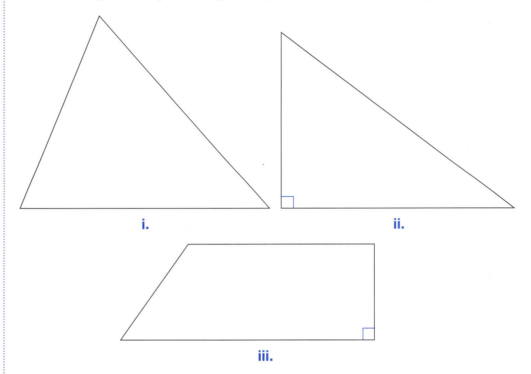

i. ii.

iii.

2. Explain what it means for two figures to be *scissors-congruent*.

3. Describe and justify each step in the dissection that will turn any parallelogram into a rectangle.

4. How does the dissection in Problem **3** lead you to the area formula for a parallelogram?

5. For each dissection in Problem **1** answer the following questions.

   a. Does the dissection algorithm reliably preserve the perimeter, sometimes preserve the perimeter, or always change the perimeter?

   b. If the algorithm reliably changes the perimeter, does it reliably increase the perimeter or decrease the perimeter?

6. In Problem **1**,

   a. will your dissection of figure **i** work for any triangle?

   b. will your dissection of figure **ii** work for any right triangle?

   c. will your dissection of figure **iii** work for any trapezoid?

7. Give two statements of the Pythagorean Theorem:

    a. one statement using sidelengths; and

    b. one statement using areas.

8. Find the area and perimeter of the following figures.

  a.

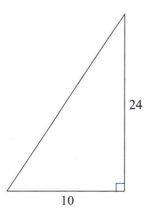

  b.

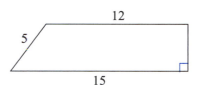

  c.

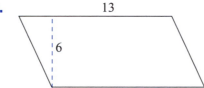

9. Quadrilateral *KITE* is a kite. Points *A, B, C,* and *D* are the midpoints of each side. Given *IE* = 12 and *KT* = 17, find the perimeter of *ABCD*.

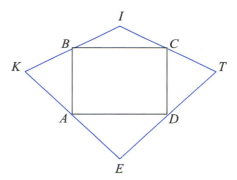

UNIT 4

A Matter of Scale
Pathways to Similarity and Trigonometry

Introduction to Maps and Blueprints

In this lesson, you will explore maps and blueprints, including what you can learn from the scales on these pictures.

What does it mean for one picture to be a good copy of another? What kinds of information do scaled drawings, like maps and blueprints, provide? How can you test to see if one drawing is a good copy of another? In this unit, you will explore these questions and learn about an important geometric idea called *similarity*.

*1 inch = 600 feet =
3-minute walk*

Explore and Discuss

The following page shows a map of downtown Seattle. The scale indicates that 1 inch on the map represents 600 feet of actual distance. It also says that an inch on the map represents about a 3-minute walk.

a What actual distance do the mapmakers claim you can walk in 1 minute? Is this reasonable?

b Other than maps, what other pictures have you seen where the images have been scaled to a smaller or a larger size?

ACTIVITY 1 ▶ **Maps**

Maps are drawn with a given scale so that a whole city, state, or even larger area can fit on a convenient-sized paper. In reading a map, it is important to keep that scale in mind.

1. Find the actual distance between the following locations (travel only on streets).

    a. The Space Needle to the intersection of Broad St. and Western Ave.

    b. Mega Mart Record Shop to the intersection of Eagle St. and 2nd Ave.

2. Lorena is at the Ramada Inn and has arranged to meet a friend at the intersection of Thomas St. and 1st Ave. N. Approximately how much time will her walk take?

3. Checkpoint Recalculate your answers to Problem **1** using the scale 1 inch = 450 feet.

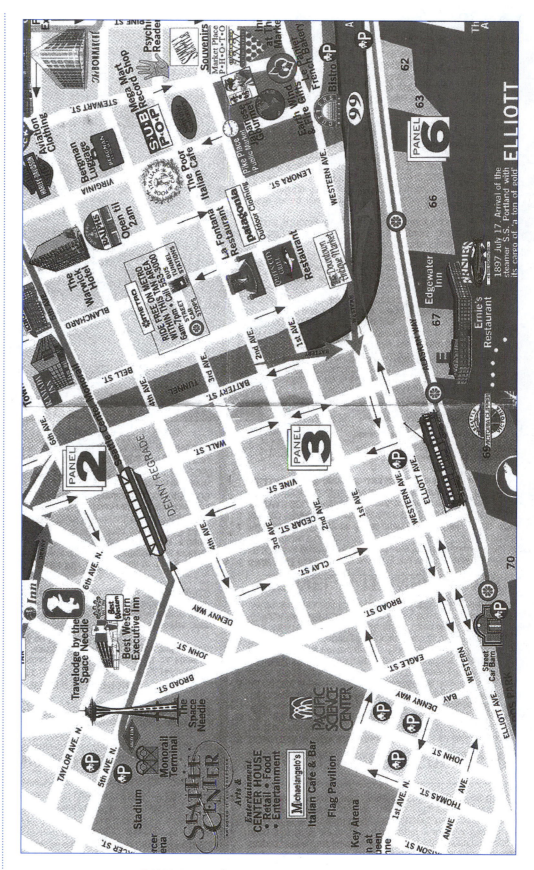

Reading a Blueprint

When architects draw plans for an apartment building, a house, a school, an office, a park, or a sports complex, they include many different sketches, each used for a different purpose. Below is an example of a *blueprint* for the outside of a house.

"Front elevation" is architect talk for "front of house."

The architect has indicated on the blueprint that the distance from the top of the roofs to the top of the chimneys is exactly 3 feet.

4. Use the blueprint, a ruler, and a calculator to find

    a. the total height of each chimney.

    b. the dimensions of the garage door.

The blueprint below shows the floor plan of the second floor of the same house.

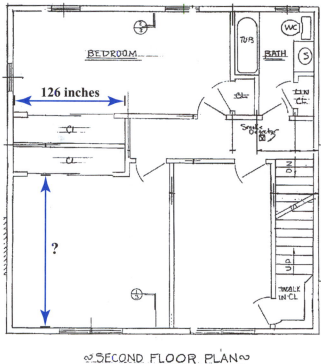

~SECOND FLOOR PLAN~

5. A measurement has been erased and replaced by "?" in this floor plan. Calculate this missing value.

6. Find the overall (full-size) dimensions of the entire second floor.

7. **Checkpoint** Make a floor plan for the house or apartment where you live or a building you visit frequently. Include measurements like the one given in the floor plan above. Make sure everything is drawn to scale, including things like sinks, bathtubs, appliances, and so on.

1. Find a new map to examine.

   a. What is the scale given by your map?

   b. Pick two points shown on the map. Decide how far apart they are based on their distance on the map.

   c. If the average person walks about 200 feet in a minute, how long would it take to walk between the two points you chose?

   d. Estimate the driving time between your two points. Does it depend just on their distance or are there other factors you should consider?

2. Often when drivers estimate the driving time between two points based on a map's scale and average driving speed, they are surprised by how much longer the trip actually takes. What are some reasons that it would take longer to drive than the length of the road shown on a map would make it seem?

3. While traveling in Wisconsin, Sydney and Mark were using a map that had a scale of 1 inch = 15 miles. Sydney's finger is 3 inches long. On the trip from Madison to Greenlake Sydney announced the following distances. What were the actual distances being estimated?

   a. At the beginning of the trip Sydney announced, "We're one finger away!"

   b. After a stop for ice cream Sydney announced, "We've got only a half of a finger to go!"

   c. When they got back into the car after taking a picture of some cows Sydney said, "Only one fingernail left in the trip!" (Her fingernail is about one eighth of her finger.)

4. If the owners of the house in Activity 2 wanted to carpet the bedroom floor with seafoam green luxury carpet, the carpet layers might use the floor plan on page 229 to plan for how much carpet they would need.

   a. What are the dimensions of the bedroom? (Do not include the closet.)

   b. How much carpet would be needed to cover the bedroom floor? Give your answer in square inches.

   c. Carpet is usually sold by the square yard. How many square yards of seafoam green luxury carpet will be needed?

5. Choose one other room from the floor plan on page 229. Complete parts **a** and **b** of Problem **4.**

What Is a Scale Factor?

In this lesson, you will develop a precise definition of scale factor and examine how to calculate it.

Most maps and blueprints provide a scale that allows you to calculate actual distances and lengths. Depending on the map, 1 inch might represent 1 mile if the map is a detailed view of a small region, or 1 inch might represent 100 miles if the map shows a larger region. The term **scale factor** describes what reduction or enlargement from the actual size was used to obtain the map, blueprint, or picture.

Explore and Discuss

Each side of this square has length 2 inches.

a What do you think it means to "scale the square by a factor of $\frac{1}{2}$"?

b Draw a figure showing what you think it would look like to scale the square by $\frac{1}{2}$. If you can think of more than one way to interpret the statement, draw a separate figure for each idea.

ACTIVITY 1 ▶ ## A Matter of Interpretation

Carlo and Amy have different interpretations of what it means to scale the square by a factor of $\frac{1}{2}$. See if you agree with either of their explanations.

Why does Carlo's folding method work? How long are the sides of his new square?

Carlo: When it said to scale by $\frac{1}{2}$, I drew a square that was half the size of the first one—you know, half the *area*. Since the area of the original square is $2 \times 2 = 4$ square inches, I needed to make a square with an area of 2 square inches. One neat way to do this is to fold all four corners of the square to the center.

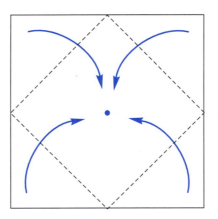

Amy: I thought that scaling by half meant we were supposed to draw the *sides* half as long. The first square has sides that are 2 inches long. So the scaled one should have sides that are 1 inch long. I drew a horizontal and vertical line on the square to divide the length and width in half. This gives me four squares, each scaled by a factor of $\frac{1}{2}$.

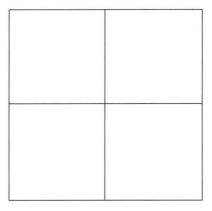

In fact, there isn't just one correct way to interpret the phrase "scale by $\frac{1}{2}$." Words can mean different things to different people. But by convention, Amy's meaning of scaling is the one that most people use.

Definition

Scaling a figure by a factor of *r*: When you scale a figure by a factor of *r*, your new figure will have lengths *r* times the corresponding lengths of the original figure.

The value of the scale factor, *r*, can be any positive number, including a fraction.

1. What features of a square are invariant when you scale it by a factor of $\frac{1}{2}$?

2. A square has a sidelength of 12 inches. How long is each side of a new square when scaling the original one by each of the following factors?

a. $\frac{1}{4}$ **c.** $\frac{2}{3}$ **e.** 1

b. $\frac{1}{3}$ **d.** 2   **f.** 1.3

3. If you scale a figure by the following values of r, will the new figure be smaller than, larger than, or the same size as the original one?

a. $r = \frac{3}{5}$ **c.** $r = 3$

b. $r = 1$ **d.** $r = 0.77$

4. Checkpoint Which of the following pairs of figures could be scaled by one half? For each pair explain why or why not.

a.  **b.** 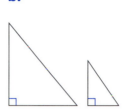 **c.**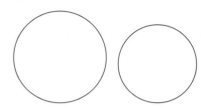

ACTIVITY 2

Calculating Scale Factors

Sometimes a scale factor isn't given but you can compare parts of figures to calculate a scale factor.

5. In each pair of pictures below, what scale factor will transform the picture on the left into the scaled picture on the right?

a.

b.

c.

d.

6. Suppose that the original pictures in Problem **5** are those on the right. By what factor would you scale them to get the pictures on the left?

7. Compare your answers to Problem **6** with your answers to Problem **5.** What is the relationship between them?

8. Many photocopy machines have a feature that allows you to reduce or enlarge (that is, to scale) a picture. Enter the amount 80% or some other percentage on a photocopy machine and copy a picture. (Some machines require the factor as a decimal; for this example, the amount would be 0.80.)

   a. By what factor have you scaled the picture?

   b. If you want to scale a picture by a factor of $\frac{3}{4}$, what percentage would you enter?

9. **Checkpoint** Label the following pairs as to whether the two scalings in the pair are the same or different.

   a. Scaling by 2 and scaling by $\frac{1}{2}$.

   b. Scaling by $\frac{1}{3}$ and scaling by 30.

   c. Scaling by $\frac{3}{5}$ and scaling by 0.6.

   d. Scaling by 1 and scaling by 100.

ACTIVITY 3 ▶ **Area and Volume**

These questions ask you to look at how the area or volume of a figure changes when you scale it.

10. **a.** Draw a square with 1-inch sides. Scale it by a factor of 2. How many copies of the 1-inch square fit inside the scaled square?

b. Start with a 1-inch square again. Scale it by a factor of 3. How many copies of the 1-inch square fit inside the scaled square?

c. If you scale a 1-inch square by a positive integer r, how many copies of it will fit inside the scaled square?

11. The equilateral triangle below has 2-inch sides.

a. Draw a scaled version of the triangle, using a factor of $\frac{1}{2}$. How many of these scaled triangles can you fit inside the original triangle?

b. Draw a scaled version of the triangle, using a factor of $\frac{1}{3}$. How many of these scaled triangles can you fit inside the original triangle?

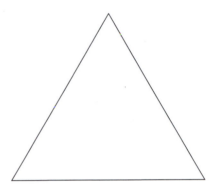

12. A cube has edges of length 1 inch.

a. If the cube is scaled by a factor of 2, how long will the sides of the new cube be? How many copies of the original cube will fit inside the scaled cube?

b. If the original cube is scaled by a factor of 3, how long will the sides of the new cube be? How many copies of the original cube will fit inside the scaled cube?

c. If you scale the original cube by a positive integer r, how many copies of the original cube will fit inside the scaled cube?

13. **Checkpoint** How many 1-inch squares will fit into a 6" × 6" square that has been scaled by a factor of

a. $\frac{1}{3}$? **b.** 3? **c.** $\frac{2}{3}$?

1. Give a scale factor that would change MEOW in the following ways.

a. Shrink it.

b. Enlarge it.

c. Shrink it very slightly.

d. Keep it the same size.

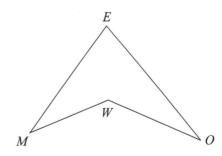

2. A rectangle that has width 12 inches and length 24 inches is scaled using the following factors. In each case, what are the rectangle's new dimensions?

a. $\frac{1}{3}$ **d.** 2.5

b. $\frac{1}{4}$ **e.** 0.25

c. 0.3

3. Use the figures below to answer these questions.

a. What is the scale factor that could be applied to figure *A* to get figure *B*?

b. What is the scale factor that could be applied to figure *B* to get figure *A*?

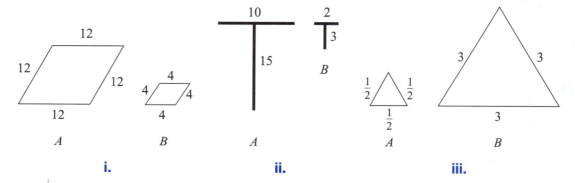

i.   ii.   iii.

4. Apply a scale factor of 4 to the figures below and answer these questions.

a. How long are the new sides of the figure?

b. How many of the original figures will fit into the new figure?

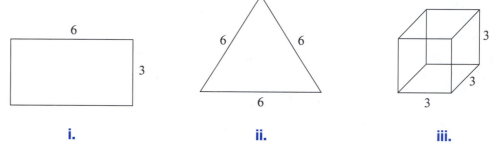

i.   ii.   iii.

5. Apply a scale factor of $\frac{1}{3}$ to the figures in Problem **4**. Then answer these questions.

a. How long are the new sides of the figure?

b. How many of the original figures will fit into the new figure?

Take It Further

6. Suppose that you take a picture and scale it by a factor of $\frac{1}{2}$. You then take your *scaled* picture and scale *it* by $\frac{1}{4}$. By how much, overall, have you scaled the original picture?

What Is a Well-Scaled Drawing?

In this lesson, you will decide if two figures are good copies of each other, and learn what tests can help you decide.

The maps and blueprints in Lesson 1 are all well-scaled drawings of cities and houses. Sometimes, though, a reduction or enlargement is not a good copy of the original. If you look at yourself in a funhouse mirror, you will see that your body has been scaled in a very bad way indeed—the mirror might stretch you to appear as skinny as a matchstick, or shrink you to appear one foot tall!

Explore and Discuss

Take a look at this picture of a horse skeleton.

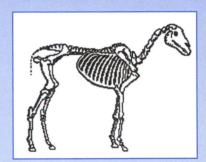

The original horse skeleton

Below are four copies of the original horse skeleton.

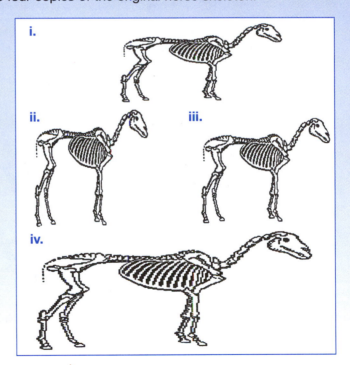

i.

ii. iii.

iv.

> **a** Decide which of the four pictures are accurate enlargements or reductions of the original horse skeleton.
>
> **b** Share your answers to part **a** with your classmates. What characteristics of the horse skeleton drawings helped you to make your decision? Did your classmates pay attention to different features?

How Can You Tell?

The next few problems will help you make better decisions about whether you are looking at a good or bad copy. You will use measurements rather than just judging by eye.

1. The picture below shows two baby chicks labeled with points *A* through *H*.

Here are the distances between some of the points:

AB = 4 cm CD = 2 cm EF = 2.5 cm GH = 1.25 cm

How can you use these measurements to help convince someone that the two chicks are well-scaled copies of each other?

2. Four locations—*A*, *B*, *C*, and *D*—are labeled on the fish below.

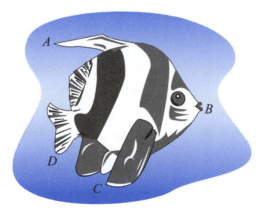

Joan measures the distance between *A* and *B*, as well as between *C* and *D*. She calculates that $\frac{AB}{CD}$ = 2.6. Michael sees her answer and asks, "But what's your unit? Is the answer 2.6 centimeters, 2.6 inches, 2.6 feet, or something else?" How would you respond to Michael's question?

3. **Checkpoint** Name at least four different measurements you could use on the following pair of figures to convince someone that the two figures are well-scaled copies of each other.

Use grid paper to scale these pictures by the given scale factor.

1.

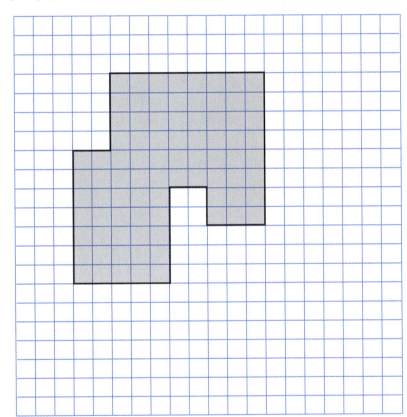

Scale factor = 2.5

2.

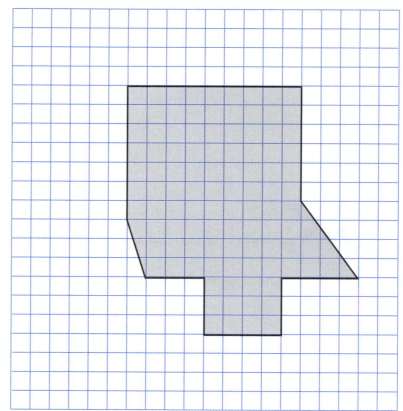

Scale factor = 0.5

3.

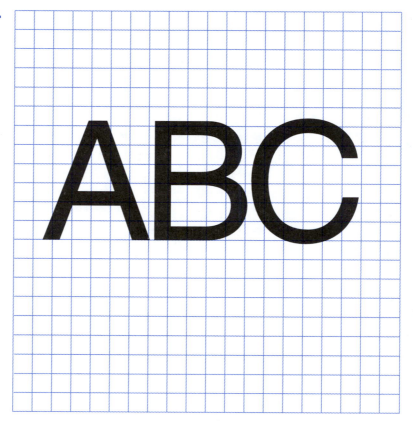

Scale factor = 2

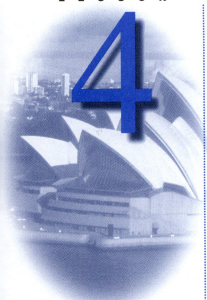

LESSON 4

Testing for Scale

In this lesson, you will find ways to test if polygons are scaled copies of each other.

In the last lesson, you thought about which characteristics of horse skeleton pictures did or did not make them well-scaled copies of each other. Now you will do the same for pairs of simple geometric figures like rectangles, triangles, and polygons.

Mathematicians usually use *scaled,* rather than *well-scaled* to refer to proportional figures. Also, the term *scale drawing* is used in architecture and other fields. From here on, we will use "scaled" rather than "well-scaled"; both terms have the same meaning.

Explore and Discuss

Below are two rectangles.

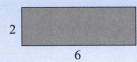

a How can you tell that the rectangles are scaled copies?

b Some people use the phrase *corresponding sides* when talking about scaled copies. What do you think this means?

c Some people say that the corresponding sides of these rectangles are *proportional.* What do you think this means?

ACTIVITY 1 ## Scaled Rectangles and Triangles

In this activity, you will learn design measurement tests to tell if two rectangles or two triangles are scaled copies of each other.

1. Below are the length and width measurements of seven rectangles. Match the rectangles that are scaled copies of each other. How did you make your decision?

a. 4" × 1"   **e.** 5" × 3"

b. 3" × 2"   **f.** 16" × 4"

c. 10" × 5"   **g.** 8" × 4"

d. 4" × 6"

2. a. The **ratio** of length to width of a particular rectangle is 1.5. A scaled copy has a width of 6. What is the length of the scaled copy?

   b. Two rectangles are scaled copies of each other. The ratio of the length of one rectangle to the length of the other is 0.6. If the smaller rectangle has a width of 3, what is the width of the larger one?

3. Take whatever measurements and do whatever calculations are necessary to check whether or not the two triangles below are scaled copies of each other. Explain your method.

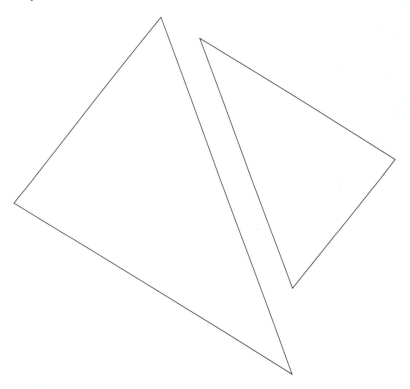

4. [Checkpoint] Kaori has two triangles with equal angle measurements. The sides of one triangle are 4, 6, and 8. The sides of the other are 9, 6, and 12. She says that the triangles are not scaled copies because

$$\frac{4}{9} = 0.44\ldots,$$
$$\frac{6}{6} = 1,$$

and

$$\frac{8}{12} = 0.66\ldots$$

Do you agree?

ACTIVITY 2 ▶ Checking for Scale Without Measuring

Think back for a minute to congruent triangles. How did you determine whether two triangles were congruent *without* taking any measurements or calculations? One way is to cut out the triangles and lay them on top of each other. If the triangles can be arranged so that they perfectly coincide, then they are congruent. Perhaps there is also a visual way to test whether two shapes are scaled copies of each other without having to take any measurements.

5. Take whatever measurements and do whatever calculations are necessary to check whether or not the two rectangles below are scaled copies of each other.

6. Below are three more pairs of rectangles. In which pair(s) are the rectangles scaled copies?

a.

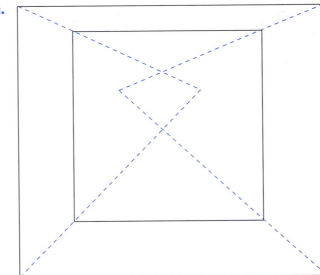

b.

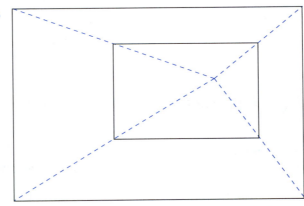

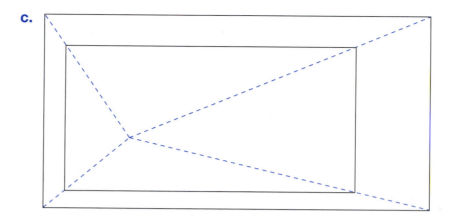

7. Your teacher will provide you with two pairs of triangles that are scaled copies of each other. Cut out each pair. Then play with the triangles and look for some visual clues that might make good tests for recognizing scaled triangles. For instance, if you place two corresponding angles on top of each other, what do you notice about the triangles' corresponding sides?

8. Trace and cut out the pair of triangles below. Based on the observations you have made in the previous problem, do you think these triangles are scaled copies of each other? Why or why not?

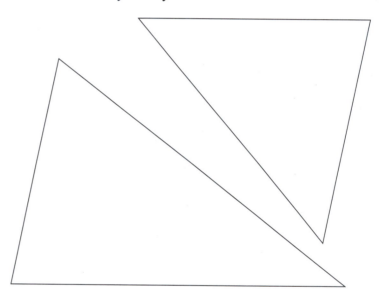

For Discussion

You have tried two approaches for checking whether or not triangles and rectangles are scaled copies: 1) you measured various parts of the triangles and did some calculations, and 2) you relied strictly on visual aspects without taking any measurements.

Discuss the various measurements, calculations, and observations you made with your classmates. Did they pay attention to the same characteristics of the triangles that you did? Make a list of all the different approaches that you and your classmates suggest.

9. a. In one classroom, students suggested that if the angles of one triangle are congruent to the corresponding angles of another triangle, then the two triangles must be scaled copies. Do you agree? Explore this conjecture with some triangles.

b. In the figure below, $\overline{DE} \parallel \overline{BC}$. Explain why, *if* the conjecture from part **a** is true, then this would prove that $\triangle ADE$ is a scaled copy of $\triangle ABC$.

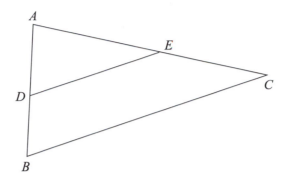

c. Martin thinks that the angle test for triangles might also be a good way to check whether other polygons, like rectangles, are scaled copies. If the angles of one polygon are congruent to the corresponding angles of another, are the polygons scaled copies? Explain.

10. Sheena has an idea for how to make a scaled copy of a triangle:

"Measure the sides of the triangle. Add the same constant value, like 1, 2, or 3, to each sidelength. Draw a triangle with these new sidelengths."

Will this new triangle be a scaled copy of the original? Try it!

11. Checkpoint Complete the following statements in as many different ways as possible. For now, write every possible test that you think might work. You will return to these ideas later in this unit and develop ways of proving or disproving these scaling tests.

a. To test if two rectangles are scaled copies of each other, ...

b. To test if two triangles are scaled copies of each other, ...

Scaled Polygons

You now have tests for scaled rectangles and triangles. What about other polygons?

12. In each figure below, decide if the pair of polygons are scaled copies. Explain how you made your decision.

a.

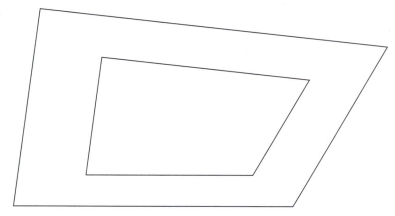

b.

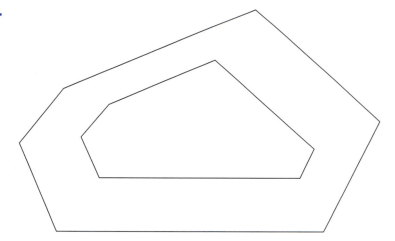

13. Trace the trapezoid below. Then draw another trapezoid inside of it that is a scaled copy (you choose the scale factor). How did you do it?

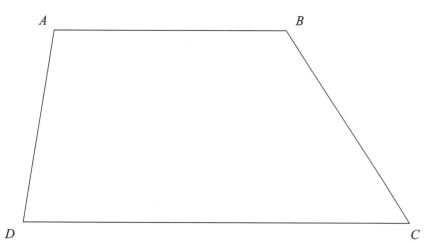

14. A polygon is scaled by a factor of $\frac{3}{4}$. The original polygon is then compared to the scaled one. Find the ratio of

    a. any two corresponding sides of the polygons.

    b. any two corresponding angles of the polygons.

15. **Checkpoint** Are either of the two statements below valid tests for checking whether or not two polygons are scaled copies? Explain your answers. If these are not valid tests, what additional requirement(s) can you add to make a test that works?

    a. Two polygons are scaled copies of each other if they can be arranged in some position so that their corresponding angles all have equal measures.

    b. Two polygons are scaled copies of each other if they can be arranged in some position so that their corresponding sides are all in the same ratio.

On Your Own

1. Can a 3-foot by 9-foot rectangle be a scaled copy of a 3-foot by 1-foot rectangle? Why or why not?

2. A square has been scaled by a factor of 2.5. It now has a sidelength of 8 inches. What was the length of the side of the original square?

3. One triangle has sidelengths of 21, 15, and 18. Another triangle has sidelengths of 12, 14, and 16. Are these triangles scaled copies? How can you tell?

4. Trace and cut out the triangles below. Use any of the methods your class has discussed to figure out whether or not any of the triangles are scaled copies of each other.

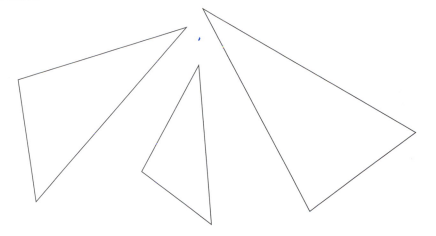

5. Two angles of one triangle measure 28° and 31°. Another triangle has two angles that measure 117° and 31°. Are the triangles scaled copies? How can you tell?

6. How can you tell if two squares are scaled copies of each other?

7. Draw two quadrilaterals so that the sides of one are twice as long as the corresponding sides of the other, but neither is a scaled copy of the other.

8. For each pair of figures listed below, explain why they must be scaled copies of each other, or give a counterexample to show why they are not always scaled copies.

   a. any two quadrilaterals

   b. any two squares

   c. any two quadrilaterals with equal corresponding angle measurements

   d. any two triangles

   e. any two isosceles triangles

   f. any two isosceles right triangles

   g. any two equilateral triangles

   h. any two rhombi

   i. any two regular polygons with the same number of sides

9. Which of the following figures must be scaled copies of each other? Explain why they are or find a counterexample.

   a. all rectangles

   b. all squares

   c. all parallelograms

   d. all trapezoids

   e. all isosceles trapezoids

   f. all regular hexagons

   g. all octagons

   h. all isosceles triangles

   i. all equilateral triangles

   j. all circles

   k. all cubes

   l. all spheres

   m. all cylinders

   n. all boxes

   o. all cones

Curved or Straight? Just Dilate!

In this lesson, you will learn one method for making scaled copies of different figures.

Testing for scale in polygons is no problem now. But how can you create scaled copies of a given figure, especially if that figure is not a polygon, but has curved sides?

Explore and Discuss

If all you had available was a ruler and pencil, how would you scale this curve by a factor of 2?

ACTIVITY

Making Scaled Copies

To scale a figure, a projector sends beams of light through it and catches those beams on a parallel surface. This is a "point-by-point" process. A 2-dimensional model might look like this:

Since the figure is enlarged by a factor of 2, point A' is twice as far from the center of dilation as point A is. Concisely put, CA' = 2CA. Likewise, CB' = 2CB.

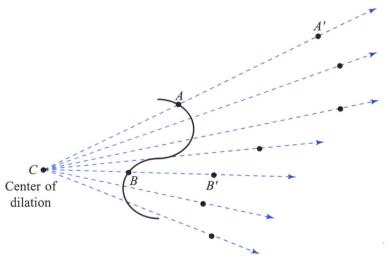

Dilating by a factor of 2

In mathematics, this process is called **dilation.** The point marked *center of dilation* represents the source of light, and the rays coming out from it represent the beams of light. The points farther out along the rays, like *A'* and *B'*, represent some of the points on the enlarged copy of the curve. To get more points, draw more rays.

You can think of dilation as a particular way of scaling a figure. So, if you are asked to "dilate a figure by 2," this means to use the dilation method to scale it by 2.

Of course, you cannot dilate *every* point on a curve, because curves have infinitely many points. The more points you choose, though, the more accurate an enlargement or reduction you will get.

If you look in a dictionary, you will find that the word dilate means "to make wider or larger," or "to cause to expand." For instance, an eye doctor might dilate your pupils. In mathematics, the word is used more generally. Dilating can refer to either expanding a figure *or* shrinking it.

Pick any center of dilation that you want.

1. **a.** Draw a circle on a piece of paper. Use the dilation method to scale it by a factor of 2. Choose enough points on the circle so that you can judge whether or not your dilation really does produce a scaled copy.

   b. Compare your result with your classmates' work. How does your choice of the center of dilation affect the result?

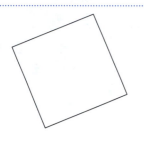

2. **a.** Draw a tilted square on a piece of paper. Dilate it by a factor of 2 (dilate at least eight points before drawing the entire dilated square).

   b. How does the orientation of the dilated square compare with the original tilted square?

3. Explain how to dilate a circle or square by a factor of

   a. $\frac{1}{2}$.

A tilted square

   b. 3.

4. The picture below shows an ornamental pattern on the left, along with a larger copy on the right that was dilated by 2. Find the location of the center of dilation.

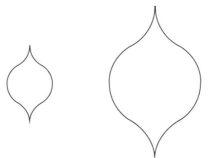

5. **Checkpoint** Make a scaled sketch of any picture you choose by dilating it by a factor of 2. One possibility is to use the picture of Trig the horse shown below. You do not need to scale all the details from your picture; a rough outline is fine, but include at least the important ones.

Below is an application of dilation with surprising results.

Stand in front of a mirror (perhaps one in the bathroom) and use some soap to trace your image. Include features like your eyes, nose, mouth, and chin.

1. Take a ruler and measure a few of the distances on your mirror picture. How far apart are your eyes? How long is your mouth? How far is it from your chin to the top of your head?

2. Compare the distances you have measured on the mirror to the actual measurements of your face. Are they the same?

3. How can the concept of dilation help to explain your results?

4. Stand in front of the mirror and have a friend trace the image of your face that she sees. Do you get the same picture as before? Why?

Dilating a curve with a ruler and pencil takes a lot of patience! You need to dilate a lot of points to get a good outline of the scaled copy. Geometry software gives a way to speed up this process.

5. Draw $\overline{AB}$ with geometry software. Construct its midpoint, M. Select points B and M. Use the software's **Trace** feature to indicate that you want the software to keep track of their paths.

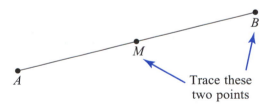

Trace these two points

a. Move point B around the screen. Use it to draw a picture or perhaps to sign your name. Compare point B's path to the path traced by point M. Are they the same? What is the relationship between them?

Point A should stay fixed as you move point B.

b. Use the software's segment tool to draw a polygon on your screen. Next, move point B along its sides. Describe the path traced by point M.

Move point B fairly quickly around the screen. If you move it slowly, the screen fills up with traced segments and the picture is hard to see.

c. In addition to tracing points B and M, also trace the entire segment AB as you move point B. How does your final picture illustrate the concept of dilation?

LESSON

6

Ratio and Parallel Methods

In this lesson, you will learn two strategies, the ratio and parallel methods, for making scaled copies of polygons.

When you scale a curve using dilation, you need to dilate a fair number of points before the scaled copy begins to even look recognizable. For simple figures like polygons, though, there are two shortcuts—the ratio and parallel methods—that make the dilation process much faster.

Try this with your own polygon.

Explore and Discuss

Suppose that you want to dilate a polygon like *ABCDE* by $\frac{1}{2}$. Draw your own polygon and pick any center of dilation (point *F* below). Draw rays from that point through every vertex of the polygon.

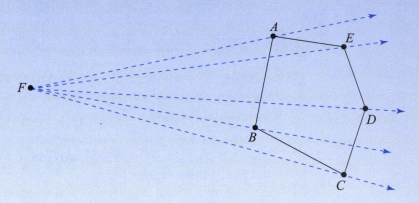

Next, find the midpoints of $\overline{FA}$ through $\overline{FE}$.

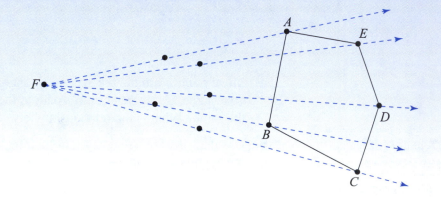

Finally, connect the midpoints to form a new polygon.

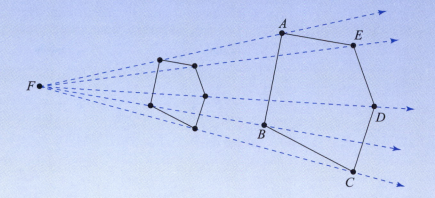

a Is your new polygon a scaled (by $\frac{1}{2}$) copy of the original? How can you tell?

b There seem to be many parallel segments in the figure above. Label the parallel segments you see on your own drawing.

c Does the ratio method work if point *F* is inside the polygon?

d Does it work if point *F* is *on ABCDE*? Try placing *F* at a vertex of the polygon.

ACTIVITY 1

The Ratio Method for Dilating a Polygon

Aside from making half-size reductions, the ratio method demonstrated above can dilate a figure by any amount you choose.

1. Start with a polygon. Use the ratio method to dilate it by $\frac{1}{3}$.

2. Use the ratio method to enlarge a polygon by a factor of 2.

3. Rosie wants to make two scale drawings of $\triangle ABC$: one that is dilated by a factor of 2, and another that is dilated by a factor of 3. She decides to make vertex *A* her center of dilation. Draw a picture like the one below. Finish her construction.

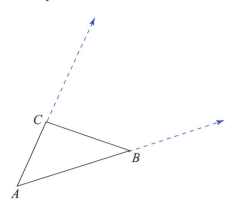

4. Steve uses the ratio method to enlarge $\triangle ABC$ below by a factor of 2. He follows this procedure:

- He measures the distance DA and finds it to be 1.0. So he moves out along $\overrightarrow{DA}$ until he finds a point A' so that $AA' = 2$ (twice as much as DA).

- He measures the distance DB and finds it to be 2.0. So he moves out along $\overrightarrow{DB}$ until he finds a point B' so that $BB' = 4$ (twice as much as DB).

- He measures the distance DC and finds it to be 1.5. So he moves out along $\overrightarrow{DC}$ until he finds a point C' so that $CC' = 3$ (twice as much as DC).

- He draws $\triangle A'B'C'$.

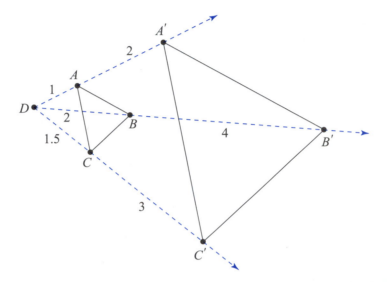

To Steve's surprise, $\triangle A'B'C'$ has sides that are proportional to $\triangle ABC$ but are not twice as long. How many times as long are they? What's the matter here?

5. Oh, no! There's been a problem at the printing press. The picture below was supposed to show $\triangle ABC$ and its dilated companion, $\triangle A'B'C'$, but an inkspot spilled onto the page. Can you salvage this disaster by calculating by how much $\triangle ABC$ has been scaled?

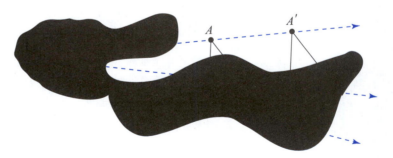

6. **Checkpoint** Copy the figure below as well as point *D*.

    a. Use the ratio method to scale the figure by $\frac{3}{4}$.

    b. Use the ratio method to scale the figure by 1.5.

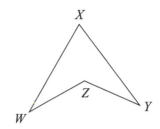

ACTIVITY 2

The Parallel Method for Dilating a Polygon

Take a look back at each picture you have drawn of a polygon and its dilated companion. You might notice something—the polygons are oriented the same way and their sides are parallel. Aha! This observation suggests another way to dilate.

The *parallel method* begins with the same setup as the ratio method.

7. To make a drawing of a quadrilateral that's dilated by a factor of 2, do the following steps.

    Step 1 Draw quadrilateral *ABCD*.

    Step 2 Pick any point *E* (the center of dilation) and draw rays from that point through all the vertices.

    Step 3 Go out along one ray ($\overrightarrow{EA}$ below) twice the distance from *E* to the polygon vertex. Mark this location *F*.

Estimate the parallel segments as best you can.

    Step 4 Starting at point *F*, move around the rays, drawing segments parallel to the sides of polygon *ABCD*.

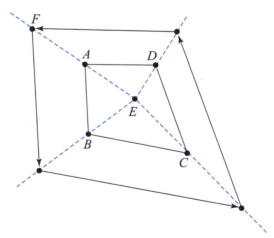

Point F is chosen so that EF = 2EA.

Start drawing parallels to the sides of *ABCD* from point *E*.

Is the new polygon a scaled (by a factor of 2) copy of the original? How can you tell?

8. Use this technique to reduce a polygon so that it's dilated by $\frac{1}{2}$.

9. Use the parallel method to enlarge a polygon by a factor of 3. Be sure to try locations for the center of dilation that are inside the polygon, on the polygon, and outside the polygon.

10. **Checkpoint** Copy the figure below as well as point *D*.

a. Use the parallel method to scale the figure by $\frac{3}{4}$.

b. Use the parallel method to scale the figure by 1.5.

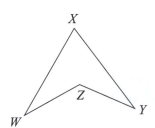

On Your Own

1. Draw any polygon.

a. Scale it by 2 using the ratio method with the center of dilation outside of the polygon.

b. Scale it by 2 using the ratio method with the center of dilation anywhere inside of the polygon.

c. Scale it by 2 using the ratio method with the center of dilation somewhere on the polygon.

d. Explain how the location of the center of dilation affects the scaled copy.

2. Draw any polygon. Create a scaled copy that shares a vertex with the original. Use any scale factor you like.

3. Draw any polygon. Create a scaled copy so that one side of the copy contains the corresponding side of the original.

4. Julia scaled a polygon three different times.

a. The first time the scaled copy was closer to the center of dilation than the original polygon. What can you say about the scale factor?

b. The second time the scaled copy was farther from the center of dilation than the original polygon. What can you say about the scale factor?

c. The third time the scaled copy was the same distance away from the center dilation as the original polygon. What can you say about the scale factor?

Using a Slider Point   If you try the parallel method with geometry software, you can create a whole series of polygons dilated to different sizes without having to start from scratch each time. Here's how:

Draw a polygon using the software. Pick a center of dilation. Draw rays from this point through the polygon's vertices. Then place a point anywhere along one of the rays. This will be your slider point. It will control the amount of dilation.

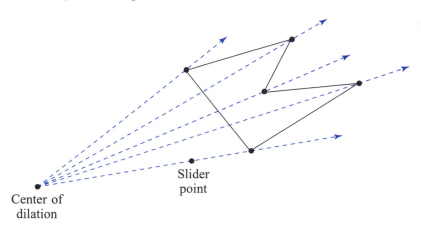

Slider
point

Center of
dilation

With the slider point as your place to begin, use the parallel method to make a dilated copy of the polygon. When your dilated polygon is complete, move the slider point back and forth along its ray. The polygon will grow and shrink, always remaining a scaled copy of the original!

5. Use the software to calculate by how much your polygon has been dilated. This dilation amount should update itself automatically as you change the location of the slider point.

   a. For what locations of the slider point is the dilation amount less than one?

   b. For what locations of the slider point is the dilation amount greater than one?

   c. Where is the dilation amount equal to one?

6. Draw $\triangle ABC$ with pencil and paper or geometry software. Your challenge is to construct a square with one side lying on $\overline{BC}$ and the other two vertices on sides $\overline{AB}$ and $\overline{AC}$. One way to get started is to draw a square with side $\overline{BC}$ facing away from the triangle; and think of A as a dilation point.

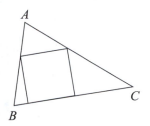

LESSON

7

The Side-Splitting and Parallel Theorems

In this lesson, you will construct proofs for the Side-Splitting Theorem and the two parts of the Parallel Theorem.

Geometry software makes it easy for you to experiment with different locations of point C.

Explore and Discuss

To prepare for the proofs, here are several questions about area to work on first. You will use these results in the proofs that follow.

a Points *A* and *B* of △*ABC* are fixed on the line below. Point *C*, which is not shown, lies on the parallel line above it and is free to wander anywhere along it. For what location of point *C* is the area of △*ABC* the largest? Why?

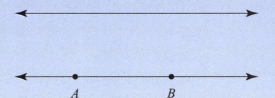

Parts **b** and **c** ask you to prove the following:

If two triangles have the same height, the ratio of their areas is the same as the ratio of their bases.

b Both △*ABC* and △*DEF* have the same height, *h*. Write an expression for the area of each triangle. Then show that the ratio of their areas is equal to $\frac{AC}{DF}$.

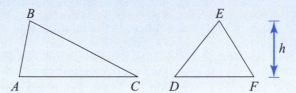

c The area of △*GEM* is 3 square inches. The area of △*MEO* is 2 square inches. What is the value of $\frac{GE}{EO}$? Why? Also find the value of $\frac{GE}{GO}$.

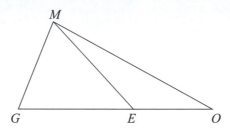

Quadrilateral *ABCD* is a trapezoid with $\overline{AB} \parallel \overline{DC}$.

d Explain why the area of △*ACB* is equal to the area of △*ADB*.

e Name two other pairs of triangles in the figure that also have equal areas.

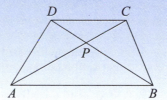

Nested Triangles

Below is a polygon and a dilated copy of it that was made using the parallel method.

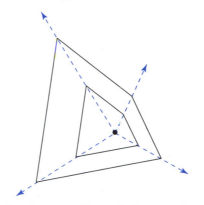

A polygon and its scaled companion

The four pictures at the right show the same polygons, only now, a pair of *nested triangles*—one triangle sitting inside another—is highlighted in each. Notice that each pair of nested triangles contains a side from the original polygon and a parallel side from the scaled polygon.

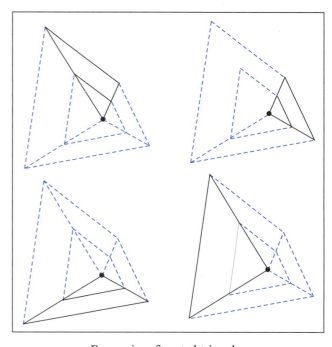

Four pairs of nested triangles

Experiment: Draw a picture like the one below that contains two parallel lines and a point, where the distance between the point and the top line is equal to the distance between the two lines.

1. Use your picture to find a super-fast way to take a ruler and draw ten segments that will each have their midpoints automatically marked.

2. Next, find a really quick way to draw ten segments that will each be divided in the ratio of 1 to 3. You will need to reposition the two parallel lines and the point to do this.

Experiment: Use geometry software to draw $\overline{XY}$. Then construct point A on the segment and point B that's not on $\overline{XY}$. Draw $\overline{AB}$ and construct its midpoint, M. Drag point A back and forth along the entire length of $\overline{XY}$ while tracing the path of point M.

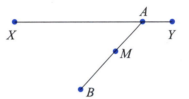

Drag point A along $\overline{XY}$.

3. **a.** Describe the path traced by M.

   b. How does the length and orientation of the path traced by point M compare to the length and orientation of $\overline{XY}$?

4. Repeat the experiment above, only this time, instead of constructing the midpoint of $\overline{AB}$, place the point M somewhere else on the segment. How does the position of M affect your answer to Problem **3**?

5. Use geometry software to draw $\overline{XY}$ and construct point A on the segment. Then draw three segments each with A as the endpoint. Construct the midpoints M_1, M_2, and M_3 of the three segments. Move point A back and forth along $\overline{XY}$ while tracing these midpoints. Describe the paths of the midpoints and anything you can say about their lengths.

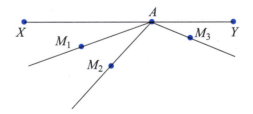

Experiment: Use geometry software to draw an arbitrary triangle *ABC*. Place point *D* anywhere on side $\overline{AB}$. Then construct $\overline{DE}$ that is parallel to $\overline{BC}$.

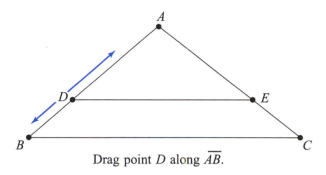

Drag point *D* along $\overline{AB}$.

△*ADE* and △*ABC* are a pair of nested triangles.

6. a. Use the software to calculate the ratio $\frac{AD}{AB}$.

   b. Find two other ratios of lengths with the same value. Do all three ratios remain equal to each other when you drag point *D* along $\overline{AB}$?

   c. Use the software to calculate the ratio $\frac{AD}{DB}$. Find any other ratios sharing this same value.

You can also think about this setup as a dilation. The center of the dilation is *A*, and the scale factor is $\frac{AB}{AD}$. You can say that $\overline{DE}$ splits $\overline{AB}$ and $\overline{AC}$ proportionally.

The following three theorems may be similar to ones that you or your classmates came up with based on the experiments. You will prove them in the next activities. First, you will need a definition.

Definition

Let *ABC* be a triangle. If *D* is a point on $\overline{AB}$ and *E* is on $\overline{AC}$, then $\overline{DE}$ splits $\overline{AB}$ and $\overline{AC}$ proportionally if

$$\frac{AB}{AD} = \frac{AC}{AE}.$$

The scale factor $\frac{AB}{AD}$ is called the *common ratio.*

Theorem 4.1   The Parallel Theorem

If a segment is parallel to one side of a triangle, then

a. it splits the other two sides proportionally, and

b. the ratio of the length of the parallel side to this segment is equal to the common ratio.

Theorem 4.2   The Side-Splitting Theorem

If a segment splits two sides of a triangle proportionally, then it is parallel to the third side.

7. **Checkpoint** In the picture below, $\overline{DE} \parallel \overline{BC}$.

The picture is not drawn to scale.

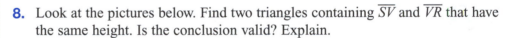

a. If $AD = 1$, $AB = 3$, and $AE = 2$, what is AC?

b. If $AE = 4$, $AC = 5$, and $AB = 20$, what is AD?

c. If $AD = 3$, $DB = 2$, and $AE = 12$, what is EC?

d. If $AE = 1$, $AC = 4$, and $DE = 3$, what is BC?

e. If $AD = 2$ and $DB = 6$, what is the value of $\frac{DE}{BC}$?

ACTIVITY 2 ▶ Proving the Parallel and Side-Splitting Theorems

Recall what the first part of the Parallel Theorem says:

> *If a segment is parallel to one side of a triangle, then it splits the other two sides proportionally.*

To prove this theorem, you need to show that $\frac{SV}{VR} = \frac{SW}{WT}$.

8. Look at the pictures below. Find two triangles containing $\overline{SV}$ and $\overline{VR}$ that have the same height. Is the conclusion valid? Explain.

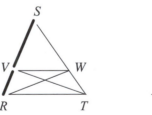

$$\frac{SV}{VR} = \frac{\text{Area } (\triangle SVW)}{\text{Area } (\triangle RVW)}$$

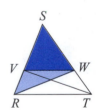

9. Look at the pictures below. Find two triangles containing $\overline{SW}$ and $\overline{WT}$ that have the same height. Is the conclusion valid? Explain.

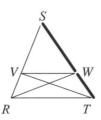

$$\frac{SW}{WT} = \frac{\text{Area } (\triangle SVW)}{\text{Area } (\triangle RVW)}$$

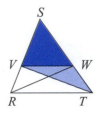

262 UNIT 4 A MATTER OF SCALE

10. In Problems **8** and **9**, you have two fractions with the same numerator.

  a. What are those fractions?

  b. The denominators of those fractions are not the same. Are they equal? Explain why.

11. Combine the results of Problems **8–10** to draw a conclusion about $\overline{SV}$, $\overline{VR}$, $\overline{SW}$, and $\overline{WT}$. Explain your work.

12. Your answers to Problems **8–11** outline a proof of the first part of the Parallel Theorem. Write up a proof using one of the methods you have learned.

13. Recall, the Side-Splitting Theorem:

> *If a segment splits two sides of a triangle proportionally, then it is parallel to the third side.*

This time, the proof is up to you! Use the same setup that you used for the Parallel Theorem proof. Write up your proof so that somebody else can follow it.

14. Checkpoint Consider these problems.

  a. In the pictures below, $\overline{AB}$ is parallel to $\overline{DE}$. Copy the pictures. Find as many lengths as you can.

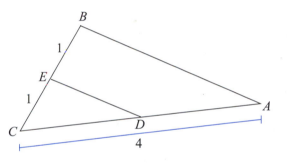

i.

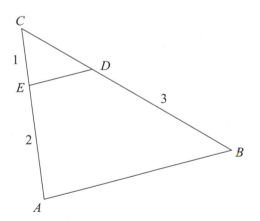

ii.

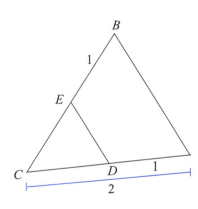

iii.

b. In the pictures below, decide if $\overline{AB}$ is parallel to $\overline{DE}$.

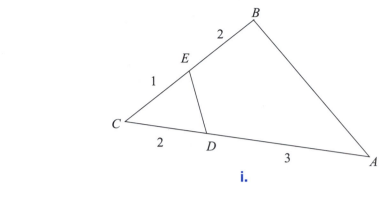

i.

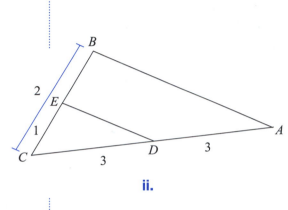

ii.

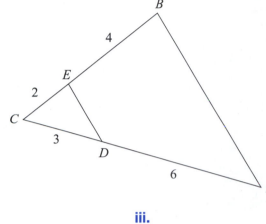

iii.

ACTIVITY 3 › **Proving the Parallel Theorem Continued**

Now you will finish proving the Parallel Theorem. Recall what this theorem says:

If $\overline{VW}$ is parallel to $\overline{RT}$, then

$$\frac{RT}{VW} = \frac{SR}{SV} = \frac{ST}{SW}.$$

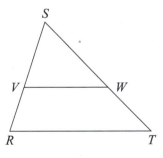

So all that remains is to prove their equality to $\frac{RT}{VW}$.

You have proved the first part of this theorem already—the Parallel Theorem guarantees that if $\overline{VW} \parallel \overline{RT}$, then $\frac{SV}{VR} = \frac{SW}{WT}$.

15. Actually, you proved that $\frac{SV}{VR} = \frac{SW}{WT}$. Use that fact to show that $\frac{SR}{SV} = \frac{ST}{SW}$.

16. Trace and cut out paper triangles of △*SVW* and △*SRT*. Then place them on top of each other like so:

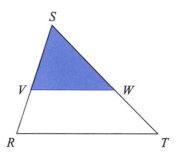

Take △*SVW* and slide it along $\overline{SR}$ until vertex *V* and *R* coincide.

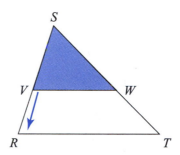

   a. Why will ∠*SVW* coincide with ∠*SRT*?

   b. Draw a picture of what the two triangles will look like after △*SVW* has been slid all the way to $\overline{RT}$.

   c. Which two segments are now parallel? Why?

17. Use this new setup to prove the remainder of the Parallel Theorem—namely, that $\frac{RT}{VW}$ equals both $\frac{SR}{SV}$ and $\frac{ST}{SW}$.

18. **Checkpoint** In the following pictures, $\overline{BC}$ is parallel to $\overline{DE}$. Copy the pictures. Find as many lengths as you can.

   a.

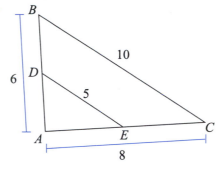

   b.

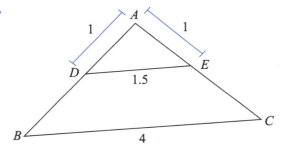

C.

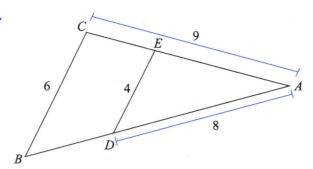

On Your Own

1. Understanding the theorems in this lesson by reading through the words can be pretty tricky. Why don't we replace some of the words by actual segment names? Rewrite the two theorems using the nested triangles below as a reference. Make each theorem as specific as possible—if a segment length or proportion is mentioned, substitute that segment or proportion in place of the words.

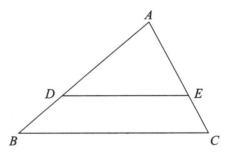

2. The Parallel Theorem says that a segment parallel to a side of a triangle "splits the other two sides proportionally." Tammy Jo has two ways to remember this. She says, "part is to part as part is to part" or "part is to whole as part is to whole." What does she mean?

3. In the picture below, the dashed lines are all parallel to $\overline{AC}$ and equally spaced. What can you conclude about how they intersect $\overline{AB}$ and $\overline{BC}$?

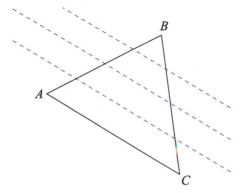

4. In the picture below, $\overline{BC}$ has been trisected.

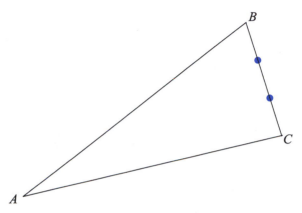

    a. Copy the picture. Use what is given to trisect $\overline{AB}$.

    b. Copy the picture. Use what is given to trisect $\overline{AC}$.

5. In the picture below, $\overline{AB}$ and $\overline{BC}$ have each been cut into six equal pieces by the dashed lines. What can you conclude?

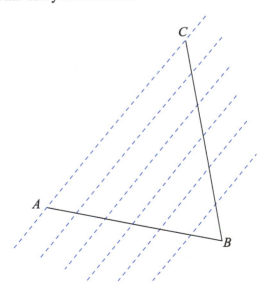

6. In each picture below, decide if $\overline{AB}$ is parallel to $\overline{DE}$. In each case, explain your decision.

a.

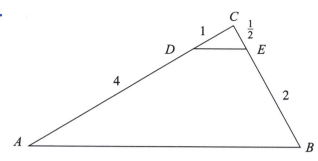

b.

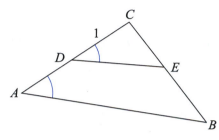

c.

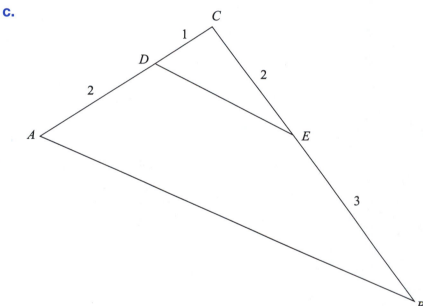

d.

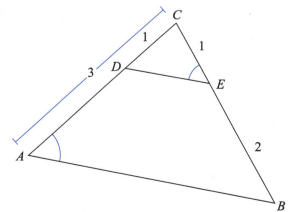

7. Use the Parallel and Side-Splitting Theorems to explain why the two rectangles in each figure below are scaled copies of each other.

a.

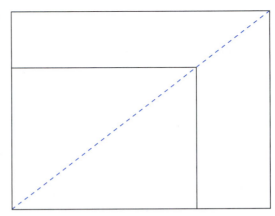

b.

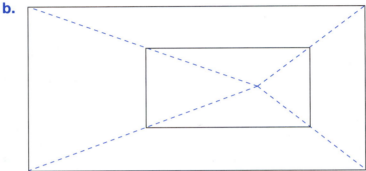

The sides of the inner rectangle are
drawn parallel to the outer sides.

8. In the figure below, $\overline{DE} \parallel \overline{BC}$. Explain why $\triangle ABC$ is a scaled copy of $\triangle ADE$.

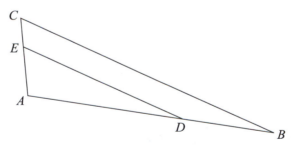

9. The sides of the outer polygon below are drawn parallel to the sides of the inner one. Explain why the polygons are scaled copies.

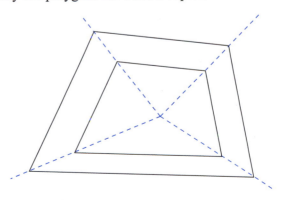

10. How can you construct, *without* taking any measurements, a triangle with the same area as quadrilateral *ABCD*?

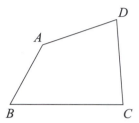

Here's one method:
First, draw diagonal $\overline{AC}$. Then draw a line through point *D* parallel to $\overline{AC}$. Extend $\overline{BC}$ to meet the parallel at point *E*.

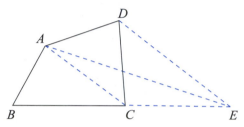

This completes the construction. The area of △*ABE* is equal to the area of quadrilateral *ABCD*. Explain why. (Hint: Find another triangle that has the same area as △*ADC*).

11. Extend the technique used in Problem **10** to construct a triangle with area equal to that of pentagon *ABCDE* below.

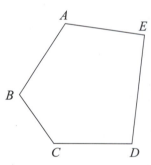

12. The picture below shows the land owned by Wendy and Juan. The border between their properties is represented by the segments *AB* and *BC*. How can you replace these segments by just a single segment that does not change the amount of land owned by either person?

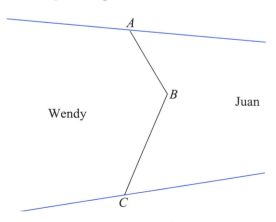

13. In pentagon *ABCDE* below, four of the sides are each parallel to a diagonal: $\overline{AB} \parallel \overline{CE}$, $\overline{BC} \parallel \overline{AD}$, $\overline{CD} \parallel \overline{BE}$, and $\overline{DE} \parallel \overline{CA}$. Prove that the remaining side $\overline{EA}$ is parallel to the diagonal $\overline{DB}$.

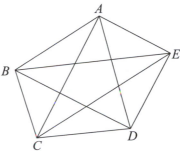

Here is an idea to get you started:

Proving that $\overline{EA} \parallel \overline{DB}$ is equivalent to showing that the area of $\triangle ABE$ is equal to the area of $\triangle ADE$. Why?

Source: This problem is from *Quantum*, November/December 1991, Volume 2, #2, p. 18. Reprinted by permission of Springer-Verlag New York, Inc. Any further production is prohibited.

By now, you have seen that there are two ways to write the first part of the Parallel Theorem. If $\overline{DE} \parallel \overline{BC}$ in the figure below, then $\frac{AB}{AD} = \frac{AC}{AE}$ and $\frac{DB}{AD} = \frac{EC}{AE}$.

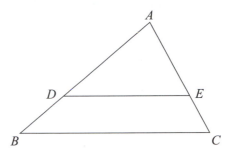

The next two problems show you how to prove that these proportions are two different ways of writing the same information.

Hint: $\frac{r-s}{s} = \frac{r}{s} - \frac{s}{s}$.

14. First, prove a related fact using algebra. Suppose that *r*, *s*, *t*, and *u* are any four numbers. If $\frac{r}{s} = \frac{t}{u}$, explain why it is also true that $\frac{r-s}{s} = \frac{t-u}{u}$.

15. Use Problem **14** as a guide to explain why the proportion $\frac{AB}{AD} = \frac{AD}{AE}$ is equivalent to $\frac{DB}{AD} = \frac{EC}{AE}$.

Midpoints in Quadrilaterals

In this lesson, you will discover an interesting fact about quadrilaterals and prove it using the Side-Splitting Theorem.

Explore and Discuss

Draw any quadrilateral *ABCD*. Construct the midpoint of each side. Then connect the midpoints to form a new quadrilateral inside of *ABCD*, which will be called a *midpoint quadrilateral*.

a Describe the features of your new quadrilateral.

b Try to classify it as a particular kind of quadrilateral.

c If you are using geometry software, experiment by moving the vertices and sides of *ABCD*. Does the *midpoint quadrilateral* retain its special features?

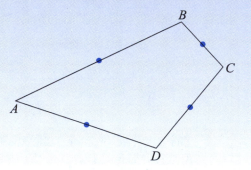

A quadrilateral and its midpoints

ACTIVITY

Why Does It Happen?

In this activity, you will use the Side-Splitting Theorem to prove and generalize your conjecture about the midpoints in quadrilaterals.

1. Prove that your conjecture about the midpoint quadrilaterals is correct. Make sure that your proof is valid for each of the three locations of *B* shown below. It may help to draw in the diagonals of the quadrilaterals.

Technically, self-crossing figures like the one on the right are not quadrilaterals, but your proof could work for these figures as well.

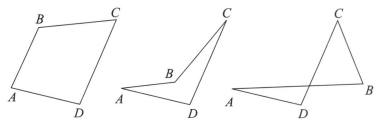

Three different locations of point *B*

2. The diagonals of a quadrilateral *ABCD* measure 8 inches and 12 inches. What is the perimeter of the midpoint quadrilateral?

3. Describe any special characteristics of the midpoint quadrilateral if the diagonals of the outer quadrilateral are

    a. congruent.

    b. perpendicular to each other.

    c. congruent and perpendicular to each other.

Juan said that he didn't use the Side-Splitting Theorem in his proof for Problem **1.** His proof is outlined below.

- In a quadrilateral *ABCD*, construct the midpoints. Connect them in order to make a new quadrilateral inside the first one.

- Draw diagonal $\overline{AC}$. Two opposite sides of the inner quadrilateral are parallel to and half as long as $\overline{AC}$ because of the Midline Theorem. You just have to look at *ABCD* as two triangles sharing a base $\overline{AC}$ to see that.

- Draw the other diagonal $\overline{BD}$. Two opposite sides of the inner quadrilateral are parallel to and half as long as $\overline{BD}$ because of the Midline Theorem.

4. Write up Juan's proof, including pictures and a conclusion about the inner quadrilateral.

Jessica said that using the Side-Splitting Theorem was better because then her proof worked for more than just midpoints. To explain her idea better, she showed the picture below.

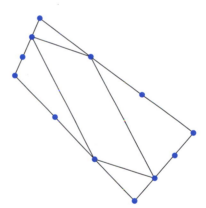

5. Explain Jessica's idea.

    a. Describe the points that Jessica connected in her figures. It may help to redraw the figure and label some points.

    b. Use the Side-Splitting Theorem to prove that the inner quadrilateral is a parallelogram.

6. Write another generalization about the midpoints in quadrilaterals. Prove your generalization using the Side-Splitting Theorem.

7. **Checkpoint** Mathematicians say that the Midline Theorem is a special case of the Side-Splitting Theorem. What does this mean? Describe that "special case" exactly.

1. Suppose a kite has diagonals with lengths 6 and 9.

   a. A quadrilateral is formed by joining the midpoints of the kite's sides. Give the perimeter of that inner quadrilateral and describe its angles.

   b. A quadrilateral is formed by joining the trisection points of the kite's sides, as Jessica did above. Give the inner quadrilateral and describe its angles.

   c. Is there only one answer to part **b?** Explain.

2. What kind of a parallelogram do you get when you connect the midpoints of a square? Explain.

3. The diagonals of a quadrilateral measure 4 and 5. What is the perimeter of the inner quadrilateral you get by joining the midpoints of the sides of the quadrilateral?

4. **a.** Draw a few quadrilaterals that have the property that when you connect the midpoints of the sides, you get a rhombus.

   b. Is there some way to tell whether a particular quadrilateral will generate a rhombus in the middle without actually testing for the rhombus? Explain.

   c. **Challenge** Suppose you were using Jessica's construction instead of the midpoint construction. What starting quadrilateral, if any, will form a rhombus inside?

5. The picture below began with an equilateral triangle whose sides measured 32 inches. The midpoints of the sides were connected over and over again to make the design.

   a. List as many facts as you can about this figure.

   b. What is the perimeter of the smallest triangle in the figure? How did you arrive at that answer?

   c. What is the length of the spiral? How did you figure it out?

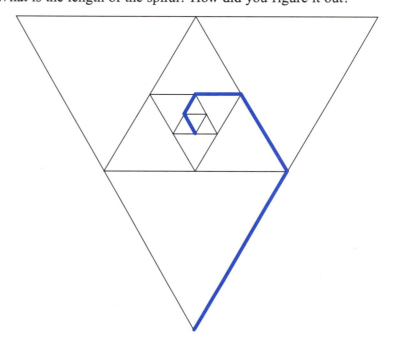

6. When the midpoints of any quadrilateral *ABCD* are connected, the inner quadrilateral (shaded below) is a parallelogram. Show that the area of the parallelogram is half the area of *ABCD*. (Hint: Think about a dissection argument.)

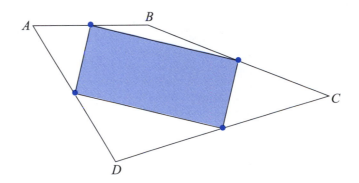

Take It Further

7. Here's a nice problem to explore with geometry software:

Draw an arbitrary quadrilateral *ABCD*. Place a point (labeled "Begin" in the following figure) anywhere you like. Construct a segment through *A* with "Begin" as an endpoint and *A* as its midpoint. Now start at the endpoint of this new segment and construct another segment, this time with *B* as the midpoint. Continue doing this until you construct a segment with *D* as the midpoint. The finishing point of this whole journey around *ABCD* is labeled "End."

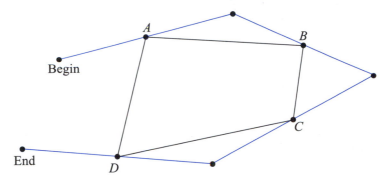

a. Draw the segment that connects "Begin" to "End." Drag the "Begin" point around the screen and watch what happens to this segment. How does its length change? How does its slope change? Make some conjectures and see if you can prove them.

b. For what kinds of quadrilaterals *ABCD* are "Begin" and "End" exactly the same point?

8. Take any two isosceles right triangles like $\triangle ABC$ and $\triangle DBE$ and place them back-to-back as shown below. Mark the midpoints of sides $\overline{AC}$, $\overline{CE}$, $\overline{ED}$, and $\overline{DA}$, and then connect them. Prove that the connected midpoints form a square.

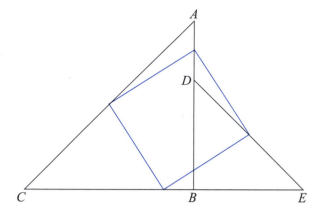

Some suggestions:

- Add $\overline{AE}$ and $\overline{CD}$ to the picture. Show that two sides of the "square" are parallel to $\overline{AE}$ and the other two are parallel to $\overline{CD}$.

- If you rotate $\triangle CBD$ by 90° clockwise about point B, where does point C land? Where does point D land? What does this tell you about the lengths of $\overline{AE}$ and $\overline{CD}$ and the angle between them?

Source: This problem is from *Quantum*, November/December 1994, Volume 8, #2, p. 9. Reprinted by permission of Springer-Verlag New York, Inc. Any further production is prohibited.

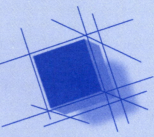

Perspective

A Method for Measuring the Earth

The Greek mathematician Eratosthenes devised an ingenious way to answer a perplexing question: How can one measure the circumference of our vast Earth?

Eratosthenes was an Alexandrian who lived around 200 B.C. He was a contemporary of Archimedes and the chief librarian at the great library in Alexandria. Eratosthenes' method for estimating the earth's circumference uses some of the facts you have proved about parallel lines. Here's what he did:

Alexandria was about 500 miles due north of another town, Syene (now called Aswan). Eratosthenes imagined slicing the Earth through the two cities, as shown in the diagram below.

These calculations are given in terms of miles and degrees. Eratosthenes used other units of measurement.

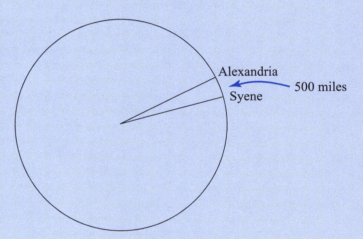

On the first day of summer each year, Eratosthenes knew that the sunlight reflecting into a well in Syene shot right back up into an observer's eyes at noon. He concluded that, on that day at that time, the center of the Earth, Syene, and the sun were in a straight line. Eratosthenes also assumed that the sun was so far away that its rays were essentially parallel by the time they hit the face of the Earth. Finally, on that day, a vertical pole erected at Alexandria cast a shadow making an angle of 7.2°, as shown in the picture below.

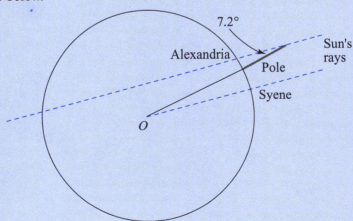

The pole at Alexandria

1. Using what you know about parallel lines, what is the measure of the central angle at *O*?

2. Using the result from Problem **1** and the fact that a circle has 360°, how much of the Earth's circumference is taken up between Syene and Alexandria? Use this to estimate the circumference of the Earth.

3. Check your estimate against a current estimate of the Earth's circumference (from an atlas or encyclopedia). How close were you and Eratosthenes?

You can read more about the connections between mathematics, the Earth, and astronomy in the book *Poetry of the Universe* by Robert Osserman (Anchor Books, 1995).

Defining Similarity

In this lesson, you will learn about an important idea called similarity. After defining similarity in terms of scaling, you will develop tests for similar triangles.

Words like *enlargements, reductions, scale factors,* and *dilations* are some of the terms you have used again and again in this unit.

Explore and Discuss

The common theme uniting them is called **similarity**. By taking a picture and enlarging or reducing it, you create another picture that is *similar* to the first. So one way to define *similar* in geometry is as follows.

Two figures are similar if one is a scaled copy of the other.

You can also use the word *dilation* to define *similar.* Here's another way to define similar.

Two figures are similar if one is a dilation of the other.

To test this definition, look at a picture of the head of Trig the horse. Trig is accompanied here by his little sister, Girt. She's smaller than Trig, but shares all of his features.

A family portrait

a Is Girt's head a dilated copy of Trig's? If so, find the center of dilation.

b Is the picture of Girt similar to the picture of Trig?

Below is another family portrait of Trig and Girt, this time in a different pose.

c Is the picture of Girt still similar to the picture of Trig?

d Can you still dilate one picture onto the other? Explain.

e Expand the dilation definition of *similarity* so that even these pictures of Trig and Girt can be called *similar.*

ACTIVITY 1 ▶ **Similarity**

Here are some suggestions for ways to define similarity using dilation terminology.

Definition

Two figures are *similar* if you can rotate and/or flip one of them so that it can be dilated onto the other.

Two figures are *similar* if one is congruent to a dilation of the other.

1. Do these definitions solve the problem you had with the second family portrait of Trig and Girt? Are these definitions equivalent?

2. a. If two figures are congruent, are they similar? Explain.

   b. If two figures are similar, are they congruent? Explain.

3. If two figures are similar to the same figure, are they similar to each other? Explain.

The symbol for similarity is the upper part of the symbol for congruence: ~ . Thus, the statement "*ABCD* ~ *EFGH*" is read "*ABCD* is similar to *EFGH*" and means that the two polygons below are scaled copies of each other.

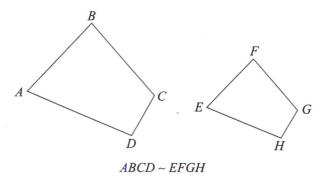

ABCD ~ *EFGH*

The similarity symbol, ~, means "has the same shape as," and the congruence symbol, ≅, means "has the same shape *and* has the same size as."

For congruent triangles, the statement $\triangle ABC \cong \triangle XYZ$ conveys specific information about their *corresponding* parts. Namely,

$$\angle A \cong \angle X, \angle B \cong \angle Y, \text{ and } \angle C \cong \angle Z$$

and

$$\overline{AB} \cong \overline{XY}, \overline{BC} \cong \overline{YZ}, \text{ and } \overline{AC} \cong \overline{XZ}.$$

For figures that are similar, the order of the letters representing their vertices also gives specific information about corresponding parts.

4. In the pictures below, $\triangle ABC \sim \triangle XYZ$ and *ABCDE* ~ *PQRST*. Copy the pictures. Label the vertices of $\triangle XYZ$ and *PQRST* correctly.

   a.

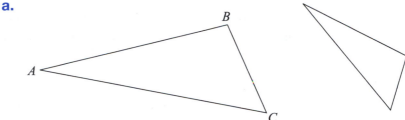

   b.

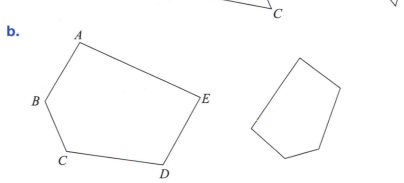

5. Checkpoint Are the following similarity statements true or false? Use measurements to decide. Explain.

a. *ABCD ~ EFGH*

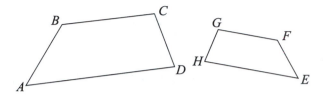

b. *ABCD ~ EFGH*

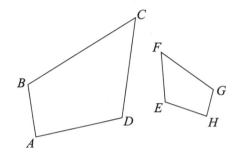

c. *ABC ~ DEF*

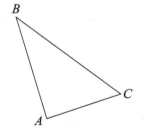

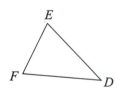

ACTIVITY 2 ▶ **Similar Triangles**

In Problem **9** of Lesson 4 on page 245, you wrote several ways to test if two triangles were similar (only then, you used the phrase "scaled copies"). One of these ways probably was:

> *Two triangles are similar if their corresponding angles are congruent and their corresponding sides are proportional.*

For future reference, refer to this test as the Congruent Angles, Proportional Sides test.

Using the word dilation, there is another test for similar triangles:

> *Two triangles are similar if one is congruent to a dilation of the other.*

Refer to this test as the Congruent to a Dilation test.

6. Suppose that △*NEW ~* △*OLD*. If *m∠N* = 19° and *m∠L* = 67°, find the measures of all the other angles.

7. If $\triangle ABC \sim \triangle DEF$, which of the following must be true?

   a. $\dfrac{AB}{DE} = \dfrac{BC}{EF}$

   b. $\dfrac{AC}{BC} = \dfrac{DF}{EF}$

   c. $\dfrac{BC}{AB} = \dfrac{DF}{DE}$

   d. $AC \times DE = AB \times DF$

8. In the figure below, $\triangle ACB \sim \triangle BAD$. Explain why $\triangle ACB$ is isosceles.

Hint: Look for congruent angles.

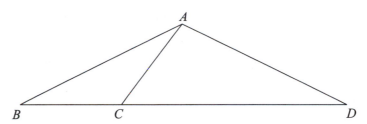

9. In the picture below, F, G, and H are midpoints of the sides of $\triangle ABC$. Show that $\triangle ABC \sim \triangle GHF$.

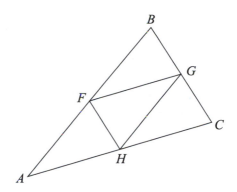

10. **Checkpoint** The picture below is the same as the one from Problem **8,** with some measurements added. Copy the picture. Find the rest of the angle measures and sidelengths.

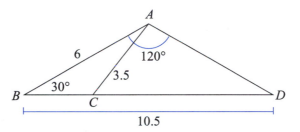

ACTIVITY 3 ▶ Tests for Similar Triangles

You know several tests to decide if triangles are congruent, for example SSS. It would be nice to have comparable tests for *similar* triangles. What do you think you will find?

11. The main tests for triangle congruence are SAS, ASA, AAS, and SSS. Are there equivalent tests for triangle similarity? Here are some possibilities to consider. For each test, draw a pair of triangles that share the attributes listed. Then check to see if they are similar. Also, see if you can find a counterexample.

a. If two triangles have all three corresponding angles congruent, the triangles are similar (AAA similarity).

b. If two triangles have a pair of corresponding sidelengths proportional and a pair of corresponding angles are congruent, the triangles are similar (SA similarity).

c. If two triangles have two pairs of sidelengths proportional and the included angles are congruent, the triangles are similar (SAS similarity).

d. If two triangles have all three corresponding sidelengths proportional, the triangles are similar (SSS similarity).

AAA Similarity Suppose two triangles, $\triangle ABC$ and $\triangle PQR$, have congruent corresponding angles. Can you prove that they are similar?

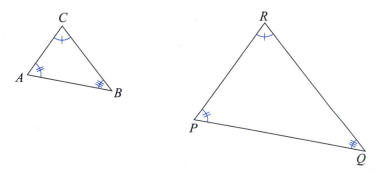

In earlier lessons, you checked to see if triangles were scaled copies (similar) by placing one inside the other to form a pair of nested triangles. Try placing $\triangle ABC$ inside $\triangle PQR$ so that $\angle C$ and $\angle R$ coincide.

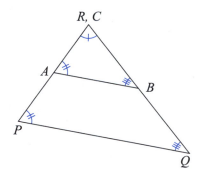

12. Do the triangles line up this way, or does one of them hang over the edges of the other? Explain.

13. From this picture, prove that $\overline{AB}$ is parallel to $\overline{CD}$.

14. Apply the Parallel Theorem to the figure above. Write several proportions between the sides of the two triangles.

15. Which definition of similarity allows you to conclude that $\triangle ABC \sim \triangle PQR$?

16. Use Problems **12–15** to prove the following theorem:

Theorem 4.3 AAA Similarity

If two triangles have corresponding angles congruent, the triangles are similar.

17. In a sense, the requirement that all three angles of one triangle be congruent to the corresponding three angles of the other is an overkill. Why?

18. Rewrite the theorem to give the *minimal* angle conditions necessary for two triangles to be similar.

19. Does it make sense to have an ASA test for triangle similarity? Explain.

20. Are two quadrilaterals similar if the angles of one are congruent to the corresponding angles of another? Either prove this as a theorem or disprove it by finding a counterexample.

SAS Similarity Is there an SAS similarity test? If such a theorem did exist, it would be stated as follows:

Theorem 4.4 SAS Similarity

If two triangles have two pairs of sidelengths proportional and the included angles are congruent, the triangles are similar.

21. You can prove this theorem by following the same basic setup used for AAA similarity. Write a proof of SAS similarity.

22. The AAA condition was too strong. Is SAS too strong as well?

23. Are two quadrilaterals similar if three corresponding sides are proportional and two included angles are congruent? Either prove this as a theorem or disprove it by finding a counterexample.

SSS Similarity Is there an SSS similarity test? If such a theorem did exist, it would be stated as follows:

Theorem 4.5 SSS Similarity

If two triangles have all three pairs of corresponding sidelengths proportional, the triangles are similar.

It may seem like cheating to change the definition. Words only mean one thing, right? Well, in mathematics, there are equivalent definitions, and sometimes one is handier than another.

24. If you try to prove this theorem using the same method as the AAA and SAS proofs, something doesn't quite work. Look back at the AAA and SAS proofs. In both of them, you placed $\triangle ABC$ inside $\triangle PQR$ so that $\angle C$ and $\angle R$ coincided. Can you do the same here?

Since the "congruent angles, proportional sides" method isn't well suited to proving SSS, you can use the "congruent to a dilation" test of triangle similarity from page 282 instead.

Suppose the corresponding sides of △ABC and △PQR are proportional.

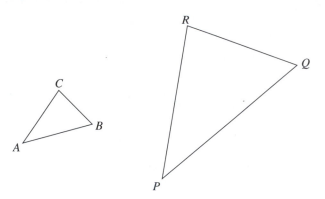

25. Write the proportionality statement for the sides of these two triangles.

26. If $\frac{PQ}{AB} = k$, where k is some number, finish each of the following statements.

a. _____ $= k(AB)$

b. _____ $= k(BC)$

c. _____ $= k(CA)$

Next, dilate △ABC by k, picking any point as the center of dilation.

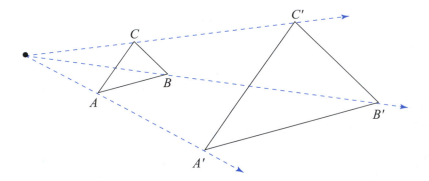

27. How do the sides of the dilated triangle △A'B'C' compare with the sides of △PQR? Justify your answer.

28. Is it true that △A'B'C' ≅ △PQR ? Which congruence postulate can you use?

29. Is △PQR "congruent to a dilation of" △ABC? Explain your answer.

30. Prove the theorem about SSS similarity. Problems **26–29** give you all the tools you need for the proof.

31. Checkpoint △JKL has JK = 8, KL = 12, and JL = 16. Points M and N are on $\overline{JK}$ and $\overline{KL}$ respectively, with JM = 6 and LN = 9.

a. Explain why $\overline{MN} \parallel \overline{JL}$.

b. Prove that △MKN ~ △JKL using each of the following theorems.

i. AA

ii. SAS

iii. SSS

1. In the picture below, $\triangle ABC \sim \triangle CDE$.

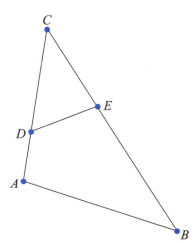

Which of the following statements are correct?

a. $ABC \sim DEC$ **d.** $CAB \sim ECD$

b. $BCA \sim DEC$ **e.** $CBA \sim ECD$

c. $BAC \sim DEC$ **f.** $CBA \sim CDE$

2. In the picture below, $\triangle QRS \sim \triangle TUV$.

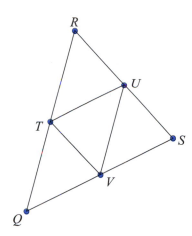

Which of the following statements are correct?

a. $\dfrac{QR}{UV} = \dfrac{SR}{TV}$ **d.** $\dfrac{QR}{QS} = \dfrac{UV}{TU}$

b. $\dfrac{QR}{SR} = \dfrac{TU}{TV}$ **e.** $\dfrac{QU}{QV} = \dfrac{RT}{RV}$

c. $\dfrac{QR}{TV} = \dfrac{SR}{TU}$ **f.** $\dfrac{QS}{UV} = \dfrac{RS}{TV}$

3. In the picture below, △*CAT* ~ △*DOT*.

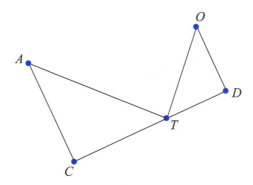

Complete the following statements.

   a. ∠*C* ≅ _____

   b. ∠*CTA* ≅ _____

   c. ∠*DTO* ≅ _____

   d. ∠*A* ≅ _____

   e. ∠*D* ≅ _____

   f. ∠*O* ≅ _____

4. The sides of a triangle are 4, 5, and 8. Another triangle is similar to it. One of its sides has length 3. What are the lengths of its other two sides? Is there more than one possible answer?

5. A triangle has sides of length 2, 3, and 4 inches. Another triangle is similar to it. Its perimeter is 6 inches. What are the sidelengths of this triangle?

6. Triangle *DEF* is similar to triangle *ABC* and has sides that are three times as long. Find the numerical values of the following ratios. Then name the triangle similarity theorem that allows you to draw your conclusion.

   a. The ratio of any two corresponding altitudes

   b. The ratio of any two corresponding angle bisectors

   c. The ratio of any two corresponding medians

7. In the figure below, $\overline{DG} \parallel \overline{BC}$ and $\overline{EF} \parallel \overline{BA}$. Prove that △*ABC* ~ △*FHG*.

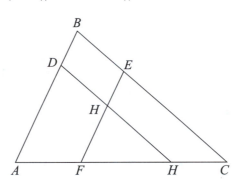

8. Segments BE and CD are altitudes of $\triangle ABC$ below. List as many pairs of similar triangles as you can find. Explain why the triangles are similar.

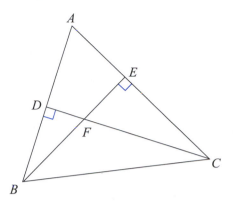

9. In the figure below, $\angle ADE \cong \angle ACB$. Explain why $\triangle ADE \sim \triangle ACB$.

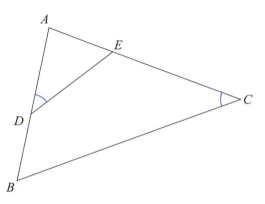

10. In the figure below, $AB = 2$ and $BC = 1$. Without making any measurements, find the values of $\frac{AD}{DE}$ and $\frac{AF}{FG}$. Explain how you got your answers.

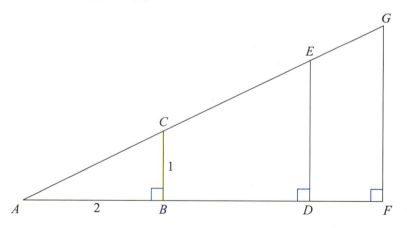

11. Draw a right triangle ABC and the altitude from the right angle to the hypotenuse. The altitude will divide $\triangle ABC$ into two smaller right triangles.

    a. There are two pairs of congruent angles, other than the right angles, in your picture. Find and label them.

    b. Make a copy of your triangle. Then cut out the two smaller right triangles. Position them in such a way as to convince yourself that they are similar to each other and to $\triangle ABC$.

    c. Explain why all three of these triangles are similar.

12. In the figure below, $AB = 4$, $BC = 5$, $AC = 6$, $DC = 2.5$, and $EC = 3$. Prove that $\triangle ABC \sim \triangle EDC$. Then find the length of segment DE.

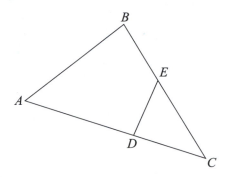

13. Quadrilateral $RATS$ is a trapezoid with $\overline{RA} \parallel \overline{ST}$ and diagonals $\overline{RT}$ and $\overline{AS}$ meeting at O.

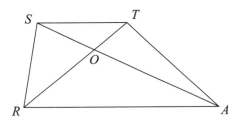

   a. Explain why $\triangle ROA \sim \triangle TOS$.

   b. From part **a**, you can say that $\frac{RO}{TO} = \frac{OA}{OS}$. Why?

   c. Given the proportion from part **b** and the fact that $\angle ROS \cong \angle TOA$, Eloida claims that $\triangle ROS \sim \triangle TOA$ by the SAS similarity test. Is this true? Explain.

14. In the picture below, F, G, and H are midpoints of the sides of $\triangle ABC$.

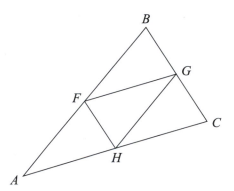

   Show that $\triangle ABC \sim \triangle GHF$.

15. In Problem **14**, there are several other pairs of similar triangles. List them and explain why they are similar.

This is a great question to spring on a friend while you're eating. Use a square napkin to pose the challenge.

16. Take a square sheet of paper and fold one fourth of it behind. You will be left with a rectangle. The challenge is to fold the paper to form a rectangle that's similar to this one, but has half its area. No fair using a ruler!

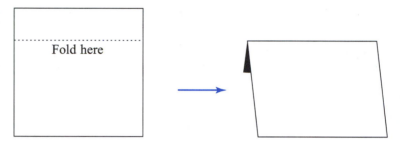

(Hint: Even though you have folded part of the square behind, you can still unfold it and work with the entire square.)

17. If the altitudes of one triangle are congruent to the corresponding altitudes of another triangle, prove that the triangles are congruent.

(Hint: Make a sketch of the two triangles and their altitudes. You should be able to find three ways to express the area of each triangle. Use this along with the SSS similarity test to first prove that the triangles are similar. Then show, using the results from *On Your Own,* Problem **6a,** that the triangles are, in fact, congruent.)

Using Similarity

In this lesson, you will explore many applications of similarity, including calculating with proportions, splitting segments, and finding invariant products.

Similarity and proportionality are important ideas in mathematics, in part because they allow you to calculate and prove many other things. This lesson explores some of the applications of similarity.

Explore and Discuss

On a sunny day, Michelle and Nancy noticed that their shadows were different lengths. Nancy measured Michelle's shadow and found that it was 96 inches long. Michelle found that Nancy's was 102 inches long.

a Who do you think is taller, Nancy or Michelle? Why?

b If Michelle is 5 feet 4 inches tall, how tall is Nancy?

c If Nancy is 5 feet 4 inches tall, how tall is Michelle?

Drawing a picture might help you answer these.

ACTIVITY 1 ▶ **Calculating Distances and Heights**

Over the ages, people have developed clever methods to answer the questions, "How far away is that?" and "How tall is that?" This activity focuses on ways that similarity can answer these distance and height questions.

A Sea Story Several years ago, two friends were sailing off Old Orchard Beach in Maine. The sky was so clear that they could see the faraway top of Mount Washington in northern New Hampshire, towering some 6600 feet above sea level. One of the two sailors held her right arm straight out in front of her in a "thumbs up" gesture to get an idea of their distance from the base of the mountain.

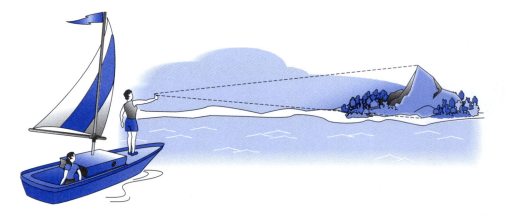

She positioned herself so that she could see how much of her thumb would cover the mountain. When the top of her thumbnail lined up exactly with the top of the mountain, one of the wrinkles on her thumb lined up with the shore at sea level—the whole 6600 feet on a thumb!

As she held her outstretched arm very still, her companion measured the distance from her eye to the place on her thumb that lined up with the edge of the shore. Then they measured the length of the part of her thumb that had covered Mount Washington. Using similar triangles, they calculated their distance to the base of the mountain. "You know," they remarked later, "that calculation turned out to be surprisingly accurate!"

1. The picture below shows a sketch of the situation.

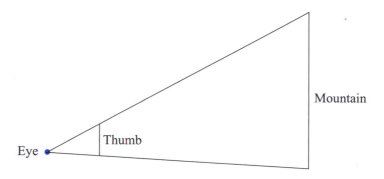

    a. What assumption is the picture making?

    b. Where is a pair of similar triangles?

    c. If the length of your thumb covering the mountain is 1 inch, and the distance from your eye to the bottom of your thumb is 14 inches, calculate your distance from the mountain.

2. Measure the length of your thumb or your index finger, and the distance from the base of the finger to your eye when your arm is fully extended. If you know the height of any object that can be covered by that finger, you can then determine your distance from that object. Pick an object and use this technique to figure out how far away it is. Check your results by measuring the actual distance.

A "Shady" Method Using shadows is a quick and accurate way to find the heights of trees, flagpoles, buildings, and other tall objects. To begin, measure the length of the shadow your object casts in sunlight. Also measure the shadow cast at the same time of day by a yardstick (or some other object of known height) standing straight up on the ground.

Remember that the height of the mountain is 6600 feet.

You can estimate what your finger covers. If your finger covers about one third of a 60-meter building, then about 20 meters is covered.

Since you know the lengths of the two shadows and the length of the yardstick, you can use the fact that the sun's rays are approximately parallel to set a proportion with similar triangles.

3. Suppose that a tree's shadow is 20 feet long and the yardstick's shadow is 17 inches.

  a. Draw a picture that shows the tree, the yardstick, the sun's rays, and the shadows.

  b. Find a pair of similar triangles. Explain why they are similar.

  c. How tall is the tree?

4. Use the shadow method to find the height of some tall object for which you can obtain the actual height. Record the details of your measurements. Prepare a presentation for your class. By how much did the result of your calculations differ from the actual height? What might cause these differences?

Tiny Planets In 1996, the very first discovery of planets that were not part of our own solar system was reported. Astronomers had good reason to believe such planets existed, but for the first time it became possible to find them. The newspapers gave accounts of the clever techniques that astronomers used to find these planets, despite the fact that the planets themselves had not been seen. No report seemed to explain why the planets were such a challenge to see, but that is something you can figure out using similarity.

As of early 1996, there were no reports of planets of the nearest star. Suppose, though, that it did have a planet the size of Earth. How large would that planet appear to the unaided eye?

First, you must figure out what that question means! Imagine looking at a person through a window and seeing how much of the window that person fills. Better yet, try tracing the outline of the person on the window.

5. If you actually make the tracing, you can measure the height of the image, but you can also figure it out if you have enough information about the height of the person, the distance of the person from you, and the distance of the window from you. Below is a picture of that situation. Carefully describe how you could use similar triangles to figure out how tall the image on the window would be.

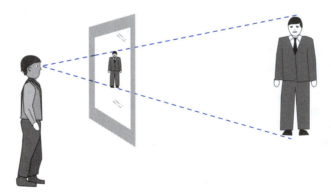

6. Apply your theory. Imagine seeing a five-foot-tall person standing outside your window about 30 feet away from you. Imagine you are about two feet from the window. About how tall will the image be?

7. Now imagine viewing the distant planet through the same window. Before you can figure out how tall the image will be, what information must you have?

8. The nearest star to us (other than our own sun) is about four light-years away. A *light-year* is a unit of distance—specifically the distance that light travels in one year. Light travels at 186,000 miles per second. How far does it travel in one year? How far away is the nearest star?

The diameter of Earth is approximately 7900 miles.

9. If a planet that is the same distance away as the star in Problem **8** has the same diameter as Earth, how large will its image appear on a window two feet from you?

You

Planet

Use $\frac{1}{20}$ of an inch for the width of an "o."

10. About how much must that image be magnified to be as big as an "o" on this page?

11. Checkpoint A child who is almost $3\frac{1}{2}$ feet tall is standing next to a very tall basketball player. The child's sister, who is studying geometry, notices that the player's shadow is about twice as long as the child's. She quickly estimates the player's height. What value does she get?

ACTIVITY 2 Segment Splitters

Below are two intriguing experiments to try, both of which depend on similarity.

Experiment One: The Projection Method Your teacher will provide you with several copies of the picture below, showing a segment *s* and ten equally-spaced points below it, all lying on a segment parallel to *s*.

12. Choose any three consecutive points on the dotted line.

a. Use a straightedge to draw a line connecting the left endpoint of segment *s* with the leftmost of the three points. Next, draw a line connecting the right endpoint of segment *s* with the rightmost point. Be careful that your lines go through the centers of the dots. In the example below, the lines meet at point *P*.

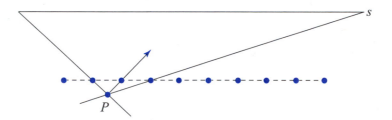

b. Draw a line through point P and the middle point. Extend your line far enough so that it intersects segment s. Again, be sure that your line goes through the center of the dot. Where does your line intersect segment s?

13. Start with a new copy of the picture. Pick three more points. This time connect the rightmost of the points to the leftmost endpoint of s. Also connect the leftmost of your points to the rightmost endpoint of s.

   a. How does this affect the location of P?

   b. Draw the third line, through the middle point and through P, extended to intersect the segment s. Does it intersect s in a different place than the previous example?

14. Start with a new copy of the picture. Split segment s into three congruent pieces without taking any measurements. Compare your work with that of your classmates to see if they did it the same way.

15. Divide segment s into five congruent pieces. Divide another segment into seven congruent pieces.

Experiment Two: The Parallels Method For this method, you will need a straightedge, a sheet of lined notebook paper, a blank overhead transparency, and an erasable marker.

16. Use the marker and the straightedge to draw a segment on the transparency. With nothing other than your lined notebook paper and marker, divide the segment in half.

17. Draw another segment on the transparency. Divide it into three congruent pieces.

18. Draw some more segments on the transparency. Divide them into five and seven congruent pieces. What is the largest number of divisions you can make with your notebook paper?

19. Explain your method to a classmate.

Why Do the Methods Work? The pictures below show examples of using the parallels method and the projection method. In both cases, segment AB is being split in half.

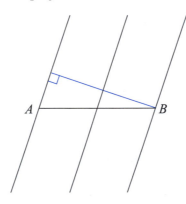 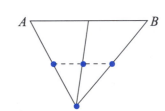

Parallels method   Projection method

20. Checkpoint Consider these problems.

    a. Use properties of similar triangles to show that both methods do indeed split segment *AB* in half.

    b. Use properties of similar triangles to show that both of your methods for trisecting segments work.

ACTIVITY 3

A Constant-Area Rectangle

Similarity can help you find the heights of tall objects and calculate distances from faraway places. But there's more to similarity than just measurement applications.

While exploring rectangles with geometry software, you may become intrigued by the following possibility: Is there a way to construct a rectangle so that its perimeter can change when you drag a vertex, but its *area must remain the same*?

If there is a way to build such a rectangle, you could call it a "constant-area rectangle."

The length and width do not have to be integers.

21. Give length and width dimensions of four rectangles that all have an area of 36 square feet.

22. If *l* represents the lengths of the rectangles in the previous problem and *w* represents their widths, what is the relationship between *l* and *w*?

23. How many rectangles are there with an area of 36 square feet?

The Power of a Point One of the special pleasures of mathematics comes from finding an unexpected connection between two topics that seem to have nothing in common. The section below contains problems that may, at first, seem like a detour from the constant-area rectangle. But as you work through it, ask yourself if any of the results can help you to construct a constant-area rectangle.

Try this with geometry software.

24. Draw a circle. Pick a point *P* anywhere inside it. Then draw a chord of the circle that passes through *P*. Point *P* divides the chord into two segments: $\overline{PA}$ and $\overline{PB}$.

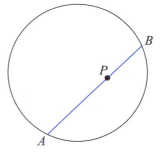

    a. Measure the lengths of these segments. Calculate *PA* × *PB*.

    b. Draw several other chords through point *P*. Calculate the same product. Record any of your observations.

These products have a special name. They are called the *power of point P.*

25. The circle below has two chords, $\overline{AB}$ and $\overline{CD}$, that intersect at point P.

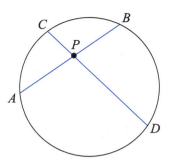

a. What can you predict about the lengths PA, PB, PC, and PD?

Hint: ∠ACD and ∠ABD intercept the same arc.

b. Add segments AC and BD to the illustration. The first step in proving your conjecture from part **a** is to show that $\triangle APC \sim \triangle DPB$. Explain why these two triangles are indeed similar.

c. Use the fact that $\triangle APC \sim \triangle DPB$ to write a proportion that includes PA, PB, PC, and PD. Rearrange the proportion to prove your conjecture from part **a.**

d. How does this result prove that the product of the chord lengths is the same for *any* chord through P?

Using your power-of-a-point findings, you can construct constant-area rectangles. To begin, the computer screen shows a rectangle on one side and the power-of-a-point construction on the other. The length PA and the width PB of the rectangle are linked to the corresponding chord segments and are equal to them. So, when the chord segment lengths change, the rectangle's dimensions will, too.

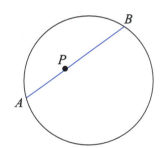

As you move point A around the circle, the chord $\overline{AB}$ spins, always passing through the stationary point P.

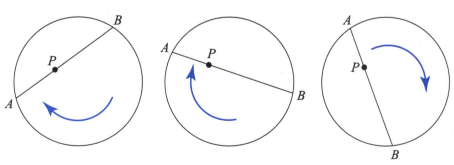

26. Use geometry software to build a construction like the one described on page 298. Your construction might include an animation feature that allows point *A* to travel automatically around the circle.

    a. As chord $\overline{AB}$ spins, describe what happens to the rectangle.

    b. The purpose of this geometry construction was to create a rectangle of constant area. Explain why you think this construction does or does not satisfy this goal.

    c. Does your construction show *all* possible rectangles that share the same area? Explain.

27. [Checkpoint] Suppose that you decide to build a collection of equal-area rectangles out of long wooden sticks. Each rectangle is to have an area of 12 square feet and no two rectangles can have the same dimensions. You can, of course, measure the sticks and then cut them at appropriate places. Measuring becomes tedious though, especially if you plan on building lots of rectangles.

How can the setup below help you to construct your rectangles?

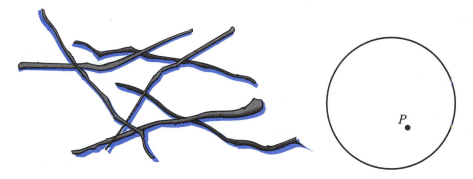

A few of your wooden sticks   Power of point $P = 12$

ACTIVITY 4

The Geometric Mean

The figure below shows a complete rectangle on the left and only the length of another rectangle on the right.

28. Use the power-of-a-point construction to construct the missing width of the rectangle so that both rectangles have the same area. The pictures below may help you think about it.

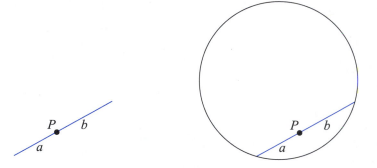

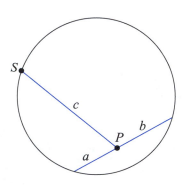

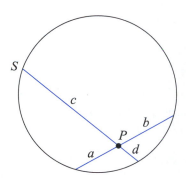

29. Once again, begin with an $a \times b$ rectangle.

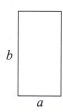

a. Use the power-of-a-point construction to draw a square with the same area as the rectangle. The pictures below may help you.

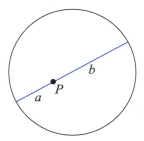

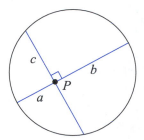

b. Explain why this construction method works.

c. What is the algebraic relationship between a, b, and c? Write a formula that relates them.

The length c has a special name.

> **Definition**
>
> If $c^2 = ab$ (or equivalently, $c = \sqrt{ab}$), then c is called the **geometric mean** of a and b.

30. Compute the geometric mean of the following numbers.

a. 2 and 8

c. 4 and 6

b. 3 and 12

d. 5 and 5

> **Ways to Think About It**
>
> The **arithmetic mean** or **average** of two numbers a and b is defined as $\frac{a+b}{2}$. So, the arithmetic mean of 6 and 24 is $\frac{6+24}{2} = 15$.
>
> The mean 15 is midway between 6 and 24:
>
> $$15 - 6 = 9$$
> $$24 - 15 = 9$$
>
> The *geometric mean* of two numbers defines a different kind of midway point between them. To get from 6 to 24, you can multiply by 2 ($6 \times 2 = 12$) and then by 2 again ($12 \times 2 = 24$). Thus 12, the midway point, is the geometric mean of 6 and 24. Another way to say that 12 is midway between 6 and 24 is to write a proportion:
>
> $$\frac{6}{12} = \frac{12}{24}$$
>
>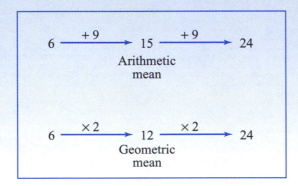

31. If you add two extra segments to the geometric mean picture on the left, then three right triangles are formed in the semicircle. Why?

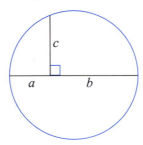
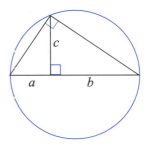

Usually, just the right triangles are shown and not the circle that is used to construct them. In the next problem, the circles were erased after constructing the right triangles.

32. Find all the unknown lengths of the segments in the two figures below.

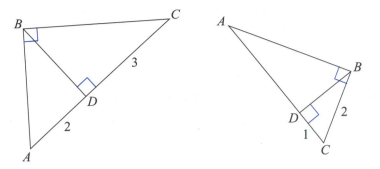

33. Construct a segment whose length is the geometric mean of a 1-inch and a 3-inch segment.

The illustration below shows four diagrams from a geometric-mean construction as point D moves to the left and point A remains stationary.

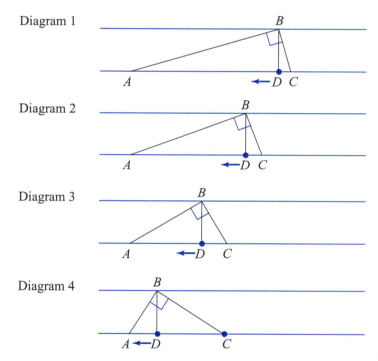

Diagram 1

Diagram 2

Diagram 3

Diagram 4

34. Use geometry software to make a sketch like the one above.

a. There are two lengths whose product remains the same throughout each of these four diagrams. Which are they?

b. Use this setup to build another rectangle-of-constant-area sketch.

c. Does your construction show *all* possible rectangles that share the same area? Explain.

35. Your constant-area-rectangle sketches show rectangles that range from narrow and tall to wide and short. Which of the two constructions—the power of a point or the geometric mean—seems to generate a larger range of constant-area rectangles? Why?

36. **Checkpoint** Consider these problems.

    a. Copy the figure below. Draw a segment whose length is the geometric mean of the segments of lengths a and b. Next draw a segment whose length is their arithmetic mean. Which is longer? Does this relationship always hold? Is the arithmetic mean ever equal to the geometric mean?

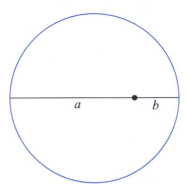

Below is a picture of a square and a rectangle.

 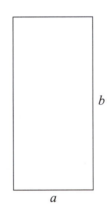

    b. Which has the larger perimeter?

    c. Which has the larger area?

1. Use the methods from this unit to calculate the height of a tree or a lightpost in your neighborhood. Show what measurements you used to find the height. Include a sketch.

2. A new Frank Lloyd Wright building is being constructed in your town. A fence surrounds the construction site, but you have found a 1-inch peephole in the fence. You look at the building so that it exactly fills the peephole, and find your eye is only $\frac{1}{2}$ inch from the hole.

    a. Draw a sketch of this situation.

    b. If the building is to have two floors and be 45 feet tall, how far is it from the fence?

    c. If you know that the building is 20 feet from the fence, how tall is it?

3. A 5-foot-tall person casts a 6-foot shadow.

    a. If a nearby flagpole casts a shadow that is 18 feet, how tall is the flagpole? Draw a sketch.

    b. Suppose the 5-foot-tall person stands so that his 6-foot shadow overlaps the flagpole's shadow, with the tips of the two shadows matching up. In this position, he is standing 6 feet from the flagpole. Draw a sketch. Find how tall the flagpole is in this case.

4. **Write and Reflect** Suppose you wanted to make a constant-area triangle instead of a rectangle. Describe at least one way to alter the methods in this lesson to make a triangle instead. Include pictures.

5. Below is an outline for a proof of a power-of-a-point conjecture. Fill in the reasons for each step. Then write up the proof.

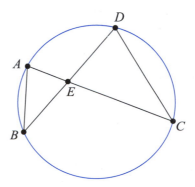

Hint: Inscribed angles.

    a. $\angle ABE \cong \angle DCE$. Why?

    b. $\angle AEB \cong \angle DEC$. Why?

    c. $\triangle ABE \sim \triangle DCE$. Why?

    d. $\frac{AE}{DE} = \frac{BE}{CE}$. Why?

    e. Write the complete proof that $(AE)(CE) = (BE)(DE)$.

6. Start with an $a \times b$ rectangle. Explain how to draw a rectangle with the same area, only this time, its length must be the very tiny segment c.

    a. Can you use the same circle as in Problem **29** to complete this construction?

    b. What difficulties do you encounter?

    c. Explain how to redraw the circle so that the construction works.

7. Construct a proof so that *BD* is the geometric mean of *AD* and *DC* by showing that △*ADB* ~ △*BDC*. Then write an appropriate proportion.

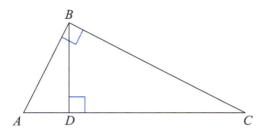

8. Below are three segments of lengths *a*, *b*, and 1. You can call a segment of length 1 a *unit segment*. For each of the problems, can you construct a segment with the given length?

All of the constructions can be done with a straightedge, a compass, and some lined notebook paper. Resist the urge to measure with a ruler. To construct some of these lengths, you will need to apply concepts that you have learned about similar triangles, segment splitters, the power of a point, and the geometric mean.

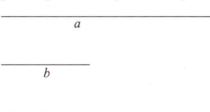

a. $a + b$

b. $a - b$

c. $3a$

d. $2a - b$

e. ab

f. $\frac{a}{b}$

g. $\frac{b}{3}$

h. a^2

i. $\sqrt{b}$

j. $\sqrt{ab}$

k. $\sqrt{a^2 + b^2}$

l. $\sqrt{a^2 - b^2}$

m. $\frac{a + \sqrt{b}}{2} + \frac{a - \sqrt{b}}{2}$

9. **Challenge** Extend your argument from Problem **20** on page 297 to show that the segment-splitting methods work for *any* division of a segment into congruent parts.

10. Draw a circle and a point *P* within it. Calculate the power of *P*.

    a. Are there other locations for *P* within the same circle that have the same power? Find a few and explain your reasoning.

    b. Find the locations of *all* points within your circle that have the same power as *P*. What does this collection of points look like?

11. Explore the power-of-a-point construction for points that lie *outside* the circle. Can you still find a constant product?

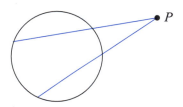

12. For a constant-perimeter rectangle, the area of the rectangle can change, but the perimeter must remain the same when you drag a side or vertex. Use geometry software to construct a constant-perimeter rectangle.

13. Rays *PA* and *PB* meet at point *P*. The segments *CD*, *DE*, *EF*, ..., form a zigzag pattern and alternate between being perpendicular to rays *PA* and *PB*.

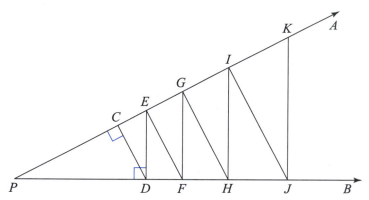

    a. If $PC = 1$ and $PD = x$, find the length of the following segments (it helps if you do them in order).

       i. $\overline{PE}$

       ii. $\overline{PF}$

       iii. $\overline{PG}$

       iv. $\overline{PH}$

    b. Do you detect a pattern here? Without doing any more calculations, give the lengths of $\overline{PI}$, $\overline{PJ}$, and $\overline{PK}$.

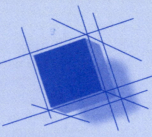

Perspective

Segment-Splitting Devices

Segment-splitting devices date back at least to the time of Euclid. In fact, London's Science Museum contains replicas of bronze splitting devices found at Pompeii. Closer to the present, the scientist Sarah Marks (1854–1923) invented and patented her own device for splitting a segment into any number of equal parts.

Sarah Marks' father died when she was seven. For many years afterward, she helped support her mother, sister, and six brothers. At the age of nine, Marks went to London to be privately educated at a school run by two of her aunts. There she was tutored in Latin, Greek, French, Hebrew, the classics, music, art, and mathematics not only by her aunts and uncles, but by several of her cousins as well.

MARKS'
PATENT LINE DIVIDER.

For Dividing any space in a number of equal parts.

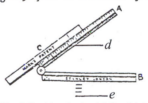

DESCRIPTION.—A B a hinged rule with firm joint, the limb A fitted to slide in an undercut groove upon the plain rule C. C has needle points on the under side to prevent it from slipping when placed in any position. The limb A of the rule is divided on both edges into eights, quarters, half-inches and inches, which are consecutively numbered so that any set may be taken.

TO USE THE LINE DIVIDER.—Suppose the space *d* to *e* is to be divided into any number of parts—say thirteen : Taking the half-inch line, hold the rule B on the line *e* and open the rule A until the division marked 13 on the inside edge is coincident with the line *d* ; now notice that the single line on rule C is opposite the 13, and in this position press it down so that the needle points on the under side get sufficient bite to prevent it slipping ; placing the fingers firmly on C, slide the part A upwards so that it may stop consecutively opposite each of the 13 divisions, as indicated opposite the line on the rule C, a pencil line drawn along B across *d e* at each stoppage opposite the numbers 12, 11, 10, 9, &c., will give the required divisions. To produce the lines in ink, the rule, after setting, may be moved to the upper line first, and the division lines be drawn downwards.

The rule may be worked in any direction for drawing line, vertical, horizontal, or oblique, and for any division of a space from 2 to 80 parts. It will be found convenient to Architects and Engineers for dividing any spaces without previous trial—such as treads and risers of stairs, joists, roof-timbers, girders, brick spaces, for drawing section line shading, &c., and will be found a saving of time for division of a space into 3 parts and upwards.

For open spaces, multiples of the space must be taken. In very close divisions, the joint may interfere with the last 2 or 3 divisions, in this case a line must be drawn at each setting and produced after the rule is removed.

SOLE AGENT—
W. F. STANLEY,
5, Great Turnstile, Holborn, London, W.C.

PRICE 5/- IN CASE.

Marks' inventions include a device for measuring a person's pulse. Her work with electrical arcs led to improved searchlights and movie projectors. She also devoted much of her time to the women's suffrage movement and to lowering the barriers that women faced in education, laboratories, and scientific societies. She was the first woman to be nominated to become a Fellow of the Royal Society of London and was awarded the society's Hughes Medal for originality in research.

Marks' segment-splitter patent is shown on page 307. See if the description given in the patent provides enough information for you to figure out how the device works.

Marks' segment splitter consists of three arms, A, B, and C. Arms A and B are hinged together, and arm C slides along a grooved track on arm A. Arm C contains just one mark, whereas arm A is marked off into 80 equal parts. (Fewer are shown in the figures below.) The underside of arm C has two protruding pins that can be pressed into a piece of paper to prevent the device from slipping.

To divide the region between lines d and e into seven equal parts, place arm B along line e and open arm A until the seventh mark touches line d. Then slide arm C until its mark coincides with this seventh mark. The pins on arm C are now pressed down to keep it in place.

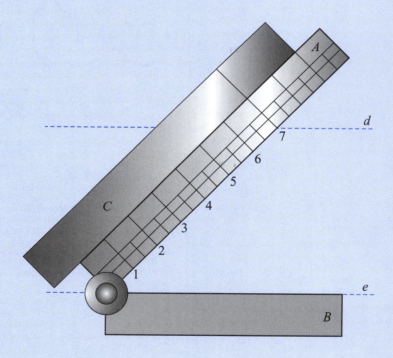

Without moving arm C, slide arm A until its sixth mark aligns with the single mark on arm C. As arm A slides up, arm B moves up also, maintaining the same angle between B and A. Draw a line along the top of arm B.

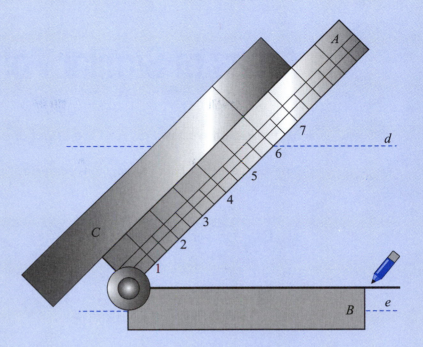

Continue sliding arm *A* up one mark at a time. For each new position, draw a line along the top of arm *B*. The picture below shows the device after the first four divisions of the region between lines *d* and *e* have been made.

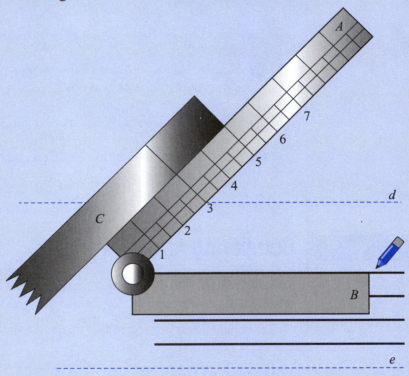

Arm C extends into the margin.

1. Explain why Marks' segment splitter works.

2. In the example above, Marks' splitter divides the region between lines *d* and *e* into seven equal parts. Explain how you would use the device to divide a *segment* into seven equal parts.

3. Does Marks' method for splitting a segment seem similar to any of the other segment-splitting methods you have worked with earlier? Explain.

4. **Project** Build a model of Marks' segment splitter.

Areas of Similar Polygons

In this lesson, you will explore the effect of scaling on the area of polygons.

How do the areas of similar polygons compare? Simple polygons, like rectangles, are a good place to start.

Explore and Discuss

Draw a rectangle. Then scale it by a factor of 2.

a How do the dimensions of the original rectangle compare with the dimensions of the scaled one?

b How many copies of the original rectangle fit into the scaled one?

c How does the area of the scaled rectangle compare to the area of the original one?

Draw a rectangle. Then scale it by a factor of $\frac{1}{3}$.

d How do the dimensions of the original rectangle compare with the dimensions of the scaled one?

e How many copies of the scaled rectangle fit into the original one?

f How does the area of the scaled rectangle compare to the area of the original one?

ACTIVITY 1 > ## Comparing Areas

In Lesson 2, you answered some questions about how many copies of scaled figures would fit inside the original. In this activity, you will formalize those ideas.

1. State and prove a theorem that starts like this:

> If rectangle $ABCD$ is scaled by a factor of r to get rectangle $A'B'C'D'$, then the area of $A'B'C'D'$... .

2. If two triangles are similar, and the scale factor is r, you know that the ratio of the lengths of two corresponding sides is r. Show that

a. the ratio of their perimeters is r.

b. the ratio of the lengths of two corresponding altitudes is also r.

c. the ratio of their areas is r^2.

3. One side of a triangle has length 10. The altitude to that side has length 12. If all the sides of the triangle are tripled, what is the area of the new triangle?

4. According to Problem **2,** if you scale any triangle by a factor of 4, then 16 copies of it should fit inside the scaled copy. Explore with paper and scissors or geometry software.

5. According to Problem **2,** if you scale a triangle by $2\frac{1}{2}$, then $6\frac{1}{4}$ copies of the original should fit inside the scaled copy Why? Check this out with paper and scissors or geometry software.

Now that you have calculated the areas of similar rectangles and triangles, take a look at similar polygons with *any* number of sides.

6. In the figure below, Polygon 1 was scaled by a factor of r to obtain Polygon 2.

Polygon 1 is divided into four triangles with areas of a, b, c, and d.

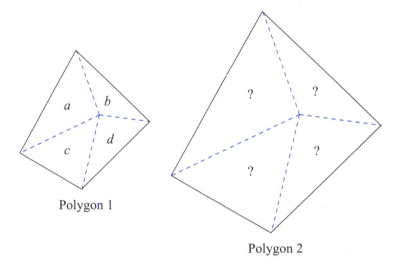

Polygon 1

Polygon 2

    a. What are the areas of the corresponding triangles in Polygon 2?

    b. What is the total area of Polygon 2?

    c. What is the total area of Polygon 1?

7. a. Use these results to complete the following theorem.

Theorem 4.6

If a polygon is scaled by some positive number r, *then the ratio of the area of the scaled copy to the original is ... .*

    b. Carefully prove Theorem 4.6, justifying each step.

8. Hans has two cornfields that he wants to plant. One measures 400' × 600' and the other measures 200' × 300'. Bessie Moonfeed, the owner of the grain store, says, "The big field will take eight bags of seed. The small field has sides half as big, so you will need four more bags for that. Will that be cash or charge?" A few days later, Hans returns to the grain store very upset. Why?

Try dividing it into
triangles.

9. a. Carefully trace the polygon below. Then estimate its area.

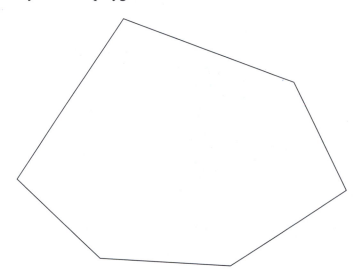

b. If you scale the polygon by a factor of 1.5, what will the new area be?

10. **Checkpoint** One square has an area that is 12 times the area of another. What is the ratio of

  a. their sides?

  b. their diagonals?

A C T I V I T Y   2 **Introducing the Apothem**

The **apothem** of a regular polygon is a perpendicular segment from the center point of the polygon to one of its sides.

11. a. Why doesn't it make a difference which side you choose when drawing an apothem?

  b. Show that the area of a regular polygon is equal to half the product of its perimeter and apothem length.

The result of the last problem is important enough to record as a theorem.

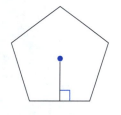

An apothem of a regular pentagon

Theorem 4.7

The area, A, of a regular polygon is equal to half the product of its perimeter, P, and the length of its apothem, a. In symbols:

$$A = \tfrac{1}{2}Pa.$$

12. What is the area of a regular hexagon whose sidelength is 8?

13. **Checkpoint** Use the area formula $A = \tfrac{1}{2}Pa$ to calculate the area of a square whose sidelength is 12. Check your result by calculating the area of the square another way.

1. A rectangle is scaled by a factor of $\frac{1}{4}$. Compare the area of the scaled rectangle to the area of the original one.

2. A triangle is scaled by a factor of 5. Compare the area of the scaled triangle to the area of the original one.

3. A polygon has an area of 17 square inches. If it is scaled by a factor of 2, what is the area of the new polygon?

4. Find the area of each figure.

   a. A stop sign with sides that are 6 inches long and an apothem that is 7.8 inches.

   b. An equilateral triangle with an apothem 1 cm and $AD = 2$ cm.

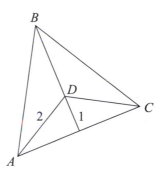

   c. A regular pentagon with the measurements shown.

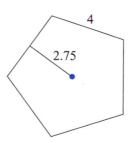

5. Can you use the formula $A = \frac{1}{2}Pa$ for irregular polygons? Why or why not?

Take It Further

As part of its new budget, your town decides to build a recreation center to benefit the community. You are elected as the student representative from your school to assist with the planning and development.

The town committee has narrowed down the choices for the recreation center to two possible locations. Both sites are located in the same neighborhood, and each spot seems appealing. Also the price of the two land parcels is roughly the same. However, you are not sure whether both spaces cover the same amount of land. You would like to pick the parcel with the larger piece of land and calculate its area so that the architect can begin to design the specifications of the recreation center.

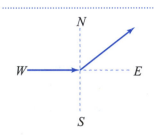

An example of a northeast turn

One possible method for finding the areas is to walk along the border of each property and record the length of each side as well as how much you turn at each corner. In fact, these are often the kinds of measurements recorded on land deeds.

Below is the description of your walk for both parcels of land.

Parcel 1 Start at one corner of the land. Walk east 798 feet, turn northeast and walk 543 feet. Head north 678 feet. Then walk along a straight line back to where you started.

Parcel 2 Start at one corner of the land. Walk 884 feet west, turn north and walk 442 feet. Then walk due east 554 feet, turn northwest and walk 61 feet. Head east 718 feet. Then walk along a straight line back to where you started.

Now it is up to you to calculate the area of each land parcel as precisely as possible. You will probably want to make a drawing of each space using either a ruler and protractor or geometry software. Here are some questions to ask yourself that might help:

- How can I draw each parcel so that its distance and angle measurements are preserved?

- How can I fit the drawing onto a single sheet of paper or computer screen?

- Can I calculate the area of the whole space at once, or should I divide it into smaller pieces?

6. Prepare a presentation for your class outlining the methods you used to calculate the land areas.

7. How close are your area calculations to those of your classmates? What might be some of the reasons for your class getting a variety of area values?

Areas and Perimeters of Blobs and Circles

In this lesson, you will learn techniques for approximating the areas and perimeters of shapes that are not polygons, leading to an area formula for circles.

Triangles and polygons are convenient shapes to study in geometry, but the truth is, many objects in our world are not composed of line segments. Circles, egg shapes, and curves of all types are just as common as polygons. How can you find the area of a shape that has curves?

Explore and Discuss

One way to find the area of a polygon is to divide it into triangles, and then find the area of each triangle. But what about a figure that isn't a polygon? For example, how can you estimate the area of this blob? Can you find it exactly?

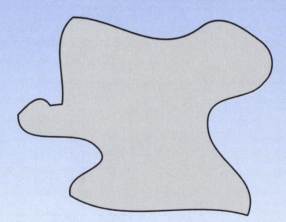

For shapes like this, usually the best you can do is to estimate the area.

a List several ways that you can estimate the *area* of an irregular shape such as the blob above.

b Try each of your methods with the blob or some other shape.

c List several ways to estimate the *perimeter* of an irregular shape like the blob.

d Try each of your methods out on the blob or some other shape.

Inner and Outer Sums

An important idea in mathematics is estimation of a value by finding upper and lower bounds, then squeezing those bounds closer together. Below is one technique that will let you do this on the blob.

1. The blob from *Explore and Discuss* has been put on $\frac{1}{2}'' \times \frac{1}{2}''$ graph paper.

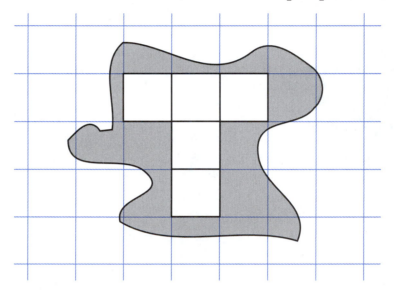

    a. Count the number of squares that are completely *inside* the figure.

    b. What is the area of each inner square?

    c. Sum the areas of these inner squares to find an area that is definitely smaller than the area of the blob.

2. Next, count all the squares that are either *inside* or *touch* the blob.

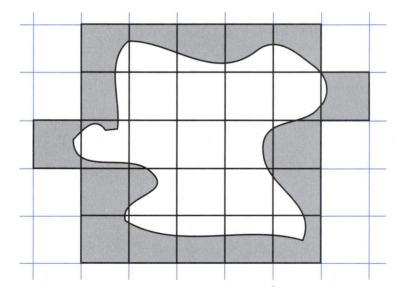

    a. How many of these squares are there?

    b. What is the area of each square?

    c. Sum the areas of these squares to find an area that is definitely greater than the area of the blob.

The inner and outer sums from Problems **1** and **2** give a pretty wide range. How can you improve on this area estimate? One way to get a closer approximation of the area of the blob is to use graph paper with smaller squares. In the picture below, the squares on the graph paper are $\frac{1}{4}" \times \frac{1}{4}"$.

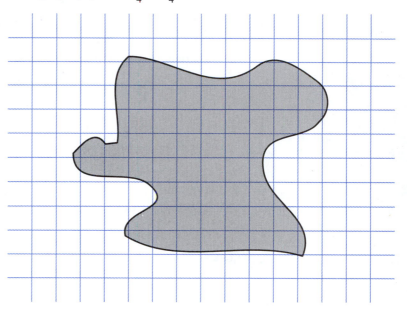

3. Use the picture above to answer these questions.

   a. What is the area of each small square?

   b. Calculate the *inner sum*—the total area of the squares that are completely inside the blob.

   c. Calculate the *outer sum*—the total area of the squares that are either inside or touch the blob.

   d. The true area of the blob is between these two values. Are these numbers closer to each other than the first estimate? Why?

Mathematicians say that you are making a "finer mesh."

In the picture below, the squares on the graph paper are $\frac{1}{8}" \times \frac{1}{8}"$.

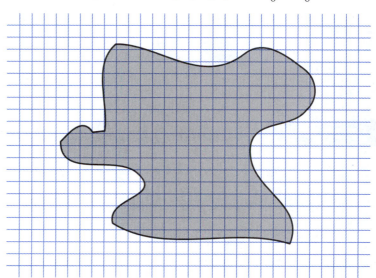

4. a. What is the area of each small square?

b. By calculating the inner sum and the outer sum, place the area of the blob between two numbers. Are these numbers closer to each other than the numbers you found in Problem **3?**

5. Give an argument to support the claim that as the number of squares per inch gets larger (that is, as the mesh of the graph paper gets *finer*), the difference between the outer sum and the inner sum gets smaller.

You now have the basic idea behind how the areas of closed curves, like the blob, are defined. In summary,

- The region is covered by graph paper, and the inner and outer sums are computed. The mesh is made finer and the process is repeated.

- This produces a sequence of inner and outer sums. If the difference between these inner and outer sums can be made as small as you want by making the mesh fine enough, then,

- ... this means the inner and outer sums get closer and closer to a single number.

- This single number, the limit of the whole process, is the area of the region.

Counting squares can be tedious, but sometimes a shape has symmetry that can make the counting easier by using shortcuts.

6. Draw a circle of radius 1 foot. Approximate its area using a mesh size of

a. 1".

b. $\frac{1}{2}$".

c. $\frac{1}{4}$".

Use shortcuts wherever possible. Describe any patterns that show up in your estimates.

7. **Checkpoint** Consider these problems.

a. Explain the method you have used to approximate the area of an irregular shape.

b. Find the area of a right triangle with sides 3, 4, and 5.

c. Suppose you didn't know the area formula for a triangle. Go through the inner and outer sums process for a 3–4–5 triangle to approximate its area. See how close you can get to the actual area.

ACTIVITY 2 ▶ **Comparing the Areas of Blobs**

You already know a way to compare the areas of two polygons when one is a scaled copy of the other. If a polygon is scaled by some positive number r, then the ratio of the area of the scaled copy to the original is r^2. Is this true for blobs, too?

Imagine that a blob and a grid of squares are drawn onto a big rubber sheet.

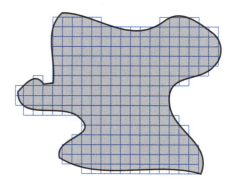

The area of these 228 squares gives a pretty good estimate of the blob's area.

Now imagine that the rubber sheet is stretched uniformly in all directions by a factor of r. This causes the blob and the squares to be scaled by r as well.

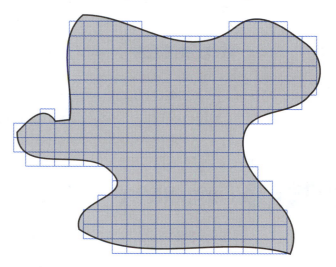

8. a. The 228 squares still give a good estimate of the area of the blob, but now the area of each square is bigger. By how much has each grown?

 b. How has the change in the squares' area affected the area of the blob?

9. The shape below on the left has an area of 4 square centimeters. The shape on the right is a scaled copy; the scale factor is 2. What is its area?

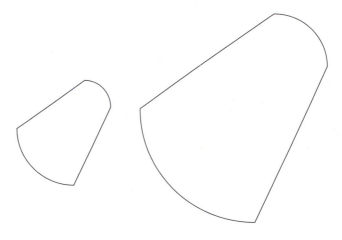

10. The two crescent moons below are scaled copies of each other. What is the ratio of their areas?

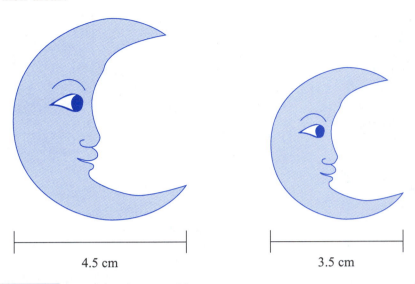

4.5 cm   3.5 cm

11. [Checkpoint] Consider these problems.

   a. If a circle of radius 2 is scaled to a circle of radius 6, how do the areas of the two circles compare?

   b. If a circle of radius 2 is scaled to a circle of radius 1, how do the areas of the two circles compare?

ACTIVITY 3 ▶ **Perimeters of Blobs and Circles**

How can you find the perimeter of a closed curve?

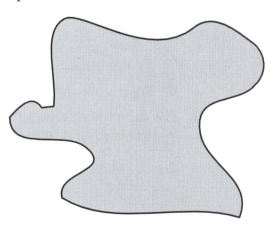

Another way to ask this is "How can you find the length of the curve?"

That is, given a blob, how can you calculate the distance around its edge?

Archimedes used a method for estimating the length of a curved path that is easy to apply. Just *approximate* the curve with line segments and add the lengths of the segments.

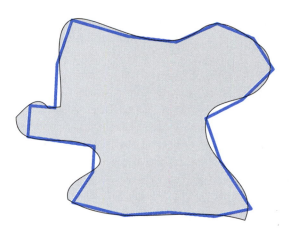

12. Approximate the length of the blob or some other curve using this technique.

13. How can you improve your estimate?

Perimeters of Circles Of all curves, perhaps the simplest is the circle. The name given to the length of a circle is one you may already know—the **circumference.** This is just another word that means *perimeter,* but it's reserved for circles and "round" 3-dimensional shapes, such as spheres and cylinders.

You may have learned a formula for the circumference of a circle in an earlier math course. You will see it again.

The *perimeter* of a circle can be found by this process.

Step 1 Inscribe a regular polygon in the circle and circumscribe a regular polygon with the same number of sides around the circle.

Step 2 Calculate the perimeter of each polygon.

Step 3 Make new regular polygons from the inscribed and circumscribed polygons by doubling the number of sides. Then calculate the perimeters of the new polygons.

Step 4 Continue this process. The *inner* and *outer* perimeters will approach a common value, and that number is what is meant by the circle's perimeter.

The following problems provide you with some practice carrying out the process described above by drawing inscribed and circumscribed polygons for a circle. You may make your own drawings, or you may fill in the table in Problem **17** by taking measurements directly from the drawings provided.

"Draw" means to use either pencil-and-paper, drawing tools, or geometry software.

14. Draw a circle. Inscribe a square in the circle. Circumscribe a square around the circle.

Calculate the perimeters of the two squares, and thus place the circumference of the circle between two numbers.

15. Use the same circle to inscribe a regular octagon in the circle and circumscribe a regular octagon around the circle.

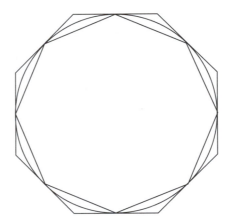

Calculate the perimeter of the two octagons, and thus place the circumference of the circle between two numbers.

16. Carry this process one step further with inscribed and circumscribed 16-gons.

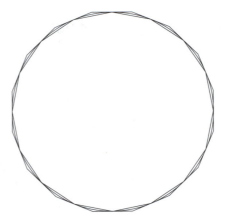

17. Copy the table. Use the data from the last three problems to complete the table.

Number of Sides	Outer Perimeter	Inner Perimeter	Difference
4			
8			
16			

18. Give an approximation for the perimeter of your circle.

19. **Checkpoint** Explain why the difference between the outer and inner perimeters gets smaller as the number of sides gets bigger.

Connecting Area and Circumference

The idea of approximating a circle with inner and outer polygons leads to a theorem that relates the area of a circle to its circumference.

Theorem 4.8

The area of a circle is one half its circumference times its radius. In symbols,

$$A = \tfrac{1}{2}Cr.$$

How can we prove this theorem? The formula, $A = \tfrac{1}{2}Cr$, looks a lot like the formula from Theorem 4.7, $A = \tfrac{1}{2}Pa$.

The first formula is about the area of a circle. The second is about the area of a regular polygon. Throughout this lesson, you have been approximating circles with polygons. That's the idea here, too. The idea is to inscribe a sequence of regular polygons in a circle and to study their areas.

20. Below are three regular polygons inscribed in a circle. The number of polygon sides increases from 4 to 8 to 16. Imagine that these pictures continue on and on with a sequence of regular polygons inscribed in this same circle.

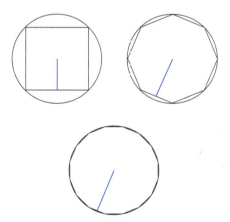

As the number of polygon sides increases, it becomes really difficult to distinguish a polygon from the circle and an apothem from a radius.

a. How many sides will each of the next three shapes have?

b. Will you ever get a polygon in this sequence whose apothem is greater than the radius of the circle? Explain.

c. Will you ever get a polygon in this sequence whose apothem is smaller than that of the previous polygon? That is, do the apothems get bigger at each step, or do they ever start getting smaller?

d. Will you ever get a polygon in this sequence whose perimeter is greater than the circumference of the circle? Explain.

e. Will you ever get a polygon in this sequence whose perimeter is smaller than that of the previous polygon?

f. Will you ever get a polygon in this sequence whose area is greater than the area of the circle? Explain.

g. Will you ever get a polygon in this sequence whose area is smaller than that of the previous polygon?

To make this more precise, you can use some notation:

- Let A, C, and r be the area, circumference, and radius of the circle.

- Number the polygons in the sequence 1, 2, 3, ...

- Let the areas of the polygons be A_1, A_2, A_3, ..., their perimeters be P_1, P_2, P_3, ..., and their apothems be a_1, a_2, a_3, ...

One more piece of notation: Instead of saying "the lengths of the apothems approach the radius," you can write this as "$a_n \rightarrow r$ as n gets larger and larger." Using this shorthand notation, you can rewrite the assumptions as:

1. $a_n \rightarrow r$

2. $P_n \rightarrow C$

3. $A_n n \rightarrow A$

This arrow notation means something quite precise in calculus: it means that you can make the difference between the length of the apothem and the radius as small as you want by making the number of polygon sides big enough.

By Theorem 4.7, for each polygon in the sequence,

$$A_n = \tfrac{1}{2} P_n a_n$$

Because of the three assumptions, as n gets larger and larger,

$$\tfrac{1}{2} P_n a_n \rightarrow \tfrac{1}{2} C r$$

and

$$A_n \rightarrow A.$$

You can think of it this way:

$$
\begin{array}{ccccc}
A_n & = & \tfrac{1}{2} & P_n & a_n \\
\downarrow & \downarrow & \downarrow & \downarrow & \downarrow \\
A & = & \tfrac{1}{2} & C & r
\end{array}
$$

To make this proof really airtight, you would need to fill in several gaps about limits and be more precise about the definitions of area and circumference.

21. A large Frisbee has area 154 square inches and diameter 14 inches. Find the circumference of the Frisbee.

22. **Checkpoint** A circular garden has radius 1 meter and circumference about 6.25 meters. What is the area of the garden?

On Your Own

1. Trace your hand on graph paper with the fingers together. Approximate the area of your palm using inner and outer sums.

2. Trace your hand on graph paper with the fingers spread. Approximate the area of your palm using inner and outer sums.

3. Imagine that a blob and a grid of squares are drawn onto a big rubber sheet. The area of these 228 squares gives a pretty good estimate of the blob's area.

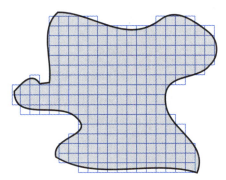

Now imagine that the rubber sheet is stretched just in one direction, so that all of the squares have one sidelength doubled and one unchanged.

a. What shape do the squares become?

b. What is the area of these new shapes compared with the areas of the squares?

c. What happens to the area of the blob?

d. Is the blob a scaled copy of the original? If yes, what is the scale factor? If no, explain why not.

For Problems **4–7,** given the sidelength and the scale factor, do the following.

a. Find the new sidelength of the cube.

b. Find the area of a face on the original cube and the area of the new face.

c. Find the volume of the original cube and the volume of the new cube.

4. A cube with sidelength 1 cm is scaled by a factor of 2.

5. A cube with sidelength 1 cm is scaled by a factor of 3.

6. A cube with sidelength 1 cm is scaled by a factor of $\frac{1}{2}$.

7. A cube with sidelength 1 cm is scaled by a factor of r.

8. You can approximate the area of 3-dimensional blobs with cubes in the same way that you approximate the area of 2-dimensional blobs with squares. How does the volume of such a blob change when it is scaled by a factor of r?

9. Tricia has a way to make the linear approximation technique easier and more accurate: She uses what she calls a *regular* approximation for a curve. She picks some length, say $\frac{1}{4}$", marks it off around the curve until she gets too close to the starting point to mark another segment. Then she just multiplies the number of segments by $\frac{1}{4}$ and adds on the last little gap. Draw a curve on your paper. Then use Tricia's method to approximate its length.

10. Many people use this *linear approximation* technique for estimating distance on road maps.

  a. Explain how this works.

  b. Using a road map and linear approximation, estimate the distance between your hometown and San Francisco. If you live in the San Francisco area, estimate the distance between your hometown and Boston.

11. Find a map of your state. Approximate the driving distance between two major cities using the scale of the map and the lengths of the roads. If your map includes a chart of driving distances between cities, compare your answer to the chart.

12. Start with a blank sheet of paper.

  Step 1   Tear the sheet of paper in half. Put one half in front of you.

  Step 2   Tear the piece you are still holding in half. Put one half on top of the sheet from Step 1.

  Step 3   Tear the piece you are still holding in half. Put one half on top of the sheets from Steps 1 and 2.

  Step 4   Continue this process until the piece you are holding is too small to tear.

  a. Let A_n be the area of the piece you are holding at Step n. Finish this statement: $A_n \rightarrow$ _____.

  b. Let the area of the original paper be 1 and A be the total area in your stack. Finish this statement: $A_n \rightarrow$ _____.

  c. Go back to your stack. If the area of the original paper is 1, what is the area of the largest sheet in your stack? The next largest? The next largest? Describe the pattern.

  d. Are your rectangles scaled copies of each other?

  e. Finish this statement: $\frac{1}{2} + \frac{1}{4} + \frac{1}{8} + \ldots + \frac{1}{2^n} \rightarrow$ _____.

Take It Further

13. Use a ruler and compass or geometry software to complete the table (all polygons are regular).

Number of Sides	Name	Length of $\frac{Perimeter}{Apothem}$
4	Square	
8		
16		

Continue the table for larger numbers of sides. Do the ratios seem to approach any particular number?

14. The Moriarty sisters, two girls who like to play jokes on their parents, decided one night to pull a prank. They jacked up their family home and slipped two logs under it, each having a circumference of 6 feet.

Then the children started up their mother's bulldozer and began to push the house. They managed to get the logs to roll one revolution before their parents awoke, very dismayed. How far did the Moriarty house move before the sisters were caught in their mischief?

15. Challenge Show that if a regular polygon with n sides and sidelength s is inscribed in a circle of radius 1, then a regular polygon with $2n$ sides inscribed in the same circle has sidelength $\sqrt{2 - \sqrt{4 - s^2}}$.

16. Project Architects, designers, and people who build swimming pools often use a device called a *planimeter* to approximate the areas of irregular shapes. Find out about planimeters and how they work.

13

Circles and 3.14159265358979323846...

In this lesson, you will formalize ideas of area and circumference for circles and learn more about the number pi.

Explore and Discuss

For starters, all circles are *similar.*

a Two circles have radii of 12 and 30. Can one of them be scaled to give a congruent copy of the other? Explain.

b Two circles have radii *r* and *R.* Can one of them be scaled to give a congruent copy of the other? Explain.

c Give a plausible argument for the following theorem:

> ### Theorem 4.9
>
> *If a circle is scaled by a positive number r, then the ratio of the area of the scaled copy to the original is r^2.*

ACTIVITY 1 ## An Area Formula for Circles

All of the problems in this unit are important, but this one is really important. Give it some time.

1. In Problem **6** in Lesson 12 on page 318, you approximated the area of a circle with radius 1 foot to be a bit more than 3 square feet. Use that result and the theorem above to find a good approximation for the area of a circle with radius

   a. 2 feet.

   b. 5 feet.

   c. 6 feet.

   d. $\sqrt{3}$ feet.

   e. $7\frac{1}{2}$ feet.

2. Give an argument to support this claim:

> ### Theorem 4.10
>
> *If the area of a circle with radius 1 is K, then the area of a circle with radius r is Kr^2.*

This theorem says that you can find the area of any circle once you know the area of a circle with radius 1. As you have calculated, the value of that area is a bit more than 3. But rather than call it "K," most people call it "pi" and represent it by the Greek letter π.

π is usually defined as the ratio of the circumference of a circle to its diameter. The two definitions give you exactly the same number.

> ### Definition
>
> Pi (π) is the value of the area of a circle whose radius is 1.

If you ask a person to tell you the value of π, he or she might say 3.14 or $\frac{22}{7}$. While these are indeed *approximations* of π, neither equals π. In fact, π cannot be represented as a ratio of two whole numbers. Its decimal representation is infinite and nonrepeating. Often, people leave the result of calculations about circles in terms of π. For the purpose of numerical calculations, though, π can be approximated to any degree of accuracy you would like.

This wedge is really called a sector of the circle.

3. The angle of the wedge in the circle at the right is 45°. The radius of the circle is 1.

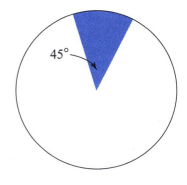

45°

    a. What fraction of the circle's area is the wedge?

    b. What is the *exact* area of the circle?

    c. What is the *exact* area of the wedge?

    d. Give two approximations for the area of the wedge.

4. Write and Reflect Henri, an inquisitive student, asks,

> "What do you mean π is the area of a circle of radius 1? One what? If you have a circle of radius 1 foot, it can't have the same area as a circle whose radius is 1 inch. This is all nonsense."

Suggest an answer to Henri's question.

5. Find the area of a circle if

    a. its radius is 10".

    b. its radius is 5 cm.

    c. its diameter is 3'.

    d. it is obtained by scaling a circle of radius 2" by a factor of 5.

6. **Checkpoint** Find the area of each shaded region.

a.

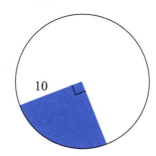

b.

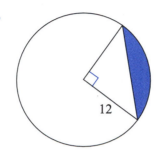

Circumference

You can use Theorem 4.8 to express the circumference of a circle in terms of its radius. Suppose a circle has radius r, circumference C, and area A. By Theorem 4.8,

$$A = \tfrac{1}{2}Cr.$$

You found that

$$A = \pi r^2.$$

So,

$$\tfrac{1}{2}Cr = \pi r^2.$$

7. Combine the two equations above.

    a. Solve for C, the circumference, in terms of r, and the radius.

    b. Rewrite this equation in terms of d, the diameter of a circle.

8. **a.** Choose the correct answer: The circumference of a circle is approximately five/six/seven times its radius.

    b. Choose the correct answer: The circumference of a circle is approximately three/four/five times its diameter.

9. In a circle of radius 2, draw a 60° central angle. The angle cuts the circle into two arcs. How long is each one?

Remember that a central angle is one whose vertex is at the center of the circle.

10. **Checkpoint** True or false? *The ratio of a circle's circumference to its diameter is the same for all circles.* Explain.

On Your Own

1. Find the area of each shaded region.

   a.
   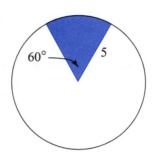

   b.

   Inner radius is 4,
   outer radius is 7.

2. Find the area of each shaded and white region.

   a.

   The circle has radius 3 cm.

   b.
   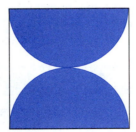

   The square has side 6 cm.

   c. Compare your answers for parts **a** and **b.** What do you notice?
   Can you explain it?

3. An equilateral triangle with sides 6.9 cm is inscribed in a circle. The length of the apothem is 2 cm.

   a. Draw a sketch.

   b. Find the area of the circle.

   c. Shade the part of the circle that is outside the triangle. Then find that area.

4. The table below gives one piece of information about four different circles. For each circle, find the missing parts.

Radius	Diameter	Area	Circumference
3			
	3		
		3	
			3

Find a tennis ball canister and check this out!

5. A canister contains three tennis balls. Which distance do you think is longer: the height of the canister or the circumference of the canister? Guess the answer. Then do the calculations to see if your guess was correct.

6. Good'n Wormy spaghetti company makes canned spaghetti. Their cans measure 5" high and 3" in diameter. What size piece of paper does the company need to make a label for the outside of its can?

Take It Further

7. The Digi-dial speedometer company makes electronic speedometers for bicycles. The device works by installing a small magnet on a spoke of your wheel and then installing another magnet on the fork of the same wheel, so that the two magnets make "contact" every time the wheel makes one revolution. A small computer converts data about the frequency of contacts into miles per hour. When you install one on your bike, you have to set it. One of the numbers you need to know is how far the wheel travels in one revolution. The instructions say to use the roll-out method: put a chalk mark on the tire where it touches the ground (and mark the ground, too), roll the bike until the mark comes back to the ground, and measure the distance between the chalk marks with a tape measure. What's an easier way to find the distance for one revolution? Try both methods with a bike.

8. The Flying Bernoulli Sisters, trapeze artists in the Italian circus, claim to have a way to calculate π. Here's what they do: They calculate two sequences of numbers, n and s_n, and then find $\frac{ns_n}{2}$.

n	s_n	$\frac{ns_n}{2}$
6	1	3
12	0.51763809	3.105828541
24		
48		

Each n is twice the one above it, and each s_n is computed from the previous one by squaring the previous one, subtracting that answer from 4, taking the square root of the result, subtracting *that* result from 2, and taking the square root of what you get.

In symbols,

$$s_n = \sqrt{2 - \sqrt{4 - \left(\frac{s_n}{2}\right)^2}} \, .$$

Use a calculator to complete the Bernoulli table. See if $\frac{n s_n}{2}$ does get close to π. Why does this work?

Perspective

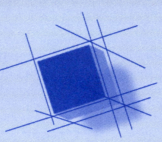

All About π

The number π has intrigued people for centuries. This essay gives a bit of history and a few notable facts about π.

The number π is not the ratio of two integers, but there are lots of ways to approximate it.

The numbers given in parentheses represent the years in which the equations were discovered.

(1579)
$$\frac{2}{\pi} = \sqrt{\frac{1}{2}} \times \sqrt{\frac{1}{2} + \frac{1}{2}\sqrt{\frac{1}{2}}} \times \sqrt{\frac{1}{2} + \frac{1}{2}\sqrt{\frac{1}{2} + \frac{1}{2}\sqrt{\frac{1}{2}}}} \times \dots$$

(1655)
$$\frac{2}{\pi} = \frac{2 \cdot 2}{1 \cdot 3} \times \frac{4 \cdot 4}{3 \cdot 5} \times \frac{6 \cdot 6}{5 \cdot 7} \times \dots$$

(1671)
$$\frac{\pi}{4} = 1 - \frac{1}{3} + \frac{1}{5} - \frac{1}{7} + \frac{1}{9} - \dots$$

(1734)
$$\frac{\pi^2}{6} = 1 + \frac{1}{4} + \frac{1}{9} + \frac{1}{16} + \frac{1}{25} + \dots$$

$$\frac{\pi}{4} = \cfrac{1^2}{1 + \cfrac{1^2}{2 + \cfrac{3^2}{2 + \cfrac{5^2}{2 + \cfrac{7^2}{\ddots}}}}}$$

The Bible contains the following description:

> *And he made a molten sea, ten cubits from one brim to the other; it was round all about, and his height was five cubits, and a line of thirty cubits did encompass it all around.* (Kings 7:23, King James Version).

According to this passage, the molten sea was round with a circumference of 30 cubits and a diameter ("from one brim to the other") of 10 cubits. Calculating the ratio of circumference to diameter gives the biblical approximation of π as $\frac{30}{10} = 3$.

This is known as the method of exhaustion.

Around 200 B.C., Archimedes found π to be between $3\frac{10}{71}$ and $3\frac{1}{7}$ (about 3.14). To obtain these values, Archimedes calculated the perimeters of 96-sided polygons inscribed in and circumscribed about a circle.

In the sixteenth century, Ludolph van Ceulen calculated π to 35 decimal places and had the result carved on his tombstone. To this day, Germans still refer to π as *die Ludolphsche Zahl* (the Ludolphine number).

An English mathematician in 1706 was the first person to use the Greek letter π to represent this number. The symbol was probably intended to stand for the word "periphery."

The English mathematician Shanks spent 20 years calculating π (without a computer) to 707 decimal places. The results were published in 1873, but sad to say, Shanks made a mistake at the 528th decimal that affected the rest of the digits. Since the mistake was not discovered until 1945, many books perpetuated this mistake by using Shanks' values.

In 1949, the computer ENIAC took 70 machine hours to calculate π to more than 2000 decimal places. By 1967, a computer had calculated π to 500,000 decimal places in 28 hours. A common way to test a computer's computational reliability is to let it crank out several thousand digits of π and then compare the result to the known value.

An account of the brothers' work appears in the article "Profiles: The Mountains of Pi" by Richard Preston, March 2, 1992 issue of The New Yorker (pages 36–67).

In 1991, two Russian computer scientists at Columbia University, David Chudnovsky and his brother Gregory, calculated π to more than 2,260,821,336 decimal places. To perform the calculation, the brothers built a supercomputer assembled from mail-order parts and placed it in what used to be the living room of Gregory's apartment. David Chudnovsky says that he and his brother undertook the project because they wanted "to see more of the tail of the dragon."

Ten decimal places of π would be enough to calculate the circumference of the earth to within a fraction of an inch if the earth were a smooth sphere. Thirty-nine decimal places are enough to calculate the circumference of a circle encompassing the known universe with an error no greater than the radius of a hydrogen atom.

How many digits per minute is this?

The world-record holder for memorizing the most digits of π recited more than 42,000 digits in 17 hours, 21 minutes, which included a break of 4 hours, 15 minutes.

There are several sites on the Internet devoted to π and its folklore. One site allows you to enter any string of digits, like 92867, and it searches for a matching string of digits in the decimal expansion of π. You might enjoy trying it with your birthdate.

You and your classmates may want to combine the data you collect.

Do an experiment, either with a computer or by polling people in the halls of your school or at lunch. Get lots of pairs of whole numbers, chosen at random. If you can, get 1000 such pairs. Count the number of pairs that have no common factor (like (5, 8) or (9, 16)). Then take this number and divide it by the total number of pairs. Your answer should be close to

$$\frac{6}{\pi^2} \approx 0.6079.$$

You can also read a wonderful book devoted just to π: Petr Beckman, *A History of Pi* (Golem Press, 1977).

An Introduction to Trigonometry

In this lesson, you will discover the basic ideas of the branch of mathematics called trigonometry.

In the final lessons of this unit, you will use similar triangles to find the measurements of unknown triangle lengths. Does this sound familiar? Well it is, with a new wrinkle added: Rather than just solve for triangle lengths, you will look for patterns and devise shortcuts to make the solution process both faster and more accurate.

Developing these kinds of improvements is immensely important. Architects, designers, engineers, and scientists use similar triangles in their work all the time and need reliable, efficient methods to deal with them.

This picture isn't exactly right. To simplify the problem, the authors overlooked something. What is it?

Explore and Discuss

Below are two different problems.

Problem 1   When you are standing at point *A*, you have to turn your head up by 27° to see the very top of a tree. The distance from you to the tree is 40 feet. How tall is the tree?

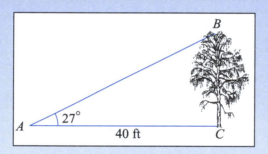

Problem 2   When you are in a boat at point *A*, you have to turn your head up by 27° to see the very top of the Statue of Liberty. If the statue, including its base, is 300 feet tall, how far are you from the bottom of the base?

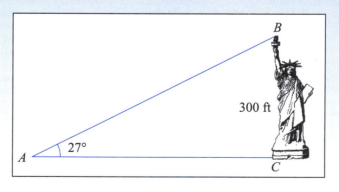

ACTIVITY 1

What Is Trigonometry?

The problems you have just solved are all examples of trigonometry in action. The word *trigonometry* can be split into three parts: *tri* (three) *gon* (angle) *metry* (measure). When put together, these words define trigonometry as "the measure of three-sided figures"—in other words, triangles. You have been measuring triangles throughout this unit, so what makes trigonometry so special? Trigonometry tables and trigonometry buttons on calculators give you accurate values of ratios like $\frac{BC}{AC}$ *without* requiring that you draw a triangle and measure its sides every single time.

For easy reference, trigonometry gives names to some of the constant ratios found in a right triangle. Specifically, in right triangle *ABC*:

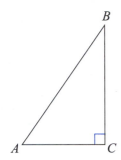

- the **sine** of $\angle A$ is defined as $\frac{BC}{AB}$,

- the **cosine** of $\angle A$ is defined as $\frac{AC}{AB}$, and

- the **tangent** of $\angle A$ is defined as $\frac{BC}{AC}$.

There are common shorthand notations for sine, cosine, and tangent:

- "sine of 27°" is abbreviated as "sin 27°"

- "cosine of 27°" is abbreviated as "cos 27°"

- "tangent of 27°" is abbreviated as "tan 27°"

You will find the values of sine, cosine, and tangent for any possible angle by entering them into a calculator or consulting a trigonometry table. For example, if you enter "tan 27°" into a calculator, it gives the value 0.5, to the nearest tenth. This value tells you that for *any* right triangle with $\angle A$ measuring 27°, the ratio $\frac{BC}{AC}$ is 0.5.

These ratios are accurate to two decimal places.

1. Rewrite these statements using the language of trigonometry.

   a. In any right triangle *ABC* with $m\angle A = 40°$, the ratio of the leg opposite $\angle A$ to the hypotenuse is 0.64.

   b. In any right triangle *DEF* with $m\angle E = 70°$, the ratio of the leg adjacent to $\angle E$ to the hypotenuse is 0.34.

   c. In any right triangle *GHI* with $m\angle H = 55°$, the ratio of the leg opposite $\angle H$ to the side adjacent is 1.43.

2. a. Find the values of sin *A*, cos *A*, and tan *A* for the triangle below.

b. Find the values of sin *B*, cos *B*, and tan *B*.

c. Which of your answers from parts **a** and **b** are the same? Can you explain why?

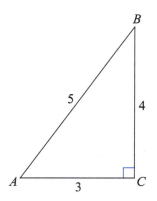

You will need to use the Pythagorean Theorem for this problem.

3. a. Find the values of sin *A*, cos *A*, and tan *A* for the triangle below.

b. Find the values of sin *B*, cos *B*, and tan *B*.

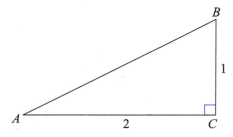

4. Find the values of sin *B* and cos *B* for isosceles triangle *ABC* below.

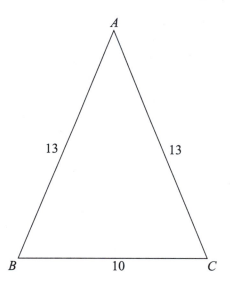

Drawing a sketch will help.

5. Triangle *RST* is a right triangle with right angle at *S*. If tan $R = \frac{2}{3}$, find the values of sin *R* and cos *R*.

6 It is possible to find the exact value of their sine, cosine, and tangent for a few special angles by using what you know about equilateral and isosceles triangles.

   a. Triangle *ABC* below is an equilateral triangle with altitude $\overline{AD}$ drawn from vertex *A*. Find the lengths of $\overline{BD}$ and $\overline{AD}$. Use the lengths to find the exact values of

     i. sin 60°, cos 60°, and tan 60°;

     ii. sin 30°, cos 30°, and tan 30°.

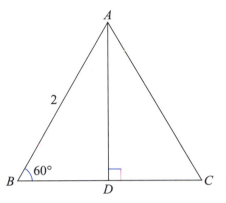

   b. Triangle *ABC* below is an isosceles right triangle with a leg of length 1. Use this triangle to find the exact values of sin 45°, cos 45°, and tan 45°.

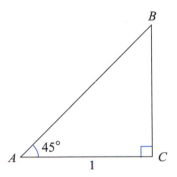

7. **Checkpoint** For each of the right triangles below, find the exact values for sine, cosine, and tangent of ∠*A* and ∠*B*.

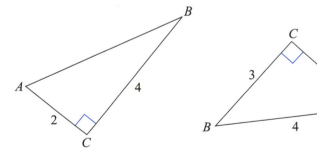

An Area Formula for Triangles

You already know the standard area formula for a triangle:

$$\text{Area} = \tfrac{1}{2}\,(\text{base})(\text{height}).$$

Using this formula, you know that the area of the triangle below is $\tfrac{1}{2}(7)(3) = 10.5$.

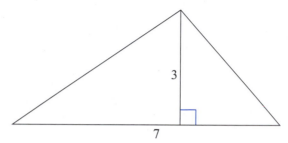

But how can you find the height of a triangle if it's not given? For example, look at the triangle below.

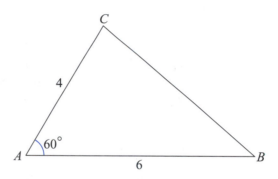

8. Find the area of $\triangle ABC$ by first calculating the length of the altitude from C. Trigonometry can help you here.

θ (the Greek letter "theta") is a variable that is often used to represent angle measures.

9. a. Find the area of the triangle below in terms of a, b, and θ.

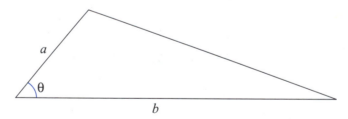

b. Use the area formula from part **a** to show that if the triangle is scaled by a factor of r, then its area is multiplied by r^2.

10. Find the area of parallelogram $ABCD$.

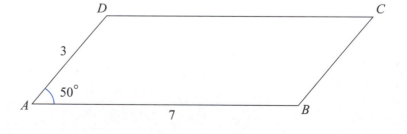

11. **Checkpoint** An airplane takes off and flies 10,000 feet in a straight line, making a 25° angle with the ground. How high above the ground does the airplane rise?

1. Define the following terms.

sine of an angle
cosine of an angle
tangent of an angle

2. The piece of paper drawn at the right originally showed a complete right triangle, $\triangle ABC$, with right angle at C. The paper was ripped, though, so that all you can see now is $\angle A$, which measures 40°.

Find as many of these values as you can using a calculator. Some might not be possible.

a. $\frac{BC}{AC}$

b. $AC + BC$

c. $\frac{BC}{AB}$

d. $AB \times AC$

e. $\frac{AC}{AB}$

f. $AB - BC$

g. $\frac{AC}{BC}$

3. A handicap access ramp slopes at a 10° angle.

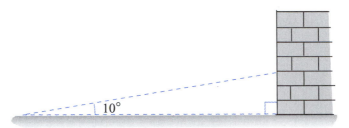

Use a calculator to solve Problems 3 and 4.

a. If the ramp meets the ground 25 feet from the base of the building, how long is the ramp?

b. Most public buildings were built before handicapped access ramps became widespread. So when it came time to design the ramps, the doors of buildings were already in place. Suppose a particular building has a door 2 feet off the ground. How long must a ramp be to reach the door if the ramp is to make a 10° angle with the ground?

4. To the nearest tenth, the value of tan 57° is 1.5. To the nearest thousandth, it is 1.540. Solve for the length of $\overline{BC}$ in the triangle below twice, using each of these values. By how much do your two answers differ?

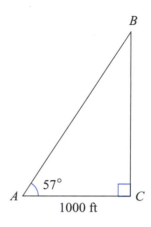

5. Triangle *JKL* is a right triangle with right angle at *K*. If side $\overline{JK}$ is three times the lenght of $\overline{KL}$, find the sine, cosine, and tangent values for ∠*J* and ∠*L*.

6. Is there an angle between 0° and 90° for which the sine of that angle equals 1.5? Explain.

Take It Further

7. Here's a clever way to find the exact value of cos 72°:

The isosceles triangle *ABC* on the left below has base angles measuring 72°, and *AB* = *AC* = 1.

The same triangle is shown on the right with segment *BD* bisecting ∠*B*. Segment *BD*'s unknown length is labeled *x*.

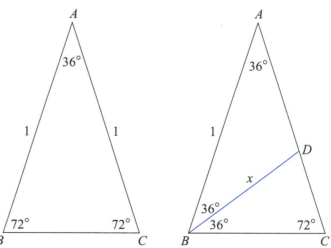

Quadratic Formula:

$$x = -b \pm \frac{\sqrt{b^2 - 4ac}}{2a}$$

a. Find the lengths of $\overline{BC}$, $\overline{AD}$, and $\overline{DC}$ in terms of *x*.

b. Explain why △*ABC* ~ △*BCD*.

c. Set up a proportion involving the lengths *AB*, *BC*, and *CD*.

d. Use this proportion and the quadratic formula to solve for *x*.

e. Divide $\triangle ABC$ into two right triangles by drawing its altitude from point A.

f. Use either of these right triangles and the value of x to find cos 72.

Now find cos 72° on a calculator. Compare the value the calculator gives to the exact value you have found.

8. By dividing $\triangle ABD$ above into two right triangles, find cos 36°. As in Problem **7,** compare the value a calculator gives for cos 36° to this exact value.

9. Find the area of a regular octagon that has a sidelength of 4 inches.

(Hint: Divide the octagon into triangles and find the area of each one. You might also consider the picture below that shows the octagon sitting in a square. If you can find the area of the square and the four triangles, how will this help?)

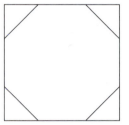

LESSON 15

Extending the Pythagorean Theorem

In this lesson, you will use ideas from trigonometry to find a formula for finding the sides of triangles that do not have a right angle.

For any right triangle, the Pythagorean Theorem gives you a way to relate the lengths of the three sides. In right triangle ABC shown at the left, $a^2 + b^2 = c^2$.

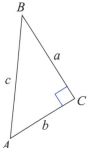

Geometrically speaking, you can think of a^2, b^2, and c^2 as the areas of squares drawn on the three sides of $\triangle ABC$.

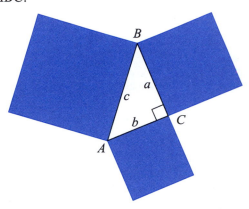

Unit 3, The Cutting Edge, gives several geometric dissection proofs of this theorem.

Since $a^2 + b^2 = c^2$, the area of the square built on the hypotenuse is equal to the sum of the areas of the squares sitting on the other two sides.

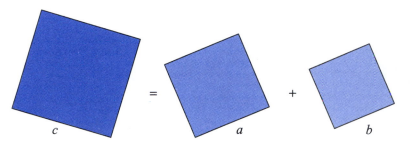

Explore and Discuss

What happens, though, for triangles with *no* right angle? Does something similar to the Pythagorean Theorem still apply?

a Use geometry software to draw an arbitrary triangle. Construct squares on its three sides and measure their areas. Add the areas of two squares to see if they total the third. Try different combinations of squares. Does the Pythagorean relationship apply? Experiment with different triangles by moving the vertices or sides of your original triangle.

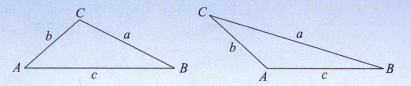

b Below are two triangles with vertices labeled *A*, *B*, and *C*. Angle *C* is obtuse in one triangle and acute in the other.

Based on your work from part a, can you predict for each triangle whether $a^2 + b^2$ will be less than, greater than, or equal to c^2?

A Recipe for Extending the Pythagorean Theorem

There is a way to extend the Pythagorean Theorem to a more general theorem that applies to *all* triangles, with or without right angles. The geometric recipe below shows you how.

Sample Recipe

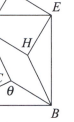

Step 1 Using geometry software, draw a triangle *ABC* with ∠*C* greater than 90°. In the picture below, the measure of ∠*C* is labeled *θ*.

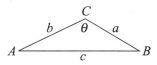

Step 2 Construct a square on side $\overline{AB}$ of your triangle so that the square covers the triangle. Compared to the squares in the Pythagorean Theorem, this one faces the opposite way.

Step 3 Construct a square on side $\overline{AC}$ of your triangle.

Step 4 Construct a parallelogram with sides $\overline{CB}$ and $\overline{CG}$.

Step 5 Construct a square on side $\overline{GH}$ of your parallelogram.

Step 6 Construct a parallelogram with sides $\overline{GF}$ and $\overline{GI}$.

Step 7 Now that your construction is complete, drag vertex *C* of △*ABC* and watch what happens.

1. a. Which features of the construction change and which stay the same?

b. Make note of any regions that appear to be congruent to each other.

Your triangle *ABC* has five quadrilaterals to accompany it. Specifically, there's a big square *ADEB*, two small squares *AFGC* and *GIEH*, and the two parallelograms *CGHB* and *FDIG*.

To help keep track of the two small squares and the two parallelograms, use the software to shade each a different color or darkness.

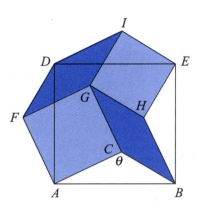

2. Experiment with different locations of point *C*, keeping θ greater than 90°, and observe how the following two regions compare in area.

   Region 1: The space occupied by the two small squares and the two parallelograms.

   Region 2: The space occupied by the large square *ADEB*.

3. Devise a way to cut up and rearrange the two small squares and two parallelograms (Region 1) so that they fit exactly within the large square (Region 2). Write your method or draw a picture to illustrate it.

There is already a lot of overlap. You only have to get those pieces that lie outside the large square to fit snugly inside it.

4. Explain what the visual equality below is saying.

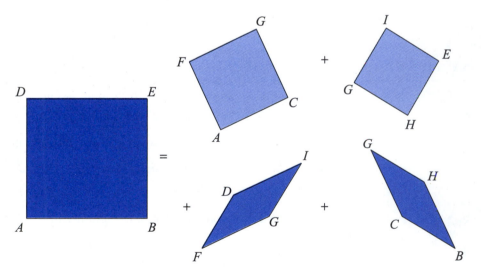

5. What is the area of each of the three squares in the picture above? Your answers should be in terms of just *a*, *b*, and *c*. Remember that in $\triangle ABC$, $BC = a$, $AC = b$, and $AB = c$.

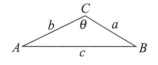

6. What are the sidelengths and angle measurements of the two parallelograms in Problem **4?** Your answers should be in terms of a, b, c, and θ.

7. Write and Reflect Explain how the theorem below applies to $\triangle ABC$. To make the theorem complete, include the sidelengths and angle measurements of the parallelogram.

Theorem 4.11

For $\triangle ABC$ with sides a, b, and c, and θ obtuse,

$$c^2 = a^2 + b^2 + 2 \left(\includegraphics{parallelogram} \right).$$

8. Use Theorem 4.11 to solve for AB in the figure below.

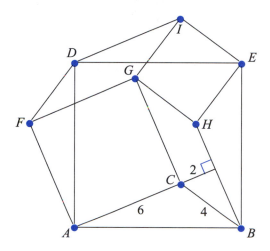

9. a. Sketch the squares and parallelograms for the triangle below.

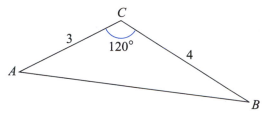

b. What are the sides and angles in the parallelogram?

c. Find the area of the parallelogram.

d. What is the length of the third side of the triangle?

10. Rewrite the formula from Theorem 4.11 in terms of the measure of angle C. That is, find a way to express the area of the parallelogram in terms of angle C.

Hint: Express the angles of the parallelogram in terms of the measure of angle C. Then use them to find the area.

11. **Checkpoint** Use theorem 4.11 to solve for the length of side $\overline{AB}$.

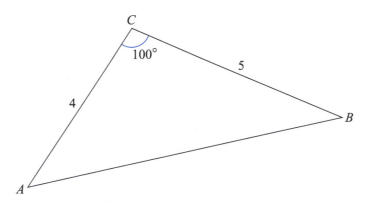

On Your Own

1. Why is the extension of the Pythagorean Theorem more powerful than the Pythagorean Theorem itself? What new kinds of problems does it allow you to solve?

2. What happens to the formula from Problem **10** if $m\angle C = 90°$?

3. Find the length of side $\overline{AC}$ in the triangle below. (Use a calculator.)

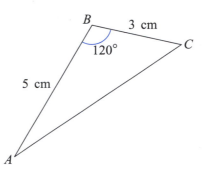

Take It Further

4. Use geometry software to see what happens to the two parallelograms in the picture created in this activity when θ equals $90°$. Can you use this setup to give a geometric dissection proof of the Pythagorean Theorem?

5. Move point C so that it lies anywhere along $\overline{AB}$. What happens to $\triangle ABC$? Write down the algebraic identity represented by this new picture.

6. The extension of the Pythagorean Theorem works when θ is greater than or equal to $90°$. How about triangles where θ is *less than* $90°$? For these triangles, the areas of parallelograms *FDIG* and *CGHB* get subtracted, rather than added, to the areas of the two small squares:

$$\text{Area}(ADEB) = \text{Area}(AFGC) + \text{Area}(GIEH) - \text{Area}(FDIG) - \text{Area}(CGHB)$$

Again, a dissection argument will help you prove this.

Move point C so that the measure of θ is now acute. Chances are, your picture has become very messy! Rather than trying to interpret all of the shading the picture already contains, it is easier to start from scratch and add and subtract the shaded regions one at a time.

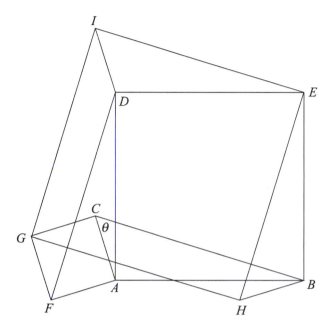

Copy the figure above. Use a pencil to shade in the areas of squares $AFGC$ and $GIEH$. You will be shading some regions twice, so perhaps write the number 2 in these places to keep track of them. Then, use your pencil's eraser to remove (subtract) the areas of parallelograms $FDIG$ and $CGHB$. If you are subtracting a region that is not already shaded, you might want to outline it and indicate that the region must be removed. When you are done, figure out how to fit everything that is left into the square $ADEB$.

Unit 4 Review

1. A map gives a scale factor of 1 inch = 5 miles = 10 minutes average driving time.

    a. How far would you travel if the route measured 3.5 inches on the map?

    b. How long would a 3.5-inch trip take?

    c. How would a 28-mile trip be represented on this map?

2. Give an approximate scale factor for the following pairs of figures. If you do not believe that the figures are scaled copies of each other explain why.

    a.
    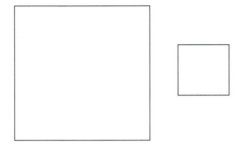

    b.

    c.
    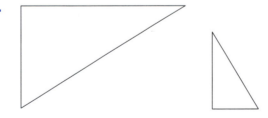

3. A figure is scaled by a factor of $\frac{3}{5}$. How do the areas of the original and the scaled copy compare?

4. A square has an area of 80 square inches. Another square has an area of 5 square inches. Can these squares be scaled copies of each other? If so, give a scale factor. If not, explain why not.

5. Complete this definition: "Two figures are similar if ..."

6. List at least three ways to test if two triangles are similar.

7. Give two more examples of statements, such as "all circles are similar" and "all squares are similar."

8. Use any method to dilate the following polygon using a scale factor of $\frac{3}{4}$.

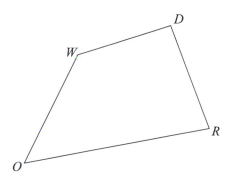

9. a. Draw a blob on a piece of grid paper. Estimate its area.

   b. Explain the process you used to estimate the area of the blob you drew for part **a.**

10. Find the area of a circle with

   a. a radius of 4 cm.

   b. a diameter of 3.5 cm.

   c. a circumference of 14π.

11. Find the area of the shaded region and the white region of the figure below.

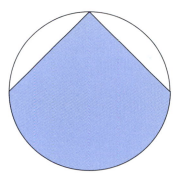

The height of the triangle is 5 cm.

12. Given $\triangle ABC$ with right angle C, define the sine, cosine, and tangent of $\angle A$.

13. Given $\triangle ABC$ with right angle C and $AC = 4$ and $BC = 3$, find the value of $\sin A$, $\sin B$, $\cos A$, and $\tan B$.

14. Explain how the Pythagorean Theorem is a special case of Theorem 4.11.

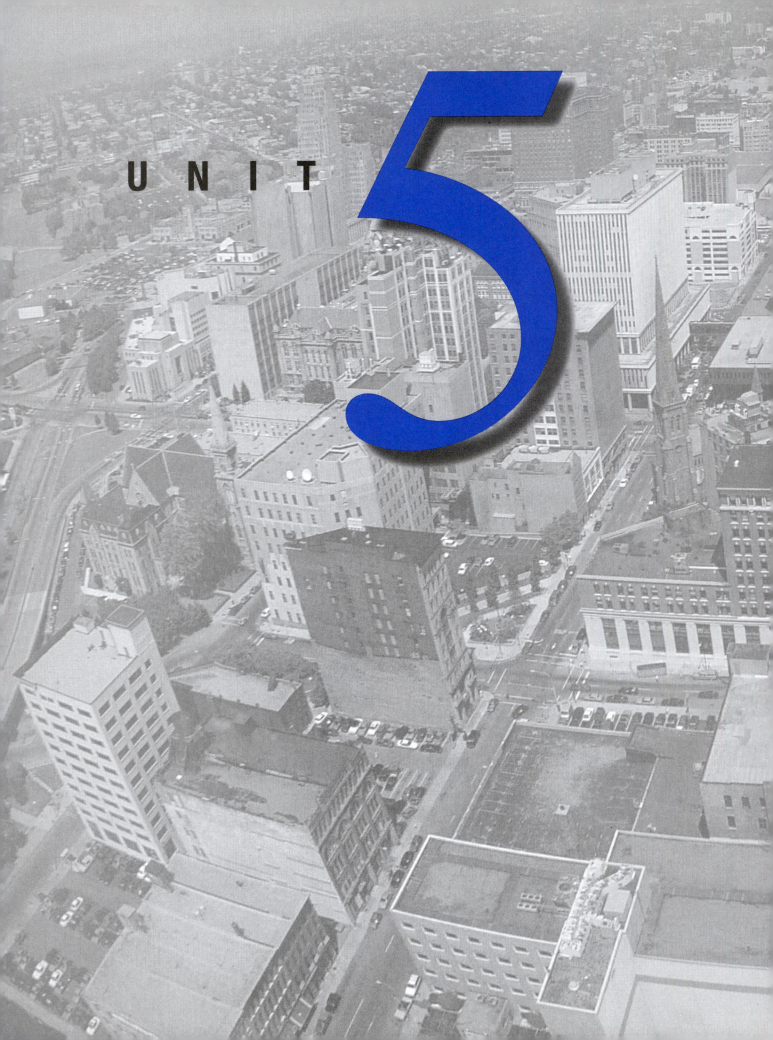

UNIT 5

The Language of Coordinates and Vectors
Connecting Geometry and Algebra

Introduction to Coordinates

In this lesson, you will learn the basics of coordinate systems and plotting points.

You have probably seen some kinds of coordinate systems before. The number line is one example, which assigns every point on the line a unique number.

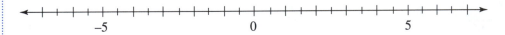

        −5   0   5

Coordinates not only tell where things are; they also identify things by where they are. In mathematics, the "things" that coordinates locate are *points,* which cannot be distinguished *except* by their location. The points are all located, at least in the Cartesian coordinate system you will be working with in this unit, with respect to a frame of reference—a point, called the **origin.**

> *Points have no size, shape, color, weight, or flavor. They have only location.*

Explore and Discuss

a On a piece of paper, draw a number line. Mark roughly where each of these numbers should go.

$$\frac{1}{2} \qquad\qquad -2.5 \qquad\qquad 1.2 \qquad\qquad \pi$$

b The picture below diagrams another "narrow path." While the number line is straight and infinite and has its origin "in the middle," this path is twisty and finite, and has its origin at one end. Explain how you could locate any point along the path with just one piece of information.

> *A, B, and the origin are not the only points in this space. But points in this space must be on the path. Point C is not in this space.*

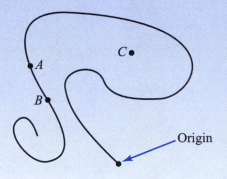

c You need to specify the exact location of a ship somewhere on a very large lake or ocean. People on the ship might have some way of figuring out their latitude and longitude and specifying their location that way, but try to find a way that people could measure from some fixed location (an *origin*) that is not on the ship. If you like, you can start by imagining that the people can *see* the ship.

One of Two Common Systems

There are two common ways to assign coordinates to points in the plane. One way uses two number lines, perpendicular to one another. Each location in the plane is described by two numbers, measured by the two number lines. The first number, often called the **x-coordinate,** comes from the horizontal number line, and says how far to the right or left of the origin the point is. The second number, often called the **y-coordinate,** comes from the vertical number line, and says how far above or below the origin the point is.

So in this system, the x-coordinate indicates a horizontal distance, and the y-coordinate a vertical distance. The order of the two numbers distinguishes pairs like (1, 3) and (3, 1). This system of assigning ordered pairs of numbers to points in the plane, called the Cartesian coordinate system, is credited to René Descartes. Read the Perspective essay at the end of this lesson for more background on Descartes and the development of the Cartesian coordinate system.

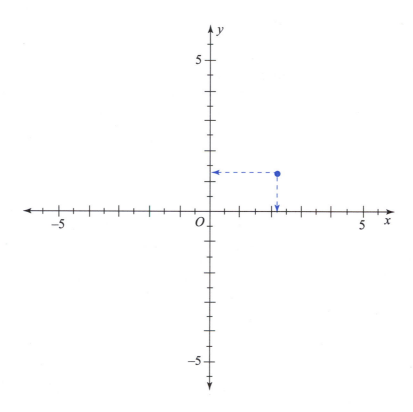

Use a Cartesian coordinate plane in working on the following problems.

Point	Coordinates
E	
O	(,)
	(0, 1)
C	(3, 2)
J	
	(2.5, −2.5)
M	
	(−2.5,)
D	
B	
G	$(-\frac{1}{2}, -3)$
	(1, 0)
A	
F	

1. a. Copy and complete the table at the left. Estimate the coordinates if necessary.

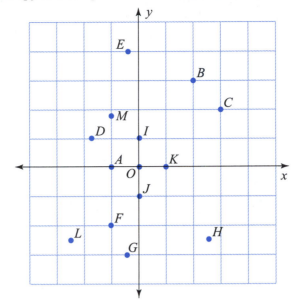

b. Where would point *N*, at (3, −3), be in comparison to point *H*: To the left or right? Above or below?

c. Locate another point, *P*, *directly* above point *F*. Is the *y*-coordinate of *P* greater than, smaller than, or the same as the *y*-coordinate of *F*? How do the *x*-coordinates compare?

2. Without drawing on paper, picture a coordinate plane in your mind. Imagine locating these points:

(2, −1) and (−2, −1)   (4, 3) and (−4, 3)

How are these two pairs of points alike?

3. On a piece of graph paper, draw and label a pair of coordinate axes. Graph the points associated with the following coordinates:

$A = (\pi, 4)$ $B = (\frac{3}{4}, -5)$ $C = (-2, \sqrt{2})$

4. Here is a sequence of coordinates: (1, 0), (1, 1), (3, 1), (3, 0), (4, 0), (4, 1), (6, 1), (6, 0), (7, 0). Extend this sequence in both directions by giving the two coordinate pairs that follow and the two coordinate pairs that precede this part of the sequence.

5. On a piece of graph paper, draw in the *x*- and *y*-axes. Pick two positive numbers, *x* and *y*, choosing values between 2 and 12. Plot the point with coordinates (*x*, *y*). Label it *A*. Point *A* will be in the first quadrant of the coordinate plane.

a. Graph point (−*x*, *y*) on your paper. Label it *B*. Point *B* is the reflection of *A* over the *y*-axis. In which quadrant did you graph *B*?

b. Graph point *C* with coordinates (−*x*, −*y*). Which quadrant contains *C*?

c. Graph (*x*, −*y*) and label it *D*. Which quadrant contains *D*?

d. Create the figure *ABCD* by connecting the points. What shape do they make? Explain your answer.

e. Compare your shape to those drawn by your classmates. What do the shapes have in common? Are they similar? Congruent? Completely different? Explain.

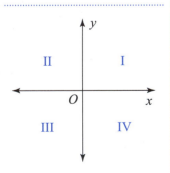

The x- and y-axes divide the plane into cuatro (Spanish for "four") quadrants.

6. a. Copy and complete the table below.

A	B	C	D	E	F	G
(x, y)	$(x + 3, y)$	$(-x, y)$	$(x, -y)$	$(2x, 2y)$	$\left(\frac{x}{2}, \frac{y}{2}\right)$	$(-y, x)$
$(2, 1)$					$\left(1, \frac{1}{2}\right)$	
$(-4, 0)$	$(-1, 0)$					
$(-5, 4)$		$(5, 4)$				$(-4, -5)$

Be sure to list any conjectures that you come up with.

b. On a piece of graph paper, plot the three points in column A. Connect them to form a triangle. Plot the three points in column B. Connect them to form a second triangle. Describe how the two triangles differ. Repeat for triangles A and D.

c. On a fresh piece of graph paper, draw triangles A and C. Describe how they differ. Repeat for triangles A and D.

d. Using fresh graph paper each time, pair triangle A with each of triangles E, F, and G. Each time describe how the two triangles differ.

7. How can you tell, by looking at the coordinates, if a point is in

a. quadrant I?

b. quadrant II?

c. quadrant III?

d. quadrant IV?

For Discussion

- Describe how a point moves if the sign of its *x*-coordinate is changed. Describe how a point moves when you change the sign of its *y*-coordinate. What happens when you change the sign of both coordinates?

- Explain this statement and then decide whether you think it is true or false: "Every ordered pair of real numbers (x, y) corresponds to a unique point in the Cartesian plane; that is, there is a one-to-one correspondence between the set of points in the plane and the set of ordered pairs of real numbers."

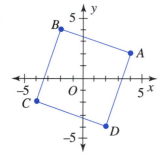

8. Checkpoint At the left is a picture of a "nonlevel square" with its center at the origin.

a. Find the coordinates of its four vertices.

b. A square has vertices at $(-3, 1)$ and $(1, 3)$ and is centered at the origin. Find the coordinates of its two other vertices.

c. Create your own nonlevel square centered at the origin. Make yours different from the two above. List the coordinates of your new square.

d. Describe the pattern in the coordinates of the vertices of such squares.

Coordinate Practice

These problems will extend your knowledge about how geometric ideas—horizontal and vertical lines, for example—are reflected in the coordinates of points.

9. The *y*-axis is the only vertical line that can pass through the point (0, 0). What do the coordinates of all the points on that line have in common? How can you tell if a point is on the *y*-axis just by looking at its coordinates?

10. Picture a horizontal line through the point (1, 1) in a coordinate plane. Name two other points on that line.

While there are infinitely many lines through (3, 7), there is only one passing through it in any given direction.

11. Suppose ℓ is the vertical line passing through (3, 7).

   a. Find the coordinates of six points that are on ℓ.

   b. Find the coordinates of six points that are not on ℓ.

   c. How can you tell if a point is on ℓ just by looking at its coordinates?

   d. Draw a different vertical line on a coordinate plane. What do all of the points on your line have in common?

12. Suppose *m* is the horizontal line passing through (3, 7).

   a. Find the coordinates of six points that are on *m*.

   b. Find the coordinates of six points that are not on *m*.

   c. How can you tell if a point is on *m* by looking at its coordinates?

13. Let *A* and *B* be two different points on the same horizontal line. If $A = (s, t)$, what coordinate of *B* do you know?

Try to visualize this or draw a rough sketch without using graph paper.

14. What are the coordinates for the intersection of the horizontal line through (5, 2) and the vertical line through (−4, 3)?

15. a. Name and plot six points whose first coordinate is the same as the second coordinate.

   b. Imagine a drawing that shows every point whose first coordinate is the same as the second coordinate. What figure would that be?

16. a. Name and plot six points whose first coordinate is the negative of the second coordinate.

   b. Imagine a drawing that shows every point whose first coordinate is the negative of the second. What figure would that be?

17. In this picture, ℓ and *m* are horizontal lines.

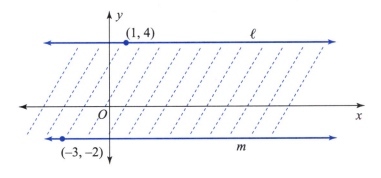

a. Find the coordinates of six points between ℓ and m.

b. Find the coordinates of six points that are not between ℓ and m.

c. How can you tell if a point is between ℓ and m just by looking at its coordinates?

18. Checkpoint Draw a set of x- and y-axes.

a. Shade in the region where the points all have x-coordinates, such that $5 \leq x \leq 8$.

b. Shade in the region where the points all have y-coordinates, such that $-3 \leq y \leq 6$.

c. Is there any overlap in the shaded regions? If so, what shape is the intersection? What is its area in square units?

d. How can you change the instructions above to make the shaded intersection take the shape of a square? What is the area of your square?

On Your Own

1. a. True or false: If point E is directly to the right of point F, then the x-coordinate of E will be greater than the x-coordinate of F.

b. If E is directly to the right of F, what can you say about the y-coordinates of the two points? Below F, and on the line through F parallel to the y-axis?

c. How do the y-coordinates compare when E is directly below F? How do the x-coordinates compare?

Estimate all coordinates to the nearest integer. Note which points may have x- or y-coordinates in common.

2. Each of the figures below is drawn on a pair of coordinate axes. Horizontal and vertical axes use the same unit of measure. The coordinates of at least one point in each figure are given to you. Make an ordered list of coordinates that might reasonably describe the points in the figure.

a.

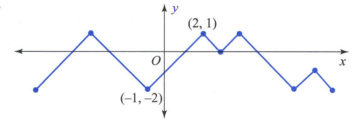

b.

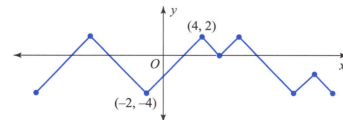

c.

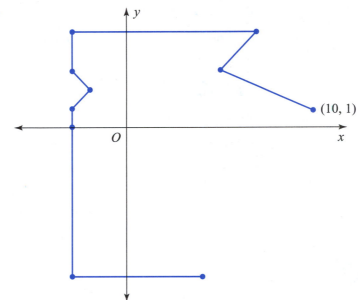

(10, 1)

3. Compare the *x*-coordinates and the *y*-coordinates of two points *C* and *D* if

   a. *D* is directly to the left of *C*.

   b. *D* is directly above *C*.

   c. *D* is to the right and below *C*.

4. On a piece of graph paper, draw three different sets of coordinate axes.

   a. On your first set of axes, shade the quadrant(s) where the points have negative *x*-coordinates.

   b. On the second set of axes, shade the quadrant(s) where the points have positive *y*-coordinates.

   c. On the third set of axes, shade the quadrant(s) where the points have negative *x*-coordinates and positive *y*-coordinates.

5. If *a* is a negative number and *b* is a positive number, state the quadrant in which each of the following points lies.

   a. (a, b)

   b. $(-a, b)$

   c. $(-a, -b)$

6. Line *a* is a horizontal line through $(-5, 12)$. Line *b* passes through the origin and makes a 45° angle with the axes as it enters quadrant I. Find the coordinates of the point where these two lines intersect.

7. Below is a rectangle.

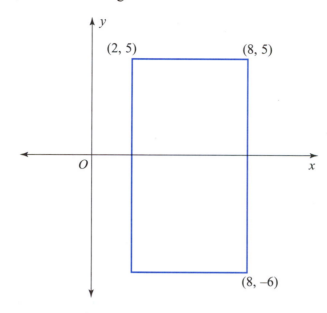

a. Find the coordinates of four points that are inside the rectangle.

b. Find the coordinates of four points that are outside the rectangle.

c. Find the coordinates of four points that are *on* the rectangle.

d. How can you tell if a point is inside the rectangle just by looking at its coordinates?

8. One side of a quadrilateral passes through (4, 6) and (4, 3). Another side passes through points $(-2, 5)$ and $(-1, 2)$. Which of the following statements are true?

a. The figure could not be a trapezoid.

b. The figure could be a trapezoid.

c. The figure must be a trapezoid.

d. The figure could not be a parallelogram.

e. The figure could be a parallelogram.

f. The figure must be a parallelogram.

9. a. How many vertical lines can be drawn through the point $(-2, 3)$?

b. Name the coordinates of the intersection of a horizontal line through $(3, -5)$ and a vertical line through $(-1, 9)$.

10. Picture or draw line ℓ through (6, 4) and $(-3, 1)$.

a. How many lines are parallel to ℓ?

b. Line m passes through (7, 4) and is parallel to ℓ. Name a point on line m other than (7, 4).

c. Line n passes through (8, 3) and is parallel to ℓ. Name a point on line n other than (8, 3).

11. a. Name four other points on the horizontal line through (3, 4).

b. Name three points that are collinear with (on the same line as) (4, 2) and (4, −3).

12. Each problem below describes a line in a coordinate plane. For each line, state how many quadrants it passes through.

a. A vertical line that passes through the origin.

b. A vertical line that does not pass through the origin.

c. A nonvertical, nonhorizontal line that passes through the origin.

d. A nonvertical, nonhorizontal line that does not pass through the origin.

13. Is it possible for a line to pass through only one quadrant? Is it possible for a line to pass through all four quadrants? Is there *any* slanting (nonvertical, non-horizontal) line that passes through only two quadrants and does *not* go through the origin?

This is the second system commonly used to assign coordinates to points in the plane. The point at the center of the system is called the pole. *And the system is called* polar coordinates.

Another way to locate a point on a plane is to describe how far it is from some central location, and in which direction. The distance might be reported in millimeters or miles (or arbitrary units), and the direction might be reported in degrees or some other appropriate unit.

14. a. Copy the table at the left. Use this distance-direction scheme to complete it. Estimate the coordinates if necessary.

Point	Coordinates
E	
	(3, 240°)
C	(4, 180°)
	(, 320°)
	(8,)
B	(3½,)
G	(, 135°)
	(1, 90°)
A	
	(7,)

b. In a coordinate scheme like this, how many points can have coordinates $(4, a°)$? Describe the location in which such points are found.

c. How many points can have coordinates $(r, 60°)$? Describe the location in which such points are found.

d. How many points can have coordinates $(0, a°)$?

Because the plane is 2-dimensional, any coordinate system must name each point by a *pair* of numbers. The numbers in the pair mean different things. In the polar coordinate system, one number indicates a distance (in some unit) from the pole, and the other indicates direction (in degrees).

Perspective

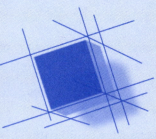

Describing Geometric Ideas with Algebraic Language

How long have people been using coordinate systems? Who was Descartes and why is he famous? This essay will help you answer these questions.

The idea of using grid systems to help specify locations is a very old one. In the West, the use of grids in map-making apparently dates back to Ptolemy, roughly the 2nd century A.D. But Western map-making declined dramatically after Ptolemy and was relatively primitive from the 13th to the 14th century. In China, the use of coordinates in map-making began equally early, but that use continued to be refined and developed through the centuries, and far surpassed Western work by the middle ages.

Even though the idea of coordinates developed early, its connection with geometric reasoning is relatively new in the history of Western mathematics. Euclid systematized geometric ideas that dated back at least as far as the 5th century B.C., but all that geometric thinking was done without coordinates. The systematic use of coordinates is usually credited to René Descartes around 1637—a rediscovery of an old idea, or perhaps a new connection between two old ideas.

> *How many years were there between 500 B.C. and 1637 A.D.? Between 1637 A.D. and now?*

In 1637, the great French philosopher, scientist, and mathematician, René Descartes (1596–1650), published a famous book called *La Géométrie*, in which he first showed how to use coordinates to translate geometry problems into algebra and vice versa. About the same time, another Frenchman, Pierre de Fermat, independently had similar thoughts, but he did not publish his ideas until 1679. In honor of Descartes, we often call coordinate geometry *Cartesian geometry,* and we call a coordinate system of axes a *Cartesian coordinate system.*

Actually, the word Cartesian comes from the Latin version of René Descartes' name: Renatus Cartesius. In the seventeenth century, it was the custom for scholars and scientists to take on Latin names. The Mercator projection is another interesting example of the use of a Latin name. The Mercator projection is the most popular kind of world map and is named in honor of Mercator, whose real name was Kramer (Mercator and Kramer are the Latin and German versions of the word *merchant*).

> *Why x and y? There is really nothing special about those letters. When you are graphing the distance traveled in a given time, the best labels for the axes might be d and t.*

If you look at Descartes' book *La Géométrie,* it would look quite unfamiliar. First, Descartes generally uses oblique axes, that is, his *x*- and *y*-axes do not meet at a right angle. Second, Descartes uses only positive numbers, which means that he does all his geometry in what we now call the *first quadrant.* This is because in Descartes' time people did not believe that there was such a thing as a negative number. (After all, how could there be anything less than nothing?)

Descartes also intentionally made his book difficult to read. The reasons for this are quite interesting. Descartes tells us that he deliberately left out a lot of explanations so that other people could not claim that he had "written nothing that they did not already know"! He also said that he didn't want "to deprive his readers of the pleasure of working things out for themselves."

Descartes made many important contributions to mathematics, but he was also an eminent scientist, and he is perhaps best known as a philosopher. In his philosophical writings he was particularly interested in the question of how we can know for certain that the things we believe are really true. His philosophical writings are still studied and admired today; you may some day have the pleasure of reading them for yourself. The most famous quotation from Descartes is the Latin sentence "Cogito, ergo sum," which means "I think, therefore I am."

Although Descartes was a Frenchman, he spent most of his adult life in Holland, where he was living when *La Géométrie* was published. You may wonder why he chose to live in a foreign country. Some of Descartes' ideas in science and philosophy were extremely controversial. In most European countries at that time, there was strict censorship of what could be published, and having unpopular ideas could be very dangerous. The notable exception to this kind of intolerance was Holland, and so Descartes went there to live so that he could say, write, and publish whatever he pleased.

Descartes was not the only one who went to Holland to enjoy freedom of thought. When the Pilgrims left England in 1608 because of religious persecution, they first went to Holland. They lived there until 1620, when they left to make a new start in America. Descartes arrived in Holland in 1629, so he and the Pilgrims missed each other by just a few years.

Descartes' contribution—the idea of describing points in terms of coordinates—opened up the possibility of describing geometric ideas in algebraic language. Objects like lines, curves, or surfaces; relationships like perpendicularity; transformations like translation, rotation, or dilation—all could be expressed in terms of functions and equations. This blending of algebra and geometry is called *coordinate* or *analytic* geometry.

Lines, Midpoints, and Distance

In this lesson, you will use coordinates to describe properties of lines, and find midpoints and lengths of segments.

Since coordinates are used to locate points, it should come as no surprise that they are helpful for locating things, such as the midpoint of a line segment. And since coordinates involve numbers, it makes sense that they might help you calculate lengths and distances better than measurement.

Notation: An easier way to write "A is at (5, 7)" is "A = (5, 7)."

Explore and Discuss

Suppose A is at (9, 5), and B is at (7, 5).

a What is the distance between A and B?

b Find the coordinates of the midpoint of $\overline{AB}$.

Suppose $C = (2, 5)$ and $D = (2, 396)$.

c Find the distance between C and D.

d Find the coordinates of the midpoint of $\overline{CD}$.

Suppose $E = (-5, -7)$ and $F = (12, -7)$.

e What is the distance between E and F?

f Find the coordinates of the midpoint of $\overline{EF}$.

g State a conjecture about the coordinates of midpoints for horizontal and vertical segments.

ACTIVITY 1 ▸ Midpoints and Distance between Points

You have found some distances and midpoints for horizontal and vertical segments. But what about segments that are not horizontal or vertical? The following problems will help you extend your methods to "slanted" segments.

1. Find the distances between the given pair of points.

    a. $I = (-110, -7)$ and $J = (-80, -7)$

    b. $K = (1, 5)$ and $L = (1, -15)$

    c. $M = (-93, 4)$ and $N = (90, 4)$

2. Here are the coordinates of four points.

    $A = (4, 2)$ $B = (8, 5)$ $C = (-4, 3)$ $D = (-7, 7)$

    Which of the following statements can you justify? How?

    a. Segments AB and CD are guaranteed congruent.

b. Segments AB and CD are approximately the same length, but are not guaranteed congruent.

3. Find the distances between the given pairs of points.

   a. $P = (a, b)$ and $Q = (a, c)$

   b. $P = (a, c)$ and $Q = (b, c)$

4. What about the distance from O to B, where $O = (0, 0)$ and $B = (1, 1)$? Solve the problem any way you can.

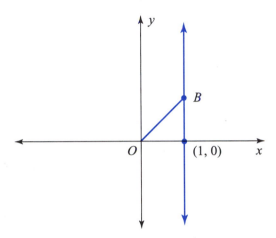

5. The following are the coordinates for the three vertices of a triangle: $E = (-1, -3)$, $F = (2, -3)$, and $G = (2, 1)$.

   a. How long are $\overline{EF}$ and $\overline{FG}$?

   b. Find the distance EG.

6. In the picture below, m is a vertical line.

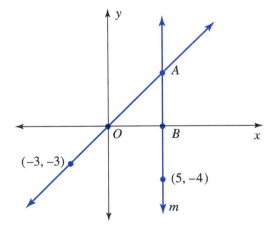

If you are not familiar with the dimensions of the particular triangles in Problems 5 and 6, use the Pythagorean Theorem.

   a. Find the coordinates of A and B.

   b. How long is $\overline{AB}$?

   c. What is the area of $\triangle ABO$?

   d. How long is $\overline{AO}$?

7. Use one set of axes for the following.

   a. Find eight points that are 5 units away from the origin.

   b. Draw the picture of all the points that are 5 units away from the origin. What shape is it? Why?

   c. Find eight more points on the figure you found in part **b.**

8. **Write and Reflect** Consider these problems.

   a. Write a set of instructions that explains how to find the distance between *any* two points if you know their coordinates.

   b. Exchange your instructions with a partner. See if you can follow your partner's instructions to find the distance between $(1, 3)$ and $(6, 15)$.

9. Devise a method to find the coordinates of the midpoint of the segment below.

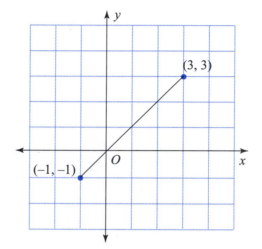

10. In the picture below, $A = (-3, -3)$, $B = (-3, 2)$, $C = (3, 3)$, and $D = (2, -3)$. Use your method from Problem **9** to find the coordinates of the midpoints of $\overline{AB}$, $\overline{CD}$, $\overline{AD}$, $\overline{BC}$, $\overline{AC}$, and $\overline{BD}$. If the method you used for Problem **9** does not work here, look at Problem **9** again. Try to develop a method that will work.

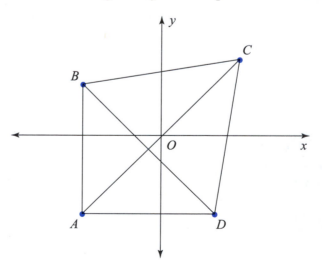

11. **Write and Reflect** Consider these problems.

   a. There are many correct ways to find the coordinates of the midpoint of a line segment when you know the coordinates of the endpoints. Describe your method precisely. Explain why it works.

   b. Exchange the methods you produced in Problems **9** and **10** with a partner. See if you can successfully follow your partner's method in finding the coordinates of the midpoint of $\overline{JK}$, where $J = (-2, 1)$ and $K = (1, 6)$.

12. **Checkpoint** Use your methods to find the midpoints and lengths of the segments whose endpoints are listed below.

   a. $(0, 7)$ and $(5, 7)$

   b. $(-3, -3)$ and $(-1, 2)$

ACTIVITY 2 # Midpoint and Distance Formulas

In this activity, you will formalize the instructions you wrote for finding the midpoint and the distance between points.

13. Assume $G = (x, y)$ and $H = (w, z)$.

   a. Use your method from Activity 1 to find the distance GH.

   b. Use your method from Activity 1 to find the coordinates of the midpoint of $\overline{GH}$.

Subscript notation is used in the standard formulas for midpoints and distance, so it is introduced here.

When only a few points need names, it is convenient enough to call them A, B, C, and so on, with coordinates (a, b), (c, d), (e, f), and so on. But it is often important to have names for many points, and then one quickly runs out of letters. Numbers *never* run out, and so the convention is to name, say, the vertices of a decagon with names like these A_1, A_2, A_3, ..., A_{10}, and the vertices of an n-gon with names like these: B_1, B_2, B_3, ..., B_n. The notation is supposed to make sense. See if you can figure out the logic of the notation in the problems below.

14. Look at the drawing of the square. Copy and complete the table below. Assume that point V_1 has coordinates (x_1, y_1) and point V_2 has coordinates (x_2, y_2), and so on.

i	Coordinates of V_i	x_i	y_i
1	(,)		
2	(,)	-2	
3	$(-4, -2)$		
4	(,)	2	

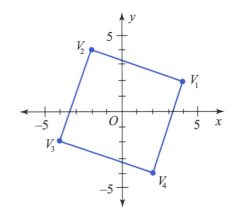

Square $V_1V_2V_3V_4$

15. Here is a claim about the coordinates of the vertices of square $V_1V_2V_3V_4$: $x_i = y_{i+1}$. Is that claim true when $i = 1$? That is, is it true that $x_1 = y_2$? Is that claim true when $i = 2$? When $i = 3$? When $i = 4$?

16. Here is another claim about the vertices of square $V_1V_2V_3V_4$: If V_i has coordinates (x_i, y_i), then V_{i+1} has coordinates $(-y_i, x_i)$.

   a. When $i = 2$, the claim says: If V_2 has coordinates (x_2, y_2), then V_3 has coordinates $(-y_2, x_2)$. Look at the table and decide whether or not this is a true statement.

   b. Pick a value of i for which the statement does not make sense.

17. Name the vertices of the square for which it is true that $y_i = \frac{1}{2}x_i$.

$P_3 = (y_2, -x_2)$

18. Start with point $P_1 = (3, 4)$ and find the coordinates of P_2, P_3, and P_4, if the points follow the rule: If P_i has coordinates (x_i, y_i), then P_{i+1} has coordinates $(y_i, -x_i)$. Plot and label the points.

19. Here is a rule for deriving a new set of points Q_i from the points V_1, V_2, V_3, and V_4: If $V_i = (x_i, y_i)$, then $Q_i = (-3 + x_i, 4 + y_i)$.

   a. The rule is written in algebraic symbols. Explain it in words.

   b. Find the new points Q_1, Q_2, Q_3, and Q_4. Plot the new points.

20. If $P_1 = (x_1, y_1)$ and you know that P_2 is a second point on the same horizontal line, how could you write its coordinates?

21. $P_i = (x_i, y_i)$, $x_i = i + 3$, and $y_i = x_i - 4$. Plot P_i as i goes from 1 to 8.

Which of these is most like your method?

There are at least two ways to write a formula for the midpoint of a segment with endpoints at (x_1, y_1) and (x_2, y_2):

(1) $M = \left(\dfrac{x_1 + x_2}{2}, \dfrac{y_1 + y_2}{2} \right)$

(2) $M = \left(x_1 - \dfrac{x_2 - x_1}{2}, y_1 - \dfrac{y_2 - y_1}{2} \right)$

22. Pick one of the formulas and translate it into a sentence in English.

Both of these formulas work; you will work through a proof of the first one, filling in the details.

The goal is to find the coordinates of M, the midpoint of the segment connecting (x_1, y_1) and (x_2, y_2). You will use what you know about horizontal lines, vertical lines, and similar triangles.

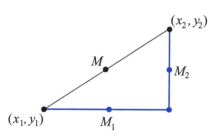

23. In the picture above, a horizontal segment is drawn from (x_1, y_1) and a vertical segment has been drawn from (x_2, y_2). What are the coordinates of the point where these segments meet?

24. Find the coordinates of M_1.

25. a. Whatever expression you used for the first coordinate of M_1, show algebraically that

$$\text{your expression} - x_1 = x_2 - \text{your expression}.$$

b. Explain what the algebra was intended to prove about the midpoint.

26. a. What are the coordinates of M_2?

b. Explain how you can be sure that the coordinates you found describe a point that is not only equidistant from endpoints (x_2, y_1) and (x_2, y_2), but also *on the line* that connects them.

Below is a new drawing. Some of the points have new labels to make it easier to talk about them.

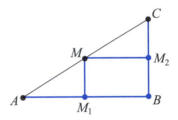

27. If a line is drawn through M_1, parallel to $\overline{BC}$, where does it intersect $\overline{AC}$? How do you know that?

28. If a line is drawn through M_2, parallel to $\overline{AB}$, where does it intersect $\overline{AC}$? How do you know that?

29. a. If you draw a vertical line through M_1, all of the points will have the same x-coordinate. What will it be?

b. If you draw a horizontal line through M_2, all of the points will have the same y-coordinate. What will it be?

c. So, what are the coordinates of M?

You have actually proved a new theorem.

Theorem 5.1

Each coordinate of the midpoint of a line segment is equal to the average of the corresponding coordinates of the endpoints of the line segment.

---

30. Use the midpoint theorem to find the midpoint between $(1327, 94)$ and $(-668, 17)$.

How many states actually comprised the U.S. in 1888?

31. What is the midpoint between $(1776, 13)$ and $(2000, 50)$?

32. Points A and B are endpoints of the diameter of a circle. Point C is the center of the circle. Find the coordinates of C given the following coordinates for A and B.

a. $A = (-79, 687)$, $B = (13, 435)$

b. $A = (x, 0)$, $B = (5x, y)$

To find a formula for the distance between two points, you can use the same right triangle that you used for midpoints.

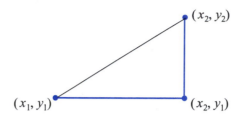

33. What is the length of the horizontal segment?

34. What is the length of the vertical segment?

35. Let d be the distance between (x_1, y_1) and (x_2, y_2). Use the Pythagorean Theorem to find d.

Theorem 5.2

The distance d between two points (x_1, y_1) and (x_2, y_2) can be found using the Pythagorean Theorem. It is the square root of the sum of the difference in the x-coordinates squared and the difference in the y-coordinates squared.

36. Find the distance between the following pairs of points.

   a. $(1, 1)$ and $(-1, -1)$

   b. $(1, 1)$ and $(4, 5)$

   c. $(2, 4)$ and $(-4, -2)$

37. a. Imagine a right triangle with its right angle sitting on the x-axis but not at the origin. Its first two vertices are at $(0, 0)$ and $(2, 0)$. Find the coordinates of the third vertex when the hypotenuse is

   i. 5 units long.

   ii. 6 units long.

   b. A different triangle has a hypotenuse 5 units long, one vertex at $(0, 0)$, and legs of equal length. Again, the right angle is on the x-axis but not at the origin. Find the coordinates of the second and third vertices of the right triangle.

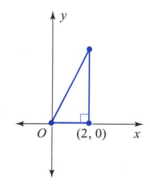

38. Find the missing coordinates for points A through G as shown in the circle with radius 10.

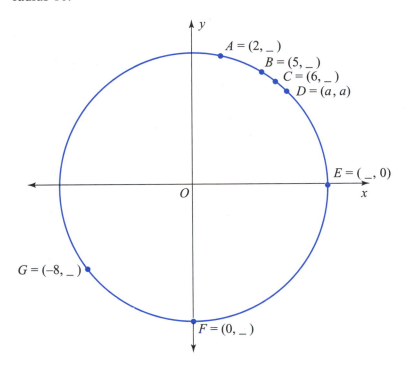

$A = (2, \_)$
$B = (5, \_)$
$C = (6, \_)$
$D = (a, a)$
$E = (\_, 0)$
$G = (-8, \_)$
$F = (0, \_)$

39. The vertices of $\triangle ABC$ are $A = (2, 1)$, $B = (4, 8)$, and $C = (6, -2)$.

  a. Find the lengths of all three sides of the triangle.

  b. Find the length of all three medians of the triangle.

40. Consider the six points $A = (5, 1)$, $B = (10, -2)$, $C = (8, 3)$, $A' = (2, 3)$, $B' = (7, 0)$, and $C' = (5, 5)$. Show that $\triangle ABC \cong \triangle A'B'C'$.

41. Consider the six points $A = (-800, -500)$, $B = (160, 12)$, $C = (-737, -484)$, $A' = (0, 0)$, $B' = (3840, 2048)$, and $C' = (252, 64)$. Show that $\triangle ABC \sim \triangle A'B'C'$.

42. Write and Reflect Consider these problems.

  a. Explain how to tell if two triangles are congruent by calculations on the coordinates of their vertices.

  b. Explain how to tell if two triangles are similar by looking at the coordinates of their vertices.

One way to show that a quadrilateral is a parallelogram is to show that the opposite sides are parallel. What is another way?

43. Checkpoint Pick four points that form a quadrilateral in a Cartesian plane. Find the midpoints of all four sides. Show that if you connect the midpoints, you get a parallelogram.

ACTIVITY 3 **Lines**

Remember Problem **15** from Lesson 1, where you drew the picture of all the points for which the x-coordinate was the same as the y-coordinate? You got a figure that looked like a line. Now you will look at some other lines in the plane, and devise a way to tell if three points are all on the same line.

44. Plot several points whose *y*-coordinates are three more than their *x*-coordinates. Is there any regularity to the points? Explain.

45. Plot several points whose *y*-coordinates are two more than their *x*-coordinates. Draw the picture of *all* the points with this property.

46. Plot several points whose *y*-coordinates are one more than their *x*-coordinates. Draw the picture of *all* the points with this property.

47. Plot several points whose *y*-coordinates are

   a. twice their *x*-coordinates.

   b. three times their *x*-coordinates.

   c. four times their *x*-coordinates.

48. Write and Reflect Consider these problems.

   a. Some of the lines you sketched had points whose coordinates were of the form (*x*, *x* + *something*). What did those lines have in common?

   b. Some other lines you sketched had points whose coordinates were (*x*, *x* × *something*). What did *those* lines have in common?

49. Find a point that is collinear with *A* = (5, 1) and *B* = (8, −3). Explain how you did it.

50. Give a set of instructions for finding points that are collinear with *A* = (5, 1) and *B* = (8, −3). Explain why your method works.

51. Write and Reflect Consider these problems.

   a. Give a set of instructions for testing points that are collinear with *R* = (−40, −30) and *S* = (80, 20). Pick some points that are collinear with *R* and *S* and some that are not. Explain why your method works.

   b. Generalize your method. Suppose *R* and *S* are two points. Give a set of instructions for testing a third point *P* to see if it is collinear with *R* and *S*. Explain why your method works.

52. **Checkpoint** Is *P* = (−4, −14) collinear with *R* = (−40, −30) and *S* = (80, 20)? Explain.

1. Three vertices of a square are (−114, 214), (186, 114), and (−214, −86).

   a. Find the center of the square.

   b. Find the fourth vertex.

2. Three vertices of a square are (−1, 5), (5, 3), and (3, −3).

   a. Find the center of the square.

   b. Find the fourth vertex.

3. Points *D* and *E* are the endpoints of one of the sides of a square. If the coordinates of the midpoint of the side, *F*, are (4.5, 17), and the coordinates of *D* are (2, 16), what are the coordinates of the other endpoint, *E*?

4. Segments AB and CD bisect each other. Find E, the point of bisection, and D, if $A = (110, 15)$, $B = (116, 23)$, and $C = (110, 23)$.

5. A segment has a length of 25 units. Give possible coordinates for the endpoints of this segment if it is

   a. a horizontal segment.

   b. a vertical segment.

   c. neither a horizontal nor a vertical segment.

6. A segment has its midpoint at $(8, 10)$. List four possibilities for the coordinates of its endpoints.

7. A segment has one endpoint at $(-7, -2)$. Its midpoint is at $(-2, 1.5)$. What are the coordinates of the other endpoint?

8. Consider point P at $(5, 0)$ and Q at $(15, 0)$.

   a. Find six points that are just as far from P as they are from Q.

   b. Find six points that are closer to P than they are to Q.

   c. How can you tell if a point is equidistant from P and Q just by looking at its coordinates?

9. Show that the length of the line segment whose endpoints are the midpoints of two sides of a triangle is one half the length of the third side of that triangle. Do this for *any* triangle. Use subscript notation.

10. Use subscript notation to prove a general version of Problem **43** on page 373.

11. Suppose $A = (4, 1)$, $B = (8, -2)$, $C = (7, 1)$, and $D = (3, 4)$.

    a. Show that the diagonals of $ABCD$ bisect each other.

    b. What does this say about the kind of quadrilateral $ABCD$ is? Why?

12. If two vertices of an equilateral triangle lie in the Cartesian plane at $(1, 0)$ and $(9, 0)$, find coordinates for the third vertex.

13. Suppose $P = (12, -5)$. How far is P from the origin? How far is P from $(15, -9)$?

14. The vertices of $\triangle DEF$ are $D = (11, -1)$, $E = (13, 10)$, and $F = (3, 5)$.

    a. Show that $\triangle DEF$ is isosceles.

    b. Call the midpoint of $\overline{FD}$ L. Find the length of the median $\overline{EL}$.

    c. Show that $\triangle ELD$ is a right triangle.

    d. Show that M, the midpoint of $\overline{DE}$, is equidistant to D, L, and E.

You have done this already for a specific right triangle in Problem 14.

15. Let the coordinates of the vertices of right triangle QRS be (x_1, y_1), (x_1, y_2), and (x_2, y_1). Show that M, the midpoint of the hypotenuse, is equidistant from the three vertices.

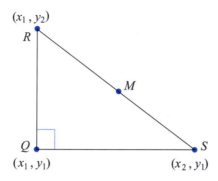

16. Group the following lines as "parallel" or "meet at the origin."

a. The y-coordinate is one half the x-coordinate.

b. The y-coordinate is three more than the x-coordinate.

c. The y-coordinate is two less than the x-coordinate.

d. The y-coordinate is negative two times the x-coordinate.

e. The y-coordinate is five times the x-coordinate.

f. The y-coordinate is one third the x-coordinate.

17. Is $(110, 9)$ collinear with $(60, 10)$ and $(10, 11)$? Why or why not?

Take It Further

For the first few problems, if you are going to add the coordinates and try to divide by 3, then you are taking the average of three points. In this case it's like a weighted average. If you are going to divide by 3, you have got to *add* three points. The question is, *which* three points?

18. Suppose $A = (2, 1)$ and $B = (32, 1)$. Find point P that is one third of the way from A to B. Explain how you did it.

19. Suppose $C = (2, 2)$ and $D = (30, 2)$. Find point Q that is one fourth of the way from C to D. Explain how you did it.

20. Consider the points $E = (5, 2)$ and $F = (11, -1)$. Find point S that is one third of the way from E to F. Explain how you did it.

21. Let $A = (-3, 5)$, $B = (5, 1)$, and $C = (7, -9)$.

a. Calculate midpoint D of $\overline{AB}$.

b. Calculate midpoint E of $\overline{BC}$.

c. Calculate F to be two thirds of the way from A to E.

d. Calculate G to be two thirds of the way from C to D.

e. Calculate midpoint H of $\overline{AC}$.

f. Calculate J to be two thirds of the way from B to H.

g. If you have not already done so, draw the picture that goes with these calculations.

h. What theorem does this remind you of?

22. Line a passes through $(0, 1)$ and $(1, 0)$. Line b passes through the origin and makes a 45° angle with the axes as it enters quadrant I. Find the coordinates where these two lines intersect.

23. Let C be the circle whose center is at the origin of the plane and whose radius is 5.

a. Find any points on C that are also on the vertical line that contains $(4, -9)$.

b. Find the intersection of C with the vertical line that contains $(3, 2)$.

c. Find any points on C that are also on the horizontal line that contains $(8, 0)$.

d. Find the intersection of C with the horizontal line that contains $(8, 3)$.

e. Find any points on C that are 13 units from the origin.

f. Challenge Find any points on C that are 13 units from $(8, 16)$.

LESSON 3

Coordinates in Three Dimensions

In this lesson, you will extend ideas of Cartesian coordinates to three dimensions.

The Cartesian coordinate plane is a system that aids mathematicians in talking about the relative positions of points, lines, and other 2-dimensional objects or drawings. Points on 3-dimensional objects may be represented by an ordered triple, (x, y, z).

In mathematics, the conventional way to extend Cartesian coordinates to three dimensions is to add a third coordinate axis, perpendicular to the other two, as shown below.

"Conventional" here means that most people do it or write it this way.

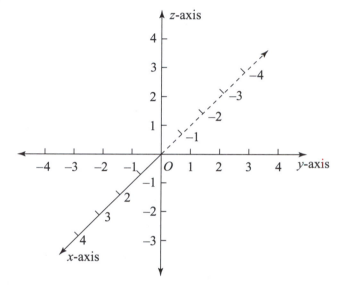

Imagine the x-axis "coming out of the page" with the positive half coming towards you and the negative half going away behind the page. The third axis is called the z-axis.

Explore and Discuss

The following mini-projects are suggestions for making one or more models of a 3-dimensional coordinate system to get a better idea of what it looks like. Try at least one of the three; you may want to keep one nearby to refer back to later.

The three axes Place one pencil on top of another so that they cross near their middles and are perpendicular to each other. Connect them with tape. Then attach the third pencil to the intersection of the first two. It needs to be perpendicular to both of them. You now have a model of 3-dimensional axes. Of course, the thickness of the pencils prevents this from being a perfect model. The third pencil can be made perpendicular to the other two, but it must be taped off to one side of their intersection.

Lattice points Begin with three or four clumps of different-colored modeling clay or gum-drops and a couple boxes of toothpicks.

Roll small, marble-sized balls of clay and connect them with toothpicks to build a cube. If you make this model with an odd number of balls to each edge you can place the origin in the center of the model. Of course, it doesn't matter which ball you designate as the origin, but it might be helpful to make it one color. Use a second color for the three axes, and a third color to designate the points with nonzero coordinates.

The three planes Start with six sheets of graph paper and three pieces of construction paper or other stiff paper. Glue a sheet of graph paper onto each side of each sheet of stiff paper. On the first sheet, cut it nearly (but not completely) in half across the width. On the second sheet, cut it nearly in half lengthwise. On the third sheet, make a slit lengthwise along the middle.

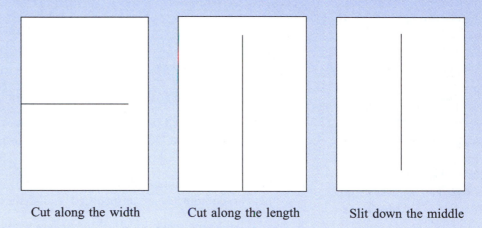

Cut along the width   Cut along the length   Slit down the middle

In the first model, the pencils depicted the three axes. Here, the pieces of graph paper represent the intersecting coordinate planes.

Connect the three sheets together: The third sheet (with the slit) fits into the cut on the first sheet. The cut on the second sheet slides over the first sheet, and the ends slide into the slit on the third sheet. You may need to wiggle the papers around a bit to make them all perpendicular to each other. You now have a model of a 3-dimensional coordinate system.

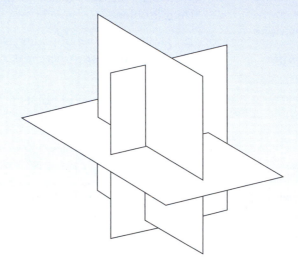

Points in Three-Dimensional Space

In three dimensions, it takes three coordinates to describe one particular point. In the picture below, B has coordinates $(3, 4, 0)$, and A has coordinates $(3, 4, 2)$.

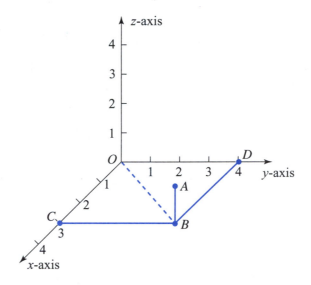

1. **a.** Name three other points where the x-coordinate is 3 and the y-coordinate is 4.

    b. How many such points are there? Picture *all* of the points whose first two coordinates are $(3, 4)$. How are those points arranged? That is, what shape is that collection of points?

2. Picture all of the points whose z-coordinate is 3. Where are these points found? The other two coordinates are unspecified, and so they can take on all possible values.

3. Name the shape made by all of the points whose first coordinate is the same as the second coordinate. The third coordinate is left unspecified.

4. Describe the set of points in three dimensions whose third coordinate is twice the second coordinate. Name the figure constructed by the graph of these points.

5. Describe all of the points whose coordinates satisfy the equation $x = y = z$. What shape do they make? Explain how you can get a solution to this problem from Problem **3**.

6. **a.** Describe the set of points in three dimensions whose coordinates satisfy the equation $x = a$, where a is a constant. What is the shape of the graph of these points?

    b. Describe the set of points in three dimensions whose coordinates satisfy the equations $x = -2$ and $y = 1$. What is the shape of the graph of these points?

You may want to experiment with a model of a 3-dimensional coordinate system if that helps you answer these questions.

No matter what point you look at, the x-, y-, and z-coordinates will always have the same value.

If you systematize how you visualize something, that can be an important habit of mind.

Ways to Think About It

There are many ways to visualize the figures in **6a** and **6b**. For **6a** you can use a 3-dimensional model, pick a particular value for a, and then imagine all possible y- and z-values to go with it. Or you can imagine the shape of $x = a$ in *two* dimensions and then extend it into three by allowing the z-coordinate to take on all possible values.

7. a. Name eight points that are 5 units from the origin in three dimensions.

   b. Picture all the points that are 5 units from the origin in three dimensions. What shape do they make?

   c. Picture all the points that are 5 units from the x-axis. How are those points arranged? That is, what shape do they make?

   d. Picture all the points that are x units from the x-axis. How are those points arranged?

8. **Checkpoint** For the following shapes in the x-y plane, add the third dimension to each by allowing z to take on all possible values. What shapes result?

   a. a point

   b. a line

   c. a circle

   d. a square

ACTIVITY 2

Midpoints and Distance in Three Dimensions

Of course, you can draw segments in three dimensions, and you might very well want to find their midpoints and lengths. Use the picture below for the next few problems.

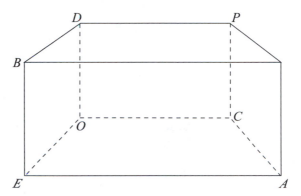

9. Point P has coordinates $(4, 5, 3)$. Figure out how far each of the following points is from P. Keep track of the thinking you use. Be sure you can explain your method in each case.

   a. $A = (4, 0, 3)$ **d.** $D = (0, 0, 3)$

   b. $B = (4, 5, 0)$ **e.** $E = (-3, 6, -3)$

   c. $C = (0, 5, 0)$ **f.** $O = (0, 0, 0)$

10. Find the coordinates of the midpoints between each pair of points.

   a. $P = (4, 5, 3)$ and $B = (4, 5, 0)$

   b. $P = (4, 5, 3)$ and $C = (0, 5, 0)$

   c. $O = (0, 0, 0)$ and $P = (4, 5, 3)$

   d. $P = (4, 5, 3)$ and $E = (-3, 6, -3)$

11. **Checkpoint** Consider these problems.

    a. Write a general rule for finding the distance between two points from their 3-dimensional coordinates.

    b. Write a general rule for finding the point midway between two points from their 3-dimensional coordinates.

Shapes in the Plane and in Space

This activity is an opportunity to get more familiar with some basic shapes, such as lines, circles, ellipses and triangles. This is a chance for you to gain a sense of some of the properties of the coordinates of these figures before moving on to operate on those coordinates.

12. Here is a recipe to perform on points in the plane: Square each coordinate and add the results.

    a. Find eight points that produce a result of 5 when you do this to them.

    b. What figure do you get if you look at all the points that produce 5?

    c. Find eight points that produce each of the following results.

        i. 13

        ii. 625

        iii. 100

        iv. 169

        v. 1

Ways to Think About It

One way to go about Problem **12** is to use a calculator. If you want a point for which the recipe produces 5, you could just *decide* that one of the coordinates of the point is 1.5, square that number, subtract the result from 5 to see what is left, and take the square root of that.

Another way is to think about sums of square integers. Can *every* integer be expressed as the sum of two squares? Can *5* be expressed as the sum of two squares?

13. **a.** If you drew the picture of all the points in the plane that are 13 units from the point (0, 0), what figure would you get?

    b. Find the coordinates of eight points that are 13 units from the point (0, 0). How can you tell if some new point is on this figure by doing a little calculation on its coordinates? In other words, how can you tell if the point (a, b) is on the figure?

    c. If you drew the picture of all the points in the plane that are 13 units from the point (4, −1), what figure would you get?

d. Find the coordinates of eight points that are 13 units from the point $(4, -1)$. How can you tell if some new point is on this figure by doing a little calculation on its coordinates? In other words, how can you tell if the point (a, b) is on the figure?

14. Suppose you perform this recipe on points in three dimensions: Take each coordinate, square it, and add the results.

    a. Find eight points that produce 81 when you perform the recipe. What do you get if you look at *all* points that produce 81?

    b. What do you get if you look at all points in three dimensions that are 7 units from the origin? Find the coordinates of two or three points that are each 7 units from the origin.

15. Let's go back to two dimensions. Here is another recipe to perform on the coordinates of points in the plane:

    Step 1 Square the x-coordinate.

    Step 2 Square the y-coordinate

    Step 3 Multiply the x-coordinate by the y-coordinate.

    Step 4 Add the results of the first three steps above.

    a. What values do you get if you perform this recipe on each of the following points?

i. $(10, 0)$	**iv.** $(6, -8)$
ii. $(0, -10)$	**v.** $(-5, 5\sqrt{3})$
iii. $(6, 8)$	**vi.** $(5, 5\sqrt{3})$

Make sure to list at least six points. Try points with coordinates of value 0, and try something like (1, −1) too.

    b. Find six points that produce 1 when you apply the recipe. What do you get if you look at *all* the points that produce 1?

16. Try this recipe:

    Step 1 Square the x-coordinate and multiply by three.

    Step 2 Square the y-coordinate.

    Step 3 Add the results.

    Find six points that produce 4 when you apply this recipe. What do you get if you look at *all* the points that produce 4?

17. Here are four related recipes with numerical outcomes. What shapes do they produce?

    a. The sum of the coordinates is 12.

    b. The difference of the coordinates is 12.

    c. The product of the coordinates is 12.

    d. The ratio of the coordinates is 12.

18. `Checkpoint` Make up your own recipe. Keep it fairly simple.

    a. See what happens when you apply that recipe to the six points listed in Problem **15.**

b. Pick what seems like a suitable goal number. Find six points that produce that number when you apply your recipe. Sketch the shape that describes *all* points that produce that number.

c. Challenge Trade sketches with someone else. See if each of you can figure out the other person's recipe, just by looking at the shape it produces.

1. Imagine this is a room in your house that has been placed on a 3-dimensional coordinate system.

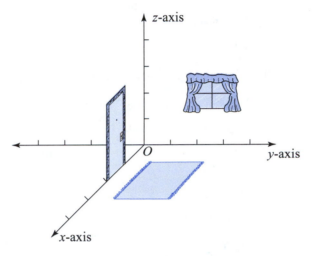

   a. Describe where the origin is.

   b. Estimate the ordered triples that describe the four corners of the door.

   c. Estimate the ordered triples that describe the four corners of the window.

   d. Estimate the ordered triples that describe the four corners of the rug.

2. The diagonal of the box below, $\overline{AH}$, is one side of $\triangle ABH$.

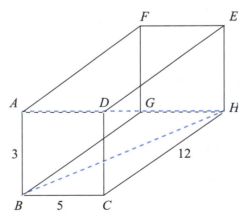

   a. What kind of triangle is $\triangle ABH$?

   b. What is the length BH?

   c. What is the length AH?

3. The diagonal of the box below, $\overline{AH}$, is one side of $\triangle ABH$.

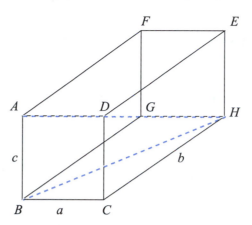

a. What kind of triangle is $\triangle ABH$?

b. What is the length BH?

c. What is the length AH?

4. Use the picture below to answer the questions.

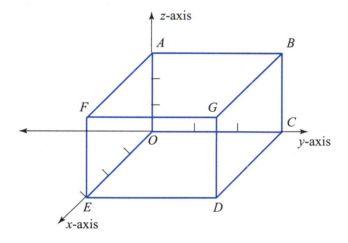

a. Find the coordinates of the vertices of the box.

b. Find the length of each segment.

   i. $\overline{OE}$

  ii. $\overline{OB}$

 iii. $\overline{AE}$

  iv. $\overline{FC}$

c. Name one other segment with the same length as each of these segments.

   i. $\overline{OE}$

  ii. $\overline{OB}$

 iii. $\overline{AE}$

  iv. $\overline{FC}$

5. A cube is sitting at the origin with sides of length 1.

    a. What are the other vertices of the cube?

    b. What is the length of a diagonal of the cube?

6. Triangle ABC has vertices $A = (2, -1, 7)$, $B = (4, 0, 5)$, and $C = (0, 0, 5)$.

    a. Find the perimeter of $\triangle ABC$.

    b. Find the midpoints of each side of $\triangle ABC$.

    c. Find the perimeter of the triangle formed by connecting the midpoints of the sides of $\triangle ABC$.

7. Name six points in three dimensions that fit the rule, "the sum of the coordinates is 30." What shape does this rule produce if you look at all the points that fit it?

8. A square with a 1-unit sidelength sits with one vertex at the origin and two of its sides lying along the axes of a 2-dimensional coordinate system. Its vertices have no negative coordinates.

    a. What are the coordinates of the farthest vertex from the origin?

    b. How far is that vertex from the origin?

9. A cube with a 1-unit sidelength sits with one vertex at the origin and three of its edges lying along the axes of a 3-dimensional coordinate system. Its vertices have no negative coordinates.

    a. What are the coordinates of the farthest vertex from the origin?

    b. How far is that vertex from the origin?

Take It Further

10. Challenge Why not extend the idea to four dimensions! A 4-dimensional *hypercube* with a 1-unit sidelength sits with one vertex at the origin and edges against the axes of a 4-dimensional coordinate system. Its vertices have no negative coordinates.

    a. What are the coordinates of the farthest vertex from the origin?

    b. How far is that vertex from the origin?

11. If two vertices of an equilateral triangle lie in the Cartesian plane at $(1, 0)$ and $(9, 0)$,

    a. find coordinates for the third vertex.

    b. embed the three points from part **a** into three dimensions, so they now look like $(1, 0, 0)$, $(9, 0, 0)$, and $(a, b, 0)$, where (a, b) is your answer to part **a.** Find the coordinates of a fourth vertex of a regular tetrahedron that has these three points as vertices.

To form a regular tetrahedron, the fourth vertex would be the same distance from each of the three points that they are from each other.

LESSON

4

Stretching and Shrinking Things

In this lesson, you will use coordinates to scale figures, and will be introduced to vectors.

Here is an experiment to try that may give you ideas about different ways coordinates can help you investigate shapes.

What does this have to do with dilations?

You now have several pentagons. Are your corresponding vertices collinear?

Explore and Discuss

On a Cartesian coordinate system, draw a pentagon whose vertices are (8, 5), (9, 9), (12, 9), (12, 6), and (10, 3). Connect each vertex to the origin. Find the midpoints of all these lines. Connect up these midpoints to make a smaller pentagon.

a How would you describe the relationship between the two pentagons?

b How would you describe the coordinates of the small pentagon's vertices in terms of the coordinates of the big pentagon's vertices?

c Use the coordinates of your original pentagon's vertices to find the vertices of a pentagon whose sidelengths are twice as long as the original's. Draw this new pentagon.

d Use the coordinates of the vertices of your original pentagon to find the vertices of a pentagon whose sidelengths are one third as long. Draw this new small pentagon.

ACTIVITY

Stretching and Shrinking with Coordinates

In Unit 4 you used the Ratio and Parallel Methods to dilate figures. Now you will do this on a coordinate plane.

1. a. Plot the four points (0, 2), (2, 3), (3, 5), and (5, 1) on a coordinate system. Connect them to make a quadrilateral. For each point, multiply both the first and second coordinates by 4. Plot the result. Connect these new points to get a quadrilateral. How is the new quadrilateral related to your original?

b. Multiply the coordinates of your original quadrilateral by −2 or −1. Plot what you get. How is this figure related to your original?

c. Multiply the first coordinates by 2 and the second coordinates by 3. How is this figure different from the figures in parts **a** and **b?**

d. Challenge Start with the four points given in part **a.** When would a new quadrilateral be entirely inside the original? When would it contain the original? When would a new quadrilateral, created by multiplying both coordinates of each point by the same number, overlap with the first quadrilateral? When would the two be disjoint?

2. In the picture below, $\overline{AB}$ is one side of a square.

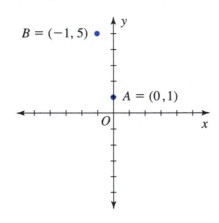

$B = (-1, 5)$

$A = (0, 1)$

a. Find two other points C and D so that $ABCD$ is a square.

b. Multiply both coordinates of each vertex of your square by $\frac{1}{2}$. Plot these new points and connect them. Describe what you get and how it is related to the original square.

c. Multiply both coordinates of each vertex of your original square by -3. Plot these new points and connect them. Describe what you get and how it is related to the original.

3. Below is a picture of the head of Trig on a coordinate system.

a. Use coordinates to scale Trig by 2.

b. Use coordinates to get a scaled copy or outline of Trig that is upside down.

You might want to scale just some important features like the nose, ears, and mouth.

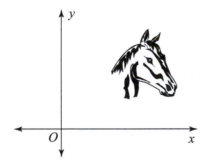

4. Plot two points X and Y. Multiply the coordinates of both points by -2. What happens to the segment XY?

5. The following points describe a triangle in space: $A = (3, 2, 6)$, $B = (0, 3, 0)$, and $C = (-1, 0, 1)$.

a. Make a sketch.

b. Scale the triangle by $\frac{1}{2}$.

c. Scale the triangle by $\frac{1}{3}$.

6. Find eight vertices that define a cube, where one of the vertices is not at the origin. How could you scale your cube by a factor of 2?

Think of a vector as an arrow. It is a line segment with a particular direction and length. You will find out more about vectors later in this unit.

A student in one of the field test sites for *Connected Geometry* came up with what she called the "vector method" for making a blow up of Trig's picture. Here's what she did:

On a copy of the picture of Trig, she drew arrows or **vectors** from the origin to interesting places on Trig's head, such as tips of the ears, end of the nose, and so on.

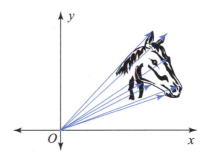

Then she *stretched* each vector, keeping its exact same direction and doubling its length. She did this by placing a ruler alongside each vector to keep the same direction, and then multiplying the vector's length by 2.

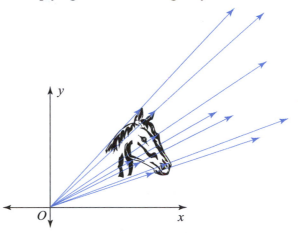

This gave an outline for "Big Trig." Then by adding more vectors, she was able to get a pretty good sketch.

She used many more vectors than this—maybe two or three times as many.

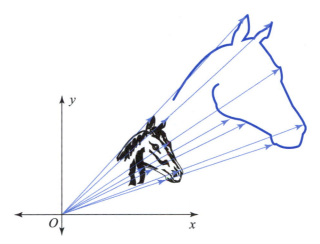

Then she added the details.

This student was a very good artist.

You may choose your own picture if you wish.

7. Try the vector method on this fish, but this time scale it down by a factor of 2, so the lengths in the new picture will be only one half as long as those in the original.

8. How would you adapt the vector method to include negative numbers? Try it. Scale one of the pictures or figures used in this lesson by −2.

9. **Checkpoint** Consider these problems.

You might want to break this down into several simpler rules: What happens when the number is ≥ 1, and so on?

   a. Write a rule that describes what happens to a figure when you multiply both coordinates of each point by some number.

   b. Compare the vector method of scaling to the coordinate method of scaling. How are the coordinates of the corresponding points in Problem **3** related to each other? How are the coordinates of the arrowhead of each original vector related to the coordinates of the arrowhead of the corresponding shrunken vector?

On Your Own

1. Go back to the *Explore and Discuss* section and plot the original pentagon on a coordinate plane.

   a. Find the midpoints of each side of the pentagon.

   b. Connect these midpoints.

   c. Is there a relationship between the new pentagon and the original? What is it?

2. Use the picture below to answer the questions.

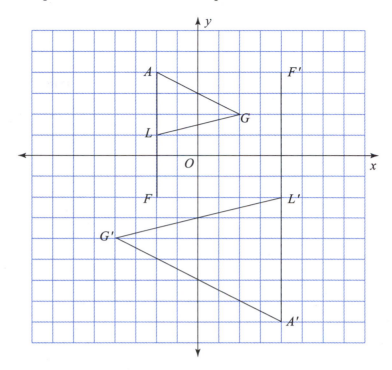

   a. What scale factor could be applied to the coordinates of *FLAG* to get *F'L'A'G'*?

   b. What scale factor could be applied to the coordinates of *F'L'A'G'* to get *FLAG*?

3. Create △*ABC* on a coordinate plane.

   a. Scale △*ABC* by 2 to create △*A'B'C'*.

   b. Scale △*A'B'C'* down by 3 to create △*A"B"C"*.

   c. Scale △*ABC* by 4 to create △*XYZ*.

   d. Scale △*XYZ* down by 6 to create △*X'Y'Z'*.

   e. How do △*A"B"C"* and △*X'Y'Z'* compare?

4. Suppose you scale figure *A* by 2 to create figure *B*. Then you scale figure *B* by 10 to create figure *C*. Then you scale figure *C* by 2 to create figure *D*. Pick a general point on figure *A*. Use operations on coordinates to answer the questions.

   a. To get figure *D* another way, you could scale figure *A* by 5 to get figure *A'* and then scale figure *A'* by what number to get figure *D*?

   b. By what scale factor could you scale figure *A* to get figure *D*?

5. The coordinates of every point on *ABCD* are multiplied by 5 to get a new figure. How do the areas of the two figures compare?

6. What is the largest scale factor you can apply to the coordinates of the figure below so that the scaled copy is still contained within the grid shown?

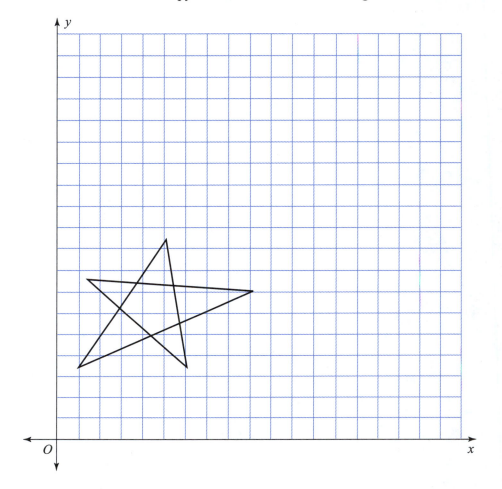

7. Use the vector method to scale the figure below by $\frac{3}{4}$ and then by $-\frac{3}{4}$.

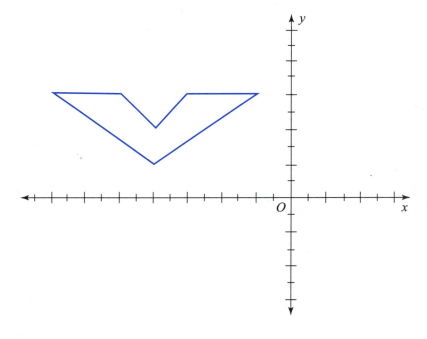

LESSON 5

Changing the Location of Things

In this lesson, you will use coordinates and vectors to translate figures about a plane.

In the previous lesson, you scaled figures by operations on the coordinates of the points on the figures. In this lesson, you will translate figures around the coordinate plane by performing different operations on the coordinates.

Explore and Discuss

On a coordinate system, draw a pentagon whose vertices are (8, 5), (9, 9), (12, 9), (12, 6), and (10, 3). Do something to the coordinates to move the pentagon 8 units to the right.

a Describe the relationship between the two pentagons.

b Describe the coordinates of the new pentagon's vertices in terms of the coordinates of the vertices of the old pentagon.

c Slide the original figure up 8 units. How would you describe the new set of coordinates?

d Slide the original figure right 6 units and up 8 units. Compare coordinates with the original.

Describe the effect on your original pentagon if you take the vertices and

e add 10 to the first coordinates and add 7 to the second coordinates.

f add 10 to the first coordinates and subtract 7 (or add −7) from the second coordinates.

ACTIVITY

Pictures from Rules, Rules from Pictures

The following problems will help you formalize the idea of translation a bit more.

1. Refer to the picture on the following page.

   a. Describe what you have to do to *AKLJ* to get *A'K'L'J'*.

   b. Describe what you have to do to the *coordinates* of the vertices of *AKLJ* to get the *coordinates* of the vertices of *A'K'L'J'*.

Mathematicians say that
AKLJ *has been*
translated to A'K'L'J'.

One easy way to say "take some points, add 10 to the first coordinates and add 6 to the second coordinates" is to say "send each point (x, y) to the point $(x + 10, y + 6)$" or even "$(x, y) \mapsto (x + 10, y + 6)$."

2. Apply the rule $(x, y) \mapsto (x + 10, y + 6)$ to the vertices of a triangle. Then connect the three new points. What figure do you get?

3. How would you write the rule "take the points and multiply the coordinates by 2" using the $(x, y) \mapsto$ notation?

4. Draw a polygon on a coordinate system. If you like, you can make your polygon look like the one below.

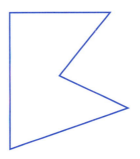

Don't forget the break "Describe ..." part of this problem.

Apply each rule below to the vertices of the polygon. Draw the resulting polygon. You might try applying the rules to more than one polygon. Describe how each resulting polygon is related to the original.

a. $(x, y) \mapsto (x + 8, y + 5)$

b. $(a, b) \mapsto (a - 8, b + 5)$

c. $(a, b) \mapsto (3a, 3b)$

d. $(x, y) \mapsto \left(\frac{x}{2}, \frac{y}{2}\right)$

e. $(x, y) \mapsto \left(\frac{x}{2} + 7, \frac{y}{2} + 10\right)$

f. $(x, y) \mapsto (-x, y + 2)$

g. $(x, y) \mapsto (2x, y + 2)$

5. For each picture, write a rule that will translate the vertices of polygon *JKLM* to polygon *J'K'L'M'*. Use the $\mapsto$ notation.

If you are having difficulty, label all the coordinates and look for a way to change the coordinates of one polygon into the coordinates of the other.

a.

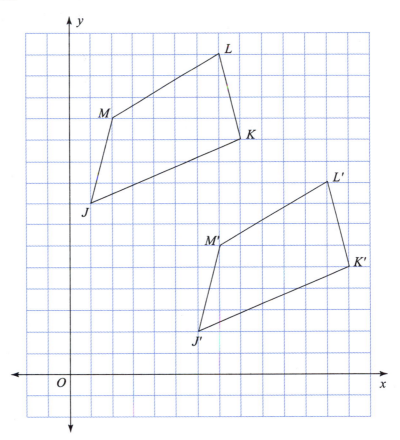

b.

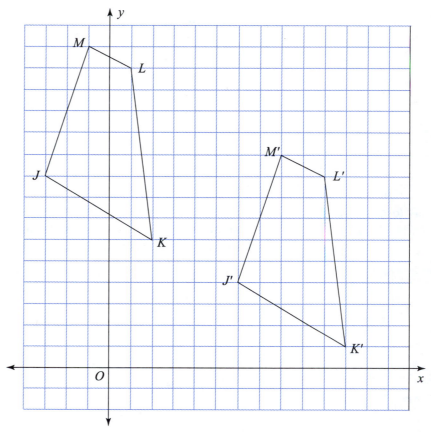

c.

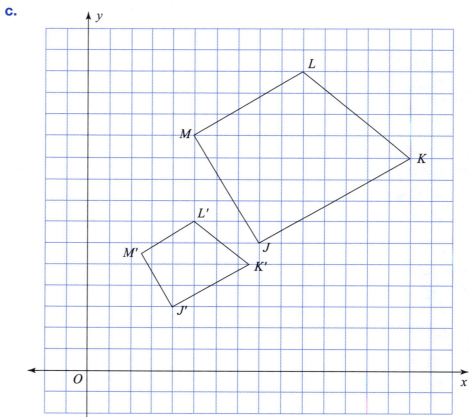

Hint: Two steps

d.

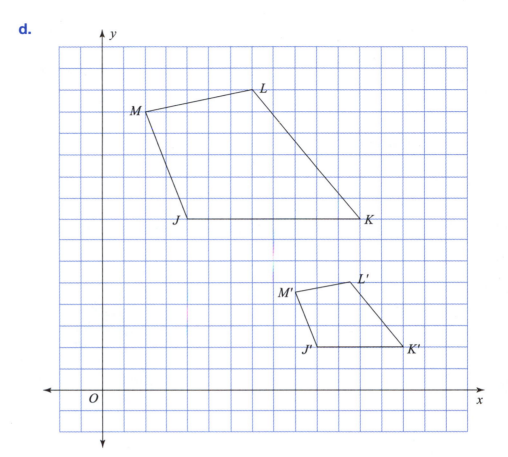

6. In $\triangle DIG$, $D = (4, 3)$, $I = (8, -2)$, and $G = (5, 7)$. In $\triangle RAT$, $R = (9, 2)$, $A = (13, -3)$, and $T = (10, 6)$. Describe the movement(s) that takes $\triangle DIG$ to $\triangle RAT$. Describe the arithmetic that changes the coordinates of the vertices from those of $\triangle DIG$ to those of $\triangle RAT$.

7. Both Jorge and Yutaka are asked to transform triangle B by scaling it by $\frac{1}{4}$ and translating it 16 units to the right, but they are not told in which order to perform the transformation. Jorge chooses to scale the triangle first and then translate it. Yutaka does the translating first and then the scaling.

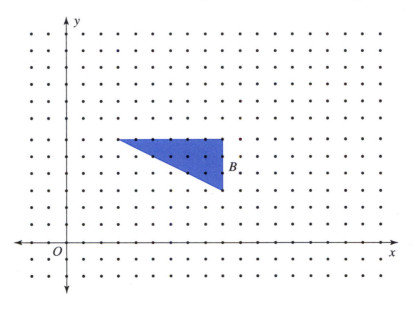

a. Pick a vertex or two of triangle *B*. Follow Jorge's and Yutaka's methods and write down where the vertices end up.

b. Do both methods end up with the same triangle in the same place?

c. Below are algebraic statements that describe Jorge's and Yutaka's solutions. Which is Jorge's? Which is Yutaka's?

$$(x, y) \mapsto \frac{1}{4}(x, y) + (16, 0)$$

$$(x, y) \mapsto \frac{1}{4}(x + 16, y)$$

What could $\frac{1}{4}(x, y)$ mean?

8. Copy the square onto dot paper. Apply each of the rules below to the vertices of the square. Draw the resulting figures. Indicate the shape of each resulting figure. Is it a square, a rectangle, or something else?

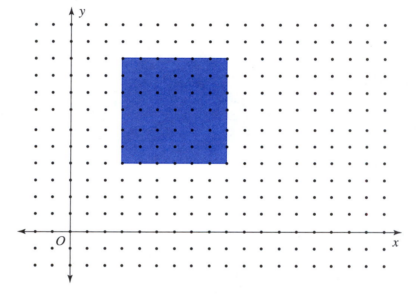

a. $(x, y) \mapsto (x + 8, y + 5)$

b. $(x, y) \mapsto (x - 3, y)$

c. $(x, y) \mapsto (3x, 4y)$

d. $(x, y) \mapsto (-x, -y)$

e. $(x, y) \mapsto (-2x, y - 2)$

f. $(x, y) \mapsto (2x + 1, -2y)$

g. $(x, y) \mapsto (x, 2(y + 1))$

h. $(x, y) \mapsto (y, x)$

9. a. Classify the rules in Problem **8** as either rules that will take a square and result in a square or that will take a square and result in a rectangle. What would be the result of applying $(x, y) \mapsto (\frac{1}{2}x, \frac{1}{2}(y + 2))$ to a square?

b. Challenge Can a rule like those above—multiplying the coordinates by a fixed number and/or adding a fixed number to the coordinates—change a square into something that is *not* a rectangle?

10. **Checkpoint** Trapezoid A is scaled and translated using two different rules. Match each of the two new trapezoids to the rule that describes it.

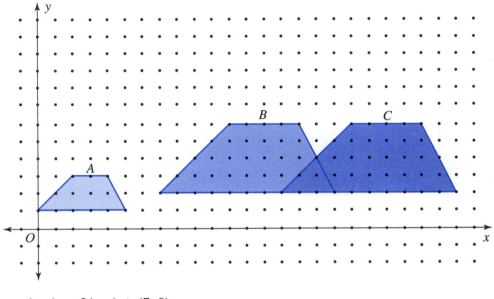

$(x, y) \mapsto 2(x, y) + (7, 0)$

$(x, y) \mapsto 2(x + 7, y)$

1. Here is a complicated rule:

$$(a, b) \mapsto \left(\tfrac{1}{2}(a-3) + 5, \tfrac{1}{2}(b + 4) + 2\right).$$

a. Describe what this rule does.

b. Simplify the algebra of the rule. Describe what your new (but equivalent) rule does.

2. Here is another complicated rule:

$$(x, z) \mapsto \left(\tfrac{1}{2}(2(x - 4) + 6), \tfrac{1}{2}(2(z + 3) - 4)\right).$$

a. Describe what this rule does.

b. Simplify the algebra of the rule. Describe what your new (but equivalent) rule does.

Check your description with some pictures: Draw a picture on a coordinate system. Apply the rule to some points on the picture. See if your description is right.

3. Here is a rule written in English:

> To move any point P, draw a line segment from P to $(3, 4)$, and plot the midpoint.

Draw a few pictures to see what the rule does. Find a way to describe the rule in the $(x, y) \mapsto$ notation.

4. Compare $(x, y) \mapsto (2x + 2, 3y + 3)$ and $(x, y) \mapsto (2(x + 1), 3(y + 1))$. Do these rules have the same result?

Use a corner of your room as the origin. Show what each rule does to a particular point.

5. So far, you have been working in two dimensions, the Cartesian plane. Describe what the following rules do to figures in space.

  a. $(x, y, z) \mapsto (3x, 3y, 3z)$

  b. $(x, y, z) \mapsto \left(\frac{1}{2}x, \frac{1}{2}y, \frac{1}{2}z\right)$

  c. $(x, y, z) \mapsto (x + 3, y + 2, z - 1)$

  d. $(x, y, z) \mapsto \left(\frac{1}{2}(2(x - 4) + 6), \frac{1}{2}(2(y + 3) - 4), \frac{1}{2}(2(z + 2) - 1)\right)$

Take It Further

6. Explain why two rules can *do* different things but have the same end result. Give an example of two rules that do different things but end up with the same result.

Adding Points and Scaling Points

In this lesson, you will investigate the effects of adding and scaling points.

In Lesson 5, you found a rule that would send *JKLM* onto *J'K'L'M'*

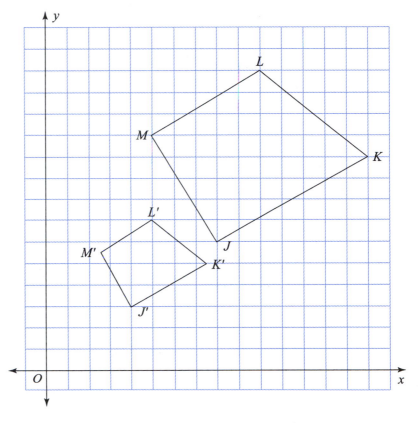

One way to do this is to multiply all the coordinates by $\frac{1}{2}$, so that your rule looks like $(x, y) \mapsto \left(\frac{x}{2}, \frac{y}{2}\right)$. There is another way to write this:

$$(x, y) \mapsto \frac{1}{2}(x, y).$$

In Lesson 5, you also found a rule that would move quadrilateral *AKLJ* onto *A'K'L'J'*.

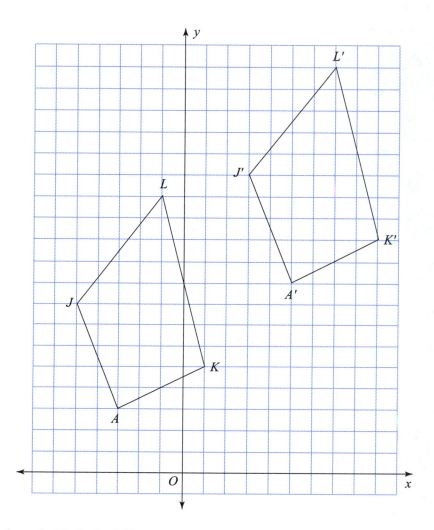

Your rule probably looked like this:

$$(x, y) \mapsto (x + 8, y + 6).$$

Another way to write the rule is

$$(x, y) \mapsto (x, y) + (8, 6).$$

Explore and Discuss

Suppose $A = (1, 3, 2)$. Use a corner of your classroom as the origin of a coordinate system. Let the unit of measure be 1 foot. Locate A by having someone in the class (with a steady hand) hold a pencil point at A. Now have other people hold pencil points at $2A = 2(1, 3, 2)$; $1.5A = 1.5(1, 3, 2)$; and $\frac{1}{2}A = \frac{1}{2}(1, 3, 2)$.

a Can someone reach $4A$?

b Where would $-1A$ be?

c Give a geometric description of all the points at which people are holding pencil points.

Let everyone sit down. Now let $A = (1, 3, 2)$ and $B = (3, 1, 4)$. Have someone in the class (with a steady hand) hold a pencil point at A. Do the same for B. Now have other people hold pencil points at $B + A = (3, 1, 4) + (1, 3, 2)$; $B + 2A = (3, 1, 4) + 2(1, 3, 2)$; and $B + \frac{1}{2}A = (3, 1, 4) + \frac{1}{2}(1, 3, 2)$.

d Can someone reach $B + 3A$? If not, where would it be?

e Where would $B - A$ be?

f Give a geometric description of all the points at which the other people (not the people at A or B) are holding pencil points.

ACTIVITY 1 ## Scaling Points

"Scaling a point" is shorthand for "dilating a point with center at the origin."

Definition

To *scale a point by a number*, multiply each of the point's coordinates by the number. For example,

$$6(4, 1) = (24, 6) \text{ and } -2(3, 1, -4) = (-6, -2, 8).$$

This problem asks you to do seven calculations. There is a reason for it.

1. For each problem, locate A and cA on a coordinate system, where cA is the point you get when you scale A by c. Use the same coordinate system for all the problems. Describe any patterns you see in your picture. How is the location of cA related to the location of A?

   a. $A = (5, 1)$, $c = 2$

   b. $A = (5, 1)$, $c = \frac{1}{2}$

   c. $A = (3, 4)$, $c = 3$

   d. $A = (3, 4)$, $c = \frac{1}{2}$

   e. $A = (3, 4)$, $c = -1$

   f. $A = (3, 4)$, $c = -2$

   g. $A = (3, 4)$, $c = -\frac{1}{2}$

So, look at 4B, 2B, −1B, $\frac{1}{2}$B, and so on. Pick eight or ten numbers, scale B by each of them, and plot the results.

2. In Problems **1c–1g** you took a single point, (3, 4), and scaled it by a bunch of numbers. Try it again with another point, say $B = (4, 6)$. Scale B by eight or ten numbers. Plot the resulting points on a single coordinate system. What do you get? Draw a picture of the set of points you would get if you looked at every possible multiple of B.

3. Write and Reflect Suppose A is a point and c is some number. Describe how cA is geometrically related to A. How would you tell someone how to locate $2A$ if that person knows where point A is? How about $-3A$? How is cA *algebraically* related to A?

4. Write and Reflect Why do you think the operation is called "scaling" instead of "multiplying"?

There is another way to scale points that is useful when you don't want to use coordinates. Here is the idea: Suppose you have a point P, and you want to scale it by some number c, say 2.

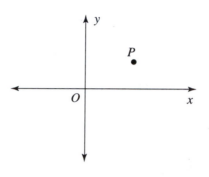

Remember, a vector is a directed line segment of a particular length, an arrow with a tail and a head.

Draw a vector from O to P. Take that vector and *stretch* it by a factor of c (in this case, a factor of 2). Now you are at cP.

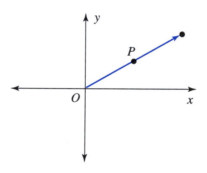

One way to stretch by 2 is to pass a ruler through O and P, and to run past P a distance equal to OP. A compass also is helpful here.

5. Suppose $A = (4, 6)$. Use vectors to locate 2A, 3A, $\frac{1}{2}A$, $\frac{2}{3}A$, and $-2A$. Check with coordinates.

Instead of taking one point and scaling it by a bunch of numbers, what happens if you take a bunch of collinear points and scale them by the *same* number? The remaining problems in this activity ask you to look at that question.

6. Draw a picture of what you get if each of the collinear points below is scaled by 2. Describe your result.

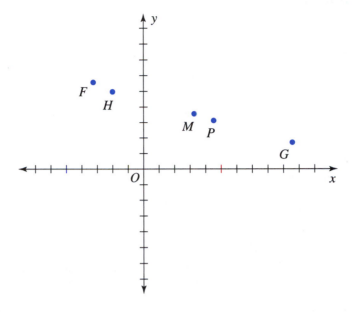

You may want to trace the following pictures on graph paper.

7. Draw a picture of what you get if each of the points below is scaled by $\frac{1}{2}$. Describe what you get.

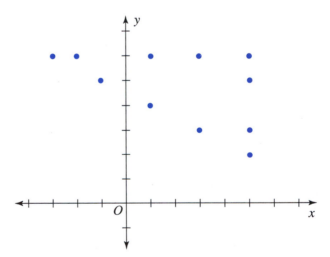

Did you need to scale
every point in order to
know what you would
get?

8. Draw a picture of what you get if each of the points below is scaled by −4. Describe what you get.

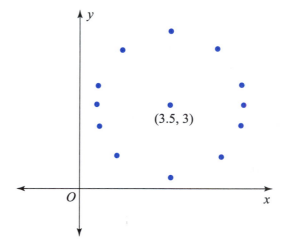

(3.5, 3)

9. Think of a circle of radius 6 centered at the origin. What do you get if you scale all the points on the circle by 2? By $\frac{1}{2}$? By $-\frac{1}{2}$?

What is the definition of
an ellipse?

10. What do you get if you scale all the points on the ellipse below by $\frac{1}{3}$? By −2?

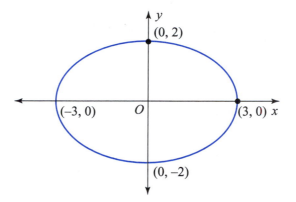

(0, 2)

(−3, 0)   O   (3, 0) x

(0, −2)

11. What do you get if you scale every point on the rectangle below by −1? By $\frac{5}{3}$?

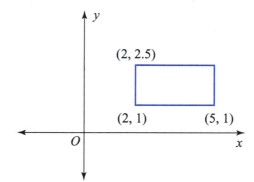

(2, 2.5)

(2, 1)   (5, 1)

12. Below is a sequence of pictures, each one drawn on a coordinate system.

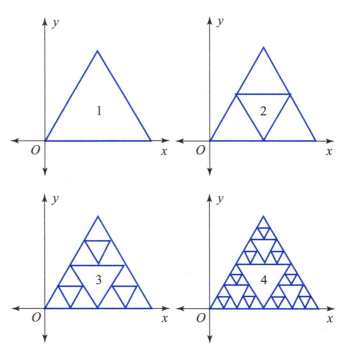

a. Describe how each picture is obtained from the previous one. Draw the next one in the sequence.

This is a job for the stretching trick.

b. What happens if you scale each picture by $\frac{1}{2}$?

13. Below is a sequence of pictures heading toward a snowflake-like figure.

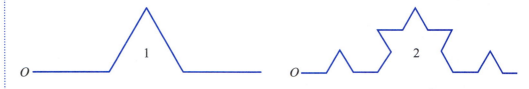

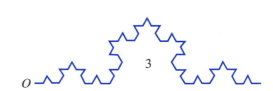

a. How is each picture obtained from the previous one? Draw the three pictures on graph paper or with a computer. Then draw the next one.

b. Place coordinate axes on each picture, with the horizontal axis running along the bottom edge of each snowflake-like figure. Describe what you get when you scale each picture by $\frac{1}{3}$.

14. Draw a picture of a pentagon that, if it is scaled by 3, produces the pentagon below.

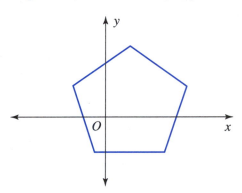

15. What would you get if you scaled the above pentagon by 1?

16. Draw a pentagon that, if it is scaled by 2 or by 3 or $n > 1$, produces an enlarged pentagon one *side* of which completely contains the corresponding side of the original.

There is more than one solution to each of these problems.

17. Draw a pentagon that when scaled has one *vertex* that always stays in the same spot.

18. Checkpoint Consider these problems.

    a. There are four triangles below. Which pairs of triangles have the property that one of the pair can be scaled to get the other? Justify your answers.

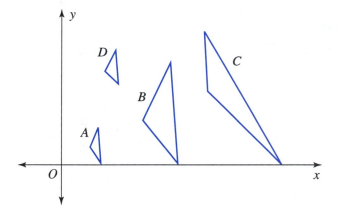

Could a life-size picture of a baby be scaled to match a life-size picture of the adult that the baby becomes?

    b. Describe some features that two figures must have in common if it is possible to scale one by a number to get the other one.

ACTIVITY 2 **Adding Points**

Definition

To *add two points*, add the corresponding coordinates. For example,

$$(-2, 7) + (4, 1) = (2, 8), \text{ and } (3, 0, 9) + (3, 1, -4) = (6, 1, 5),$$

which can be written simply as

$$A + B = (A + B).$$

So, for example, you can write the formula for the midpoint of $\overline{AB}$ as $M = \frac{1}{2}(A + B)$

19. Draw a separate coordinate system for each problem. Then locate O (the origin), A, B, and $A + B$.

    a. $A = (5, 1)$, $B = (3, 6)$

    b. $A = (4, -2)$, $B = (0, 6)$

    c. $A = (4, -2)$, $B = (-3, -5)$

    d. $A = (4, -2)$, $B = (-4, 2)$

    e. $A = (4, -2)$, $B = (-1, 4)$

    f. $A = (3, 1)$, $B = (6, 2)$

    g. Describe anything you see in your pictures. How is the location of $A + B$ related to the locations of A, B, and O? Describe how to locate $A + B$ in the picture below.

Try connecting A, B, O, and A + B in some order.

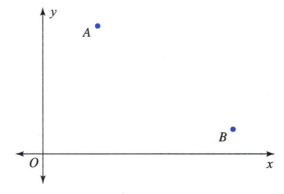

Search for an invariant!

In the previous problem, you took several pairs of points and added them. That was just to give you an idea of what the sum of two points looks like. Now keep one of the points constant and let the second point vary along some familiar sets.

20. Suppose $A = (3, 5)$ and $B = (6, -1)$.

    a. Plot B, $2B$, $0.5B$, $-1B$, $4B$, $-\frac{18}{5}B$, and three or four other multiples of B.

    b. Add A to each of the points you plotted in part **a.** Plot the results on the same coordinate system.

    c. What would the picture look like if you plotted the sum of A and *every possible* multiple of B?

21. Let $P = (1, 4)$. Plot the sum of P with each of the points on the rectangle below.

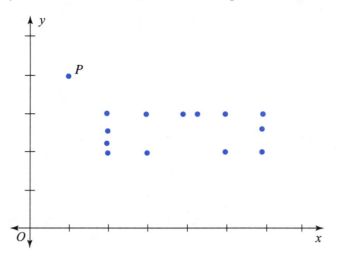

Of course, you cannot add P to every point on the rectangle, but what would you get if you could?

22. Let $P = (1, 4)$. Plot the sum of P with each of the points on the rectangle below.

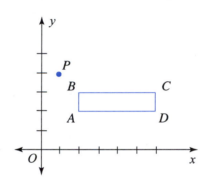

Problems 23 and 24 you are asked to do something you have not done before: to add points without using coordinates.

23. Add Z to each of the collinear points and plot the sums.

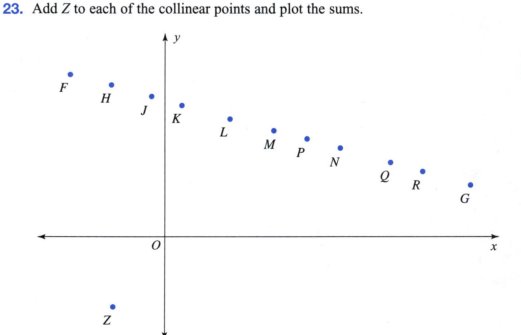

24. Plot the points you get when you add P to each point on the triangle. Do the same for Q.

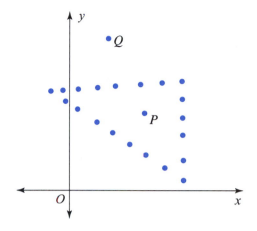

25. Think of a circle of radius 6 centered at the origin. Describe what you get if you add all the points on the circle to $(8, -1)$. To $(-4, 2)$. To $(0, 6)$. To $(0, 0)$.

Draw a picture if it helps to solve the problem or explain the answer.

26. Look at the picture below.

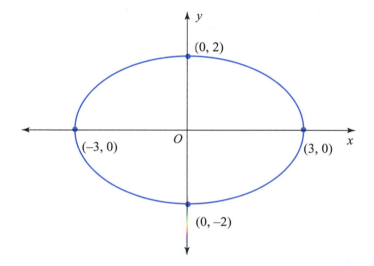

What would you get if you added $(5, 4)$ to all the points on the ellipse?

The last few problems ask you to describe the set of points you would get if you add a point P to *every* possible point on a shape—a circle or an ellipse, for example. You can add P to a *few* points on the shape, and that probably gives you an idea of what to expect. Geometry software can help you plot many more points than you would be able to plot by hand.

27. Below is a picture of Trig, drawn on a coordinate system. Sketch the picture that results from adding P to every point on Trig.

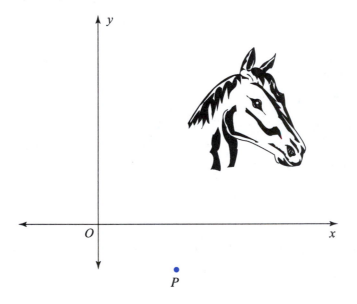

Being able to reverse directions, as you are asked to do in Problems 27 and 28 helps to make your understanding deeper and more flexible.

28. Draw a rectangle such that, if you add A to each of its points, you will get a rectangle like the one below.

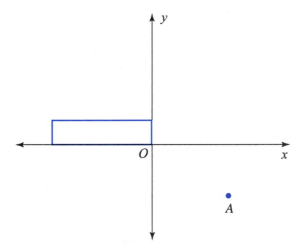

29. Draw a pentagon such that, if you add A to each of its points, you will get a pentagon like the one below.

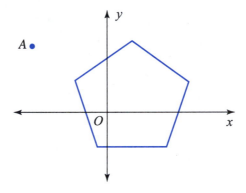

30. What would you get if you added O to every point on the above pentagon?

31. **Checkpoint** Consider these problems.

a. There are five triangles below. Which pairs of triangles have the property that one can be moved onto the other by adding a point to every point on one triangle? Justify your answers.

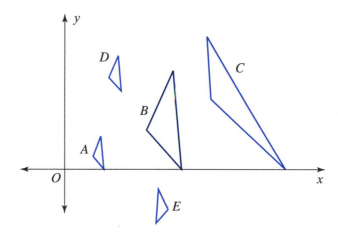

b. Describe some features that two figures on a coordinate plane must have in common if it is possible to add a point to every point on one figure to get the other one.

ACTIVITY 3 # Vectors

What really happens when you add (2, 3) to every point in a figure? The figure gets moved to the right 2 units and up 3 units.

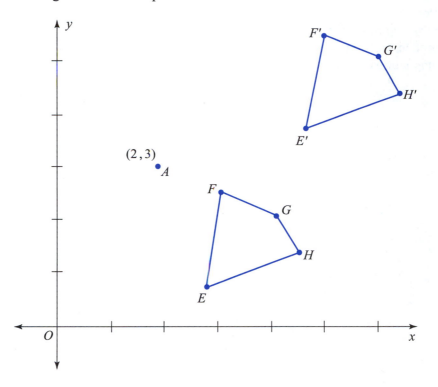

One way to think of the diagram on page 285 is that the whole quadrilateral is translated to another location by first moving over 2 and then up 3. The picture below tries to suggest this by showing what happens to the vertices and one point along one of the sides.

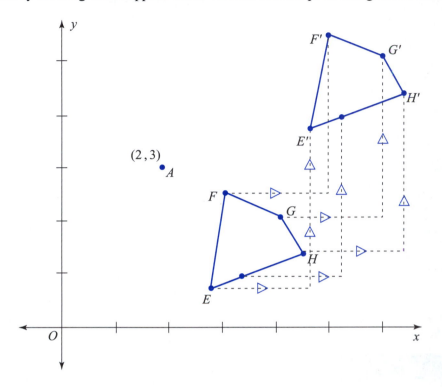

You can also think of adding (2, 3) another way. Suppose you want to add $A = (2, 3)$ to every point in $EFGH$. Think of a vector from O to A:

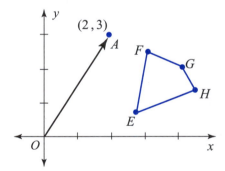

Now imagine that the quadrilateral gets moved once, this time in the direction of A and as far as the length of A.

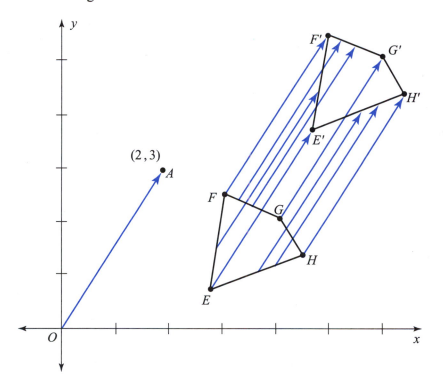

32. Make a copy of the picture below. Translate the figure by $(-2, -4)$.

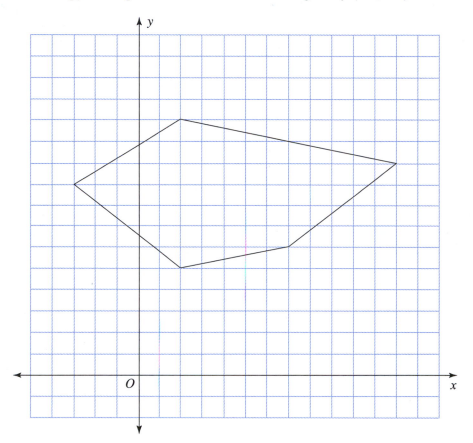

33. a. By what point should you translate the bottom figure to get the top one?

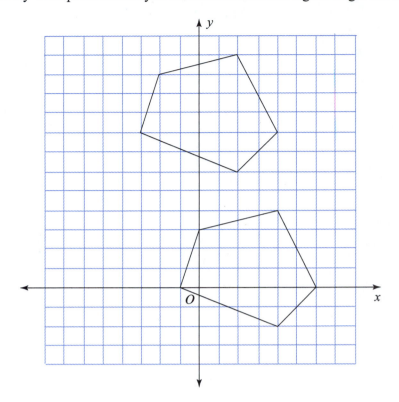

b. By what point should you translate the top figure to get the bottom one?

34. Triangle *ABC* has been moved four different times. Match each triangle with the point by which △*ABC* was translated.

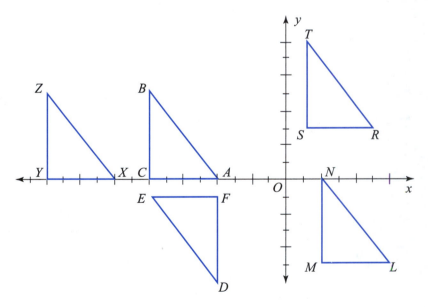

a. $(-6, 0)$

b. $(10, -5)$

c. $(9, 3)$

d. not a translation

35. Describe what each of the following will do to a figure:

    a. translate by $(0, 4)$

    b. translate by $(-1, 3)$

    c. translate by $(2, -2)$

    d. translate by $(-2, 0)$

36. **Checkpoint** Copy the picture below.

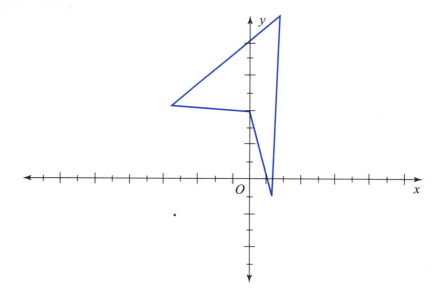

    a. Use vectors to translate the figure by $(3, -1)$.

    b. By what point would you have to translate the new figure to move it back to the original position?

    c. If a figure A is translated to figure B by the point (x, y), what point will translate figure B to figure A? That is, write a general rule for reversing a translation.

1. Write everything you know about scaling points. What is the definition? Describe shortcuts you have developed. How do you scale a complicated shape? What happens when you scale by a negative number?

2. True or false: "If the distance from A to B is 10, then the distance from $\frac{1}{2}A$ to $\frac{1}{2}B$ is 5." If the statement is true, explain or show why. If it is false, try to fix it.

3. Below is a picture of point A and the origin O.

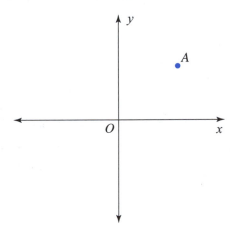

   a. Locate 2A, $\frac{1}{2}A$, and $-2A$.

   b. What can you say about the number c if A is between O and cA? If cA is between O and A? If O is between A and cA?

4. Write everything you know about adding points. Include the definition. Describe shortcuts you have developed. What do you get when you add a point to every point on a complicated shape, like a circle, a pentagon, or a picture of Trig?

5. Suppose $A = (-1, 4)$ and $B = (3, 2)$. Locate the following points on a coordinate system.

   a. $A + 2B$

   b. $3A + 2B$

   c. $-2A + B$

   d. $-2A + \frac{1}{2}B$

   e. $4A + \frac{1}{2}B$

   f. $kA + jB$

6. Suppose $A = (-1, 4)$ and $B = (3, 2)$.

   a. Find numbers a and b so that
   $$aA + bB = (-6, 10).$$

   b. Can every point on the plane be written as $aA + bB$ for some choice of a and b? Explain.

> You can do this problem either by drawing a picture and estimating or by setting up some equations.

7. Suppose $A = (6, 4)$ and $B = (3, 2)$. Can you find numbers a and b so that $aA + bB = (9, 6)$? How about $(-9, 4)$? Describe, with a picture perhaps, the set of points on the plane that can be written as $aA + bB$ for all possible choices of a and b.

8. Copy the grid below.

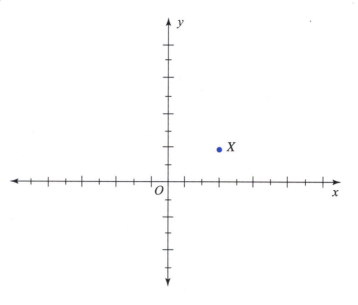

a. Use vectors to

   i. translate X by $(-3, 4)$. Call the new point A.

   ii. translate A by $(5, 0)$. Call the new point B.

   iii. translate B by $(3, -4)$. Call the new point C.

   iv. translate C by $(-5, 0)$. Where is the final point?

b. What type of figure is $ABCX$? Can you prove it?

Take It Further

9. True or false? If true, give a proof or justify it with an argument and include a picture. If false, give a counterexample.

a. If A is a point and c is a number, then cA is $|c|$ times as far from the origin as A is from the origin.

b. If C is somewhere on the segment from A to B, then $2C$ is somewhere on the segment between $2A$ and $2B$.

Choose numbers to use for some of these if you need to.

c. If M is the midpoint of the segment from A to B, then $3M$ is the midpoint of the segment between $3A$ and $3B$.

d. If the measure of $\angle ABC$ is $45°$, then the measure of $\angle(2A)(2B)(2C)$ (the angle that goes from $2A$ to $2B$ to $2C$) is $90°$.

10. Design a geometry software sketch that has coordinate axes, two points A and B, and their sum $A + B$. Put a trace on $A + B$.

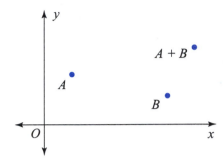

a. Animate A along a segment or a circle. Find a simple description for the path taken by $A + B$.

b. Do the same with B.

11. Suppose $A = (5, 12)$. Prove that $2A$ is collinear with A and the origin.

Ways to Think About It

There are several ways to approach Problem **11**. You can use the Triangle Inequality. Calculate three distances: from O to A, O to $2A$ and A to $2A$. Show that two of the distances add up to the other one.

7

The Algebra of Points

In this lesson, you will formalize ideas about adding and scaling points and prove three theorems.

Now that you have a good feel for working with coordinates, you can start proving some theorems about adding and scaling points.

Convention: Single lower-case letters, like c or k, will stand for numbers. Single upper-case letters, like A or X, will stand for points.

Explore and Discuss

There are a lot of letters flying around here, and keeping track of what they mean takes a little practice. For example, if A is a point and c is a number, then cA is a *point*, but $c(OA)$ is c times the distance from O to A, so it is a number. One way to keep things straight in your head is to substitute numerical values for the letters and see what *kind* of thing you get. For example, to figure out what kind of thing $c(OA)$ is, say to yourself, "If $A = (3, 4)$ and $c = 2$, then $OA = 5$ (do what is in parentheses first). So $c(OA) = 10$, a number." Try it. Classify each of these expressions as a *point*, a *number*, or just plain *meaningless*.

- **a** cB
- **b** $c + A$
- **c** AB
- **d** $c(AB)$
- **e** $k(OA)$
- **f** $A(BC)$
- **g** x

- **h** xX
- **i** XX
- **j** Bc
- **k** $A + B$
- **l** $c(A + B)$
- **m** $A(c + B)$

A Scaling Theorem

When you scale a point by a number c, cA is on the line between the origin and A. Find A in the picture below.

A, 2A, $\frac{1}{3}$A, and -2A all lie on the same line here. What other familiar point is on that line?

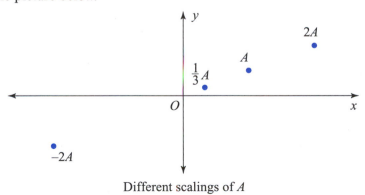

Different scalings of A

1. Describe the location of cA, in relation to A and the origin, when

   a. $c > 1$.

   b. $c < 1$.

   c. $c = 1$.

2. Suppose $A = (7, 1)$, $B = (-2, 4)$, and $C = (-3, -6)$. On a coordinate system, draw $\triangle ABC$. Then draw the triangle whose vertices are

Use a fresh coordinate system for each problem.
"$-A$" means $-1A$.

   a. $2A$, $2B$, $2C$.

   b. $\frac{1}{3}A$, $\frac{1}{3}B$, $\frac{1}{3}C$.

   c. $-A$, $-B$, $-C$.

3. Silly Fran had five points, scaled them, and forgot to keep track of the originals. Help Fran find A, B, C, D, and E.

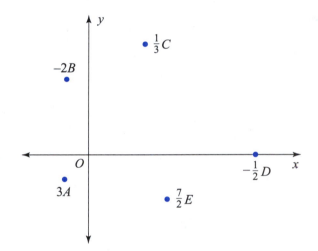

When there are several cases to consider, it is often a good idea to take one case at a time. So take one of the cases, say $c \geq 1$ (this is the easiest one). Here is the way the theorem might look for that case:

Theorem 5.3

Because $c \geq 1$, part 2 of the theorem implies that A is between O and B.

Suppose A is a point, c is a number greater than or equal to 1, and let $B = cA$.

1. B is collinear with A and the origin.

2. B is c times as far from the origin as A is.

The Triangle Inequality says that A is between O and B if and only if $OA + AB = OB$. So one way to prove the collinearity part of the theorem is to show that $OA + AB = OB$.

To prove the second part, show that $OB = c(OA)$.

This problem set asks you to implement the third item above. It asks you to prove exactly the same statement for several different points. The most important part of the problem is to construct in your mind a process for proving the statement that works for all the given cases and that will, in fact, work for *any* case.

Can you draw a picture of what you are asked to show?

4. Let $A = (3, 4)$, $c = 3$, and $B = cA$. Show that

$$OA + AB = OB$$

and that $OB = c(OA)$.

5. Let $A = (-5, 12)$, $c = 2$, and $B = cA$. Show that

$$OA + AB = OB$$

and that $OB = c(OA)$.

6. Let $A = (6, 8)$, $c = 1.5$, and $B = cA$. Show that

$$OA + AB = OB$$

and that $B = c(OA)$.

7. Let $A = (3, 8)$, $c = 4$, and $B = cA$. Show that

$$OA + AB = OB$$

and that $OB = c(OA)$.

Below is a proof of a general case of Theorem 5.3. Suppose $A = (a_1, a_2)$, c is a number greater than or equal to 1, and $B = cA$. This means that $B = (ca_1, ca_2)$. There are two things to prove:

Thing 1: $OA + AB = OB$.

$$OA + AB = \sqrt{a_1^2 + a_2^2} + \sqrt{(ca_1 - a_1)^2 + (ca_2 - a_2)^2}$$

$$= \sqrt{a_1^2 + a_2^2} + \sqrt{[a_1(c-1)]^2 + [a_2(c-1)]^2}$$

$$= \sqrt{a_1^2 + a_2^2} + \sqrt{a_1^2(c-1)^2 + a_2^2(c-1)^2}$$

$$= \sqrt{a_1^2 + a_2^2} + \sqrt{(c-1)^2(a_1^2 + a_2^2)}$$

$$= \sqrt{a_1^2 + a_2^2} + \sqrt{(c-1)^2}\sqrt{a_1^2 + a_2^2}$$

$$= \sqrt{a_1^2 + a_2^2} + (c-1)\sqrt{a_1^2 + a_2^2}$$

$$= (1 + (c-1))\sqrt{a_1^2 + a_2^2}$$

$$= c\sqrt{a_1^2 + a_2^2}$$

and

$$OB = \sqrt{(ca_1)^2 + (ca_2)^2}$$

$$= \sqrt{c^2 a_1^2 + c^2 a_2^2}$$

$$= \sqrt{c^2(a_1^2 + a_2^2)}$$

$$= \sqrt{c^2}\sqrt{a_1^2 + a_2^2}$$

$$= c\sqrt{a_1^2 + a_2^2}$$

So, $OA + AB = OB$, as desired.

8. Write and Reflect There are many details that need to be filled in here. Go through the proof, giving reasons for each of the steps.

9. Write and Reflect The assumption in the theorem is that $c \geq 1$. Is that ever used? Trace the above proof using several different numbers for c; try 3, 2, $\frac{1}{2}$, -1 and -2. Where does the proof fail for the numbers that are less than 1? Why?

Thing 2: $OB = c(OA)$. If you look carefully at the proof of Thing 1, you will see that both OA and OB were calculated in the course of proving $OA + AB = OB$. It turned out that

$$OA = \sqrt{a_1^2 + a_2^2}$$

and

$$OB = c\sqrt{a_1^2 + a_2^2}.$$

So, by substitution, $OB = c(OA)$.

Once you do Problem 10, you will know what scaling by c does for any c ≥ 0.

10. The proof for Theorem 5.3 assumes that $c \geq 1$. Modify the proof so that it works for $0 \leq c < 1$.

11. Now you know that Theorem 5.3 works for any $c \geq 0$. Prove that it also holds for negative c. (Hint: If $c < 0$, then $-c > 0$. How is $-cA$ related to cA?)

12. Checkpoint Theorem 5.3 is still worded in a way that assumes that $c \geq 1$. But, in Problems **10** and **11,** you described what scaling does for any kind of c. State a new and improved version of Theorem 5.3 that tells the whole story in one or two sentences.

ACTIVITY 2 ## An Adding Theorem

Now for adding. If A and B are points, you want a theorem that tells you how to locate $A + B$ in terms of A and B. A little experimenting is in order.

13. Plot each pair of points as well as their sum. Describe how the sum is located with respect to the points and the origin. Draw segments connecting the origin, O, to A and B, and segments connecting $A + B$ to A and B. What kind of figure do you get in each case?

   a. $A = (-3, 5)$, $B = (5, 1)$

   b. $A = (2, 3)$, $B = (5, -1)$

   c. $A = (-2, -2)$, $B = (0, 6)$

   d. $A = (-3, 3)$, $B = (5, -5)$

   e. $A = (-3, 5)$, $B = 2(5, 1)$

   f. $A = (-3, 5)$, $B = \frac{1}{2}(5, 1)$

The last problem suggests a theorem:

Theorem 5.4

If $A = (a_1, a_2)$ and $B = (b_1, b_2)$, then $A + B$ is the fourth vertex of the parallelogram that has A, O, and B as three of its vertices and $\overline{OA}$ and $\overline{OB}$ as two of its sides.

14. This seems like a pretty wordy way of stating a theorem. Do you really need the "and $\overline{OA}$ and $\overline{OB}$ as two of its sides" part? Try to make the theorem clearer by stating it in your own words. Use a picture if you think it helps.

Now for the proof. Basically, you want to show that the quadrilateral whose vertices are O, A, $A + B$ and B is a parallelogram. The strategy is *to show that the opposite sides have the same length.*

The next problem set asks you to prove exactly the same statement for several different points. The most important part of the problem is to construct in your mind a process for proving the statement that works for all the given cases and that will, in fact, work for *any* case.

Can you draw a picture of what you are asked to show?

15. Let $A = (3, 4)$, $B = (5, -12)$, and $P = A + B$. Draw a picture of O, A, B, and P. Show that $OA = BP$ and $OB = AP$.

16. Let $A = (8, 15)$, $B = (-4, 3)$, and $P = A + B$. Draw a picture of O, A, B, and P. Show that $OA = BP$ and $OB = AP$.

17. Let $A = (8, 6)$, $B = (3, 1)$, and $P = A + B$. Draw a picture of O, A, B, and P. Show that $OA = BP$ and $OB = AP$.

The previous problem set was designed to help you identify the key calculations.

Below is a proof of Theorem 5.4 that captures what is common to all the calculations in the problem set above. Let $A = (a_1, a_2)$, $B = (b_1, b_2)$. Let $P = A + B$, so that

$$P = (a_1 + b_1, a_2 + b_2).$$

The situation looks like this:

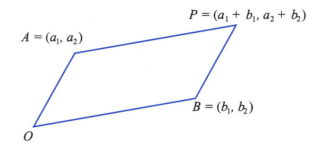

To prove the opposite sides have the same length, you need to show two things:

Thing 1: $OA = BP$. You have

$$OA = \sqrt{a_1^2 + a_2^2}$$

and

$$BP = \sqrt{[(a_1 + b_1) - b_1]^2 + [(a_2 + b_2) - b_2]^2}$$

$$= \sqrt{a_1^2 + a_2^2}.$$

So $OA = BP$.

Thing 2: $OB = AP$

18. Prove Thing 2.

The segment goes from B + A to C + A, and is obtained from $\overline{BC}$ by "translating by the vector from O to A."

19. Does Theorem 5.4 hold if O, A, and B are collinear? For example, where is the parallelogram if $A = (9, -1)$ and $B = (18, -2)$?

20. **Checkpoint** If A, B, and C are points, show that the quadrilateral whose vertices are B, C, $C + A$, and $B + A$ is a parallelogram by showing the opposite sides are congruent. Draw a picture.

ACTIVITY 3 ▶ **More on Scaling Points**

Theorem 5.3 tells you what happens if you scale a point by a number. If $B = cA$, then B is collinear with A and O. But is *every* point on $\overleftrightarrow{OA}$ a multiple of A?

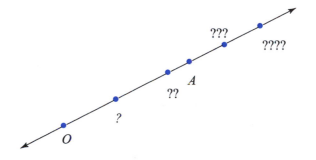

Are they all multiples of A?

Yes indeed.

The proof makes use of what you know about similar triangles. Just as in the proof of Theorem 5.3, you have to worry about where B is located in relation to the origin and A. Look first at the special case where A is between the origin and B.

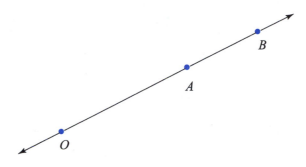

Is B a multiple of A?

As usual, assign generic coordinates to A and B. Let $A = (a_1, a_2)$ and $B = (b_1, b_2)$. Then, on a coordinate system, the measurements look like the diagram below.

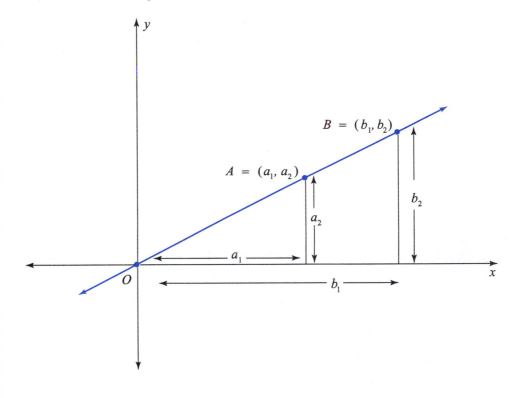

There are two right triangles in the picture (what are they?) and, *because OAB* is a straight line, the triangles are similar. Since corresponding sides of similar triangles are proportional,

$$\frac{b_2}{a_2} = \frac{b_1}{a_1}.$$

An important habit of mind: Give things names so you can more easily deal with them.

Give this common ratio a name; call it "*c*." So,

$$\frac{b_2}{a_2} = \frac{b_1}{a_1} = c.$$

But then, $b_2 = ca_2$ and $b_1 = ca_1$ (why?). So,

$$B = (b_1, b_2) = (ca_1, ca_2) = c(a_1, a_2) = cA,$$

and that's what was to be proved. The result can be summarized in a theorem:

Theorem 5.5

Suppose A and B are points so that B is collinear with A and the origin. Then there is a number c so that B = cA.

21. Write out the proof of Theorem 5.5 in your own words.

22. Prove Theorem 5.5 in the case where *B* is between *O* and *A*.

23. Prove Theorem 5.5 in the case where *O* is between *B* and *A*.

Theorems 5.3 and 5.5 can be combined to give a complete picture.

Theorem 5.6

If A is a point, the set of all multiples of A is the line through the origin and A.

24. **Checkpoint** Name 6 multiples for each point.

    a. (7, 3)

    b. (3, 7)

    c. $\left(\frac{1}{2}, -3\right)$

ACTIVITY 4 **Parallel Lines**

You can use results of the scaling and adding theorems to find tests for parallel lines. In the following problems, think about parallelograms formed by the points, and about what you know about points on the same line through the origin.

25. Suppose $\overleftrightarrow{AB}$ and $\overleftrightarrow{CD}$ are parallel lines. Show that there is a number k so that $B - A = k(D - C)$. Draw a picture. If you are having difficulty, try it with actual coordinates.

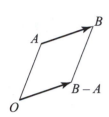

One way to think about this is that the vector from A to B is equivalent to the vector from O to B − A.

Theorem 5.7

If A and B are points, then the segment from O to B−A is parallel and congruent to $\overline{AB}$.

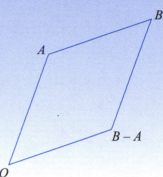

26. Show that Theorem 5.7 holds if $A = (3, 5)$ and $B = (7, 9)$.

27. Prove that Theorem 5.7 holds for any points A and B.

28. Show that if P and A are points and t is a number, the segment from P to $P + tA$ is parallel to the segment from O to A. Draw a picture. If you are having difficulty, try it with actual coordinates.

29. Suppose $A = (4, 1)$, $B = (9, 5)$, $C = (3, -1)$, and $D = (13, 7)$. Show that $\overleftrightarrow{AB} \parallel \overleftrightarrow{CD}$.

30. Suppose that A, B, C and D are points and t is a number so that $B - A = t(D - C)$. Show that $\overleftrightarrow{AB} \parallel \overleftrightarrow{CD}$. Draw a picture. If you are having difficulty, try it with actual coordinates.

You must show two things: MA = MB and that M is on $\overline{AB}$.

31. If A and B are points, show that the midpoint of $\overline{AB}$ is

$$\tfrac{1}{2}(A + B).$$

Draw a picture. If you are having difficulty, try it with actual coordinates.

This is just the Midline Theorem.

32. Checkpoint Let $A = (4, 4)$ and $B = (8, -4)$. Copy the figure from Problem **25.** Draw $\overline{OB}$. Look at $\triangle OAB$. Let M be the midpoint of $\overline{OA}$ and N be the midpoint point of $\overline{OB}$. Show that $\overline{MN} \parallel \overline{AB}$ and $MN = \tfrac{1}{2}AB$.

A C T I V I T Y 5 ▶ ## Properties of Points

Is it obvious to you that algebra with points behaves like algebra with numbers? There are many properties of point addition and scaling that look like familiar number properties. Here are eight properties that are often used in a branch of mathematics called *linear algebra*. They are given here in a theorem.

Theorem 5.8

If $A = (a_1, a_2)$, $B = (b_1, b_2)$, and $C = (c_1, c_2)$ are points and d and e are numbers, then

1. $A + B = B + A$

2. $A + (B + C) = (A + B) + C$

3. $A + O = A$

4. $A + (-1A) = O$

5. $(d + e)A = dA + eA$

6. $d(A + B) = dA + dB$

7. $d(eA) = (de)A$

8. $1A = A$.

These look like statements about numbers, but they are really about points.

33. Prove each part of the theorem. You will use properties of numbers (the coordinates) to prove properties of points.

34. Use adding and scaling points to show that the line joining the midpoints of two sides of a triangle is parallel to the third side and half as long.

35. Suppose A and B are points. Explain how to locate each of the points below.

Draw a picture. If you are having difficulty, try it with actual coordinates.

    a. $\frac{1}{3}A + \frac{2}{3}B$

    b. $\frac{2}{3}A + \frac{1}{3}B$

    c. $\frac{1}{4}A + \frac{3}{4}B$

    d. $\frac{3}{4}A + \frac{1}{4}B$

    e. $\frac{3}{5}A + \frac{2}{5}B$

    f. $kA + (1 - k)B$ (here, $0 \leq k \leq 1$)

36. **Checkpoint** Let $A = (4, 2)$, $B = (5, -3)$, $C = (6, 4)$, and $P = \frac{1}{3}(A + B + C)$.

Show that, in $\triangle ABC$, P is $\frac{2}{3}$ of the way from any vertex to the midpoint of the opposite side.

On Your Own

1. Classify each expression as a *point*, a *number*, or *meaningless*.

    a. cZ

    b. $c(XY)$

c. $c(X + Y)$

d. $A(X + Y)$

e. AX

f. $X + Y$

g. $M(C + X)$

2. There is another way to think about Theorem 5.4.

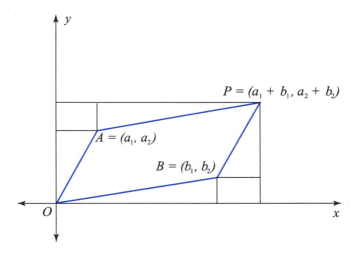

$P = (a_1 + b_1, a_2 + b_2)$

$A = (a_1, a_2)$

$B = (b_1, b_2)$

a. Label lengths in the above figure. Use congruent triangles and corresponding parts to show that the opposite sides of $OBPA$ are parallel.

b. Draw a picture that shows the situation if A is in the second quadrant. Develop a proof for this case.

c. How many cases would you need to claim a complete proof?

What is $-1A + A$*?*

3. Explain how to locate $-1A$ on the plane if you know A.

4. Copy each of the pictures below.

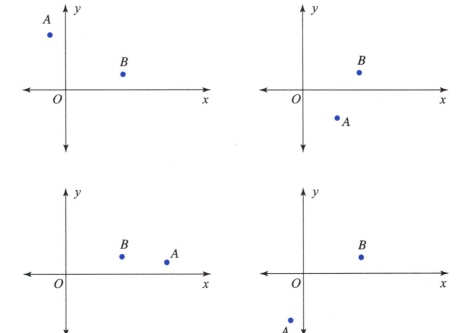

B − A means B + (−1A).

a. Locate $A + B$.

b. Locate $B - A$.

c. Find a way to describe the position of $B - A$ in terms of the origin, A, and B.

5. Copy the picture below. Locate

a. $2A + 3B$.

b. $3A - B$.

c. $-2A - \frac{1}{2}B$.

d. $0.7A + 3.6\,B$.

Think of 2A + 3B as "two
As and three Bs." Go out
two As from the origin,
go out three Bs from the
origin, and complete the
parallelogram.

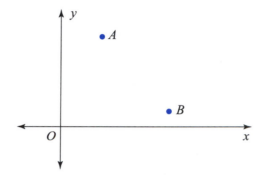

6. Points A and B are marked with open circles. For each of the other points in the picture, estimate numbers x and y so that $xA + yB$ will get you to the point.

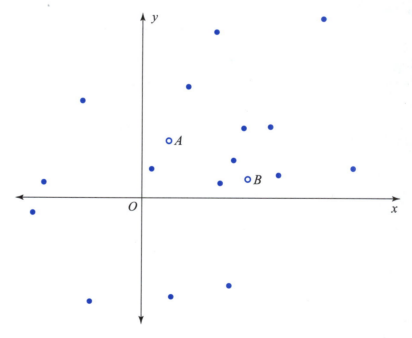

7. Suppose P and A are points. If A is the midpoint of $\overline{PX}$, express X in terms of P and A.

8. Suppose A is a point. If O is the midpoint of $\overline{AC}$, express C in terms of A.

9. If $A = (3, 2)$, $B = (-4, 5)$, and $C = (7, 4)$, show that $\triangle ABC \cong \triangle O(B-A)(C-A)$. Draw a picture.

10. If A, B, and C are *any* three points, show that $\triangle ABC \cong \triangle O(B-A)(C-A)$.

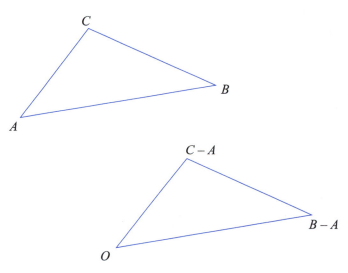

Move it down to O

Notice that O = A − A.

Sometimes people say that "triangle ABC has been translated to the origin by moving A to O."

11. Draw three points A, B, and C. Draw the pictures of the triangles you get by moving

Put them all on the same set of axes for a nice effect.

   a. A to the origin.

   b. B to the origin.

   c. C to the origin.

12. If A, B, and C are points, show that the quadrilateral whose vertices are B, C, $C + A$, and $B + A$ is a parallelogram by showing the opposite sides are parallel. Draw a picture.

13. Given two points A and B, how can you find a point that is one third of the way from A to B? Explain.

Take It Further

14. Let A, B, and C be points, and let

$$P = \tfrac{1}{3}(A + B + C).$$

Show that, in $\triangle ABC$, P is $\tfrac{2}{3}$ of the way from any vertex to the midpoint of the opposite side.

15. Extend the results of this lesson to three dimensions. In particular, how do Theorems 5.3 and 5.5 play out in three dimensions?

Vectors and Geometry

In this lesson, you will learn more about vectors, and use them to solve problems.

In the last few lessons, you studied the geometry behind adding and scaling points. There's a very convenient language for using what you have learned. It involves the notion of *vector*.

mph = miles per hour

In physics, people distinguish a *speed* (like 30 mph) from a *velocity*. A velocity has a direction as well as a size, so that 30 mph northeast is different from 30 mph due south. This idea can be modeled on a coordinate system by thinking about line segments that have a *direction*.

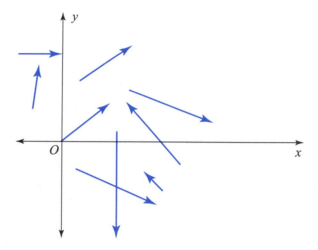

Some directed segments

Each arrow in this figure can stand for a velocity. The length of the segment is the speed, and the direction of the segment tells you which way the velocity is headed. The arrows are often called *vectors*.

For navigators, vectors might stand for trips or currents or wind velocities. A wind velocity of 10 mph at a bearing of 40° from north could be represented like this:

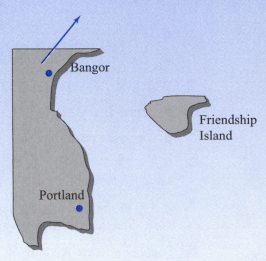

The *length* of the arrow represents the speed of the wind, and the *direction* of the arrow shows the direction of the wind.

a On a copy of the picture, draw a vector that represents a 20 mph wind with the same direction as the one in the picture.

b On a copy of the picture, draw a vector that represents a 40 mph wind with the opposite direction from the one in the picture.

ACTIVITY 1

Vector Algebra

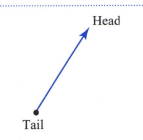

The kind of vectors used in this unit are called "located vectors." The navigator's vectors are called "free vectors."

When speaking of winds, navigators might be most interested in the magnitude of the wind (10 mph, for example) and the direction of the wind (40° from north, for example). It makes much less sense to talk about the "place from where the wind starts." In other words, the tail of the vector really doesn't matter. A 10 mph wind at 40° from north is the same whether it blows while you are sitting in Bangor or Portland or on a pier on Friendship Island. So, the vectors in this situation are somehow the same.

In geometry, it turns out to be useful *not* to call them the same. They are not "equal," because they are not absolutely identical. For us, vectors will have a definite location. A vector requires an *initial point* (a tail) and a *terminal point* (a head). Instead of saying that certain vectors in the situation are the same, you could say that they are *equivalent*. So, if A and B are points, the "vector from A to B" can be thought of as the arrow that starts at A and ends at B. The notation for this vector is $\overrightarrow{AB}$. Notice that this is also the notation for the *ray* that starts at A and goes through B. For the rest of this unit, forget about rays; $\overrightarrow{AB}$ will stand for the vector with tail A and head B. Here is some practice that will help you get used to the idea.

1. Let $A = (5, 1)$, $B = (-2, 5)$, $C = (9, -2)$, $D = (16, -6)$. On a coordinate system, draw each of the following vectors.

   a. $\overrightarrow{AB}$

   b. $\overrightarrow{CD}$

   c. $\overrightarrow{AC}$

   d. $\overrightarrow{BA}$

   e. $\overrightarrow{OA}$

   f. $\overrightarrow{OB}$

   g. $\overrightarrow{B(A + B)}$

   h. $\overrightarrow{A(A + B)}$

   i. $\overrightarrow{O(B - A)}$

2. Which of the vectors in Problem **1** have the

   a. same direction?

   b. opposite direction?

3. If $A = (5, 3)$ and $B = (8, 7)$, find a vector that starts at the origin and has the same length and direction as $\overrightarrow{AB}$.

4. If $A = (5, 3)$ and $B = (8, 7)$, find a vector that starts at the origin and has the same direction as $\overrightarrow{AB}$, and is twice as long as $\overrightarrow{AB}$.

5. Suppose you are working on a coordinate system. If $A = (3, 5)$ and $B = (8, 1)$, find two points C and D (neither at the origin) so that the vector from A to B is equivalent to the vector from C to D.

6. Below is a pair of sentences:

   $\overrightarrow{A(A + B)}$ is equivalent to $\overrightarrow{OB}$, and $\overrightarrow{O(A + B)}$ is one diagonal of the parallelogram whose vertices are O, A, $A + B$, and B. In addition, $\overrightarrow{OA}$ is equivalent to $\overrightarrow{B(A + B)}$.

   The sentences are pictured in the figure below. Copy the picture. Label the five vectors that are mentioned in the sentences above.

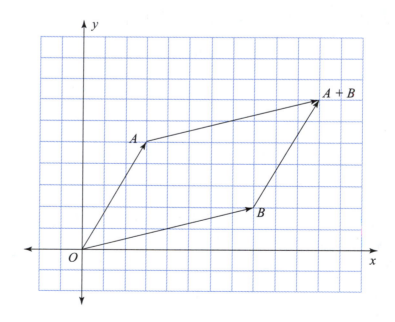

7. If $R = (7, 2)$ and $S = (15, 6)$, find the head of a vector that starts at $(4, -3)$ and is equivalent to the vector from R to S.

8. Copy the picture below. Draw some equivalent vectors that suggest the movement of one polygon onto the other. Using coordinates, how could you show that all your vectors are equivalent?

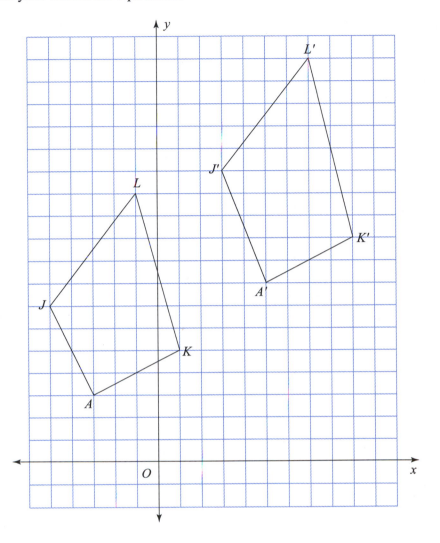

9. True or false? If A and B are any points, the vector from O to B is equivalent to the vector from A to $A + B$. Explain.

10. Suppose that T is the function that moves points according to the rule $T(X) = X + (10, 5)$. Pick two points A and B. Draw the vectors from A to $T(A)$ and from B to $T(B)$. Show that these vectors are equivalent.

11. Plot a dozen or so random points. Translate each of them by $(8, 5)$. Draw a vector from each of your points to the corresponding translated points. Are all of the vectors you drew equivalent? Explain your answer.

Remember, this means that, for example, $T(4, 2) = (4, 2) + (10, 5) = (14, 7)$.

This picture will give you a feel for the function T *defined by the rule* $T(X) = X + (8, 5)$.

12. If A and B are any two points, there is a vector that starts at O and is equivalent to the vector from A to B. Find a way to calculate the coordinates of the head of the vector with tail O below.

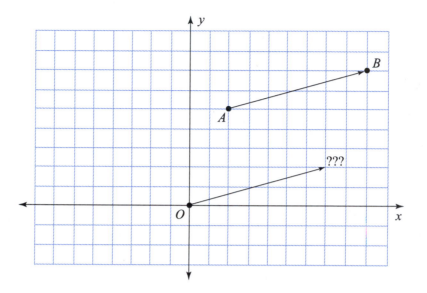

13. Find a way to tell if the two vectors below are equivalent just by looking at their coordinates and doing some calculations with them.

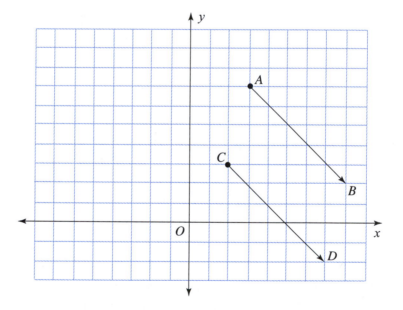

14. Prove the following theorem:

Theorem 5.9

If A and B are points, the vector from O to $B - A$ is equivalent to the vector from A to B.

Ways to Think About It

Because of Theorem 5.9, many people think of vectors starting at the origin as "representatives" for whole classes of vectors, all equivalent.

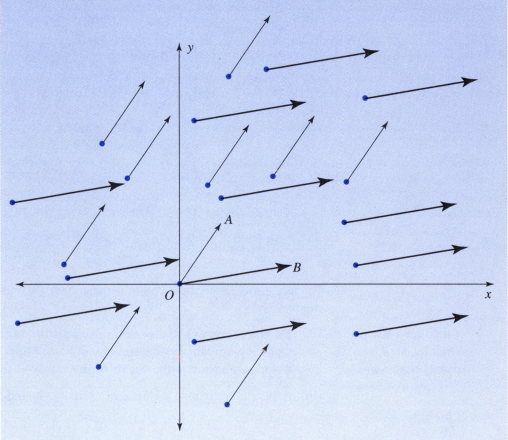

Two families of vectors represented by $\overrightarrow{OA}$ and $\overrightarrow{OB}$

It sometimes brings clarity to a problem if you draw vectors from the origin to all the important points. For example, you discovered the parallelogram law for adding by drawing vectors from the origin to A, B, and $A + B$. You learned how to scale a point A by drawing a vector from the origin through A.

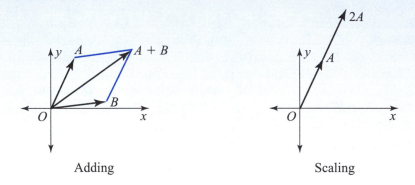

Adding   Scaling

With these two properties in place, forming various combinations of points can be pictured as stretching and adding vectors.

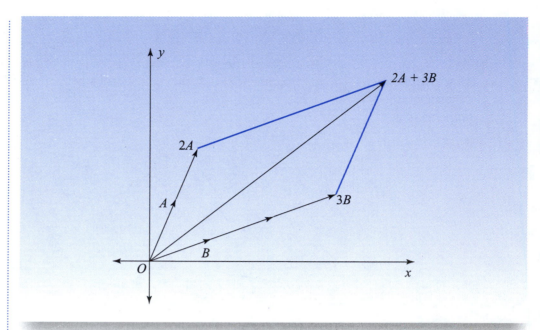

The next problem set asks you to try visualizing the algebra of points as "vector algebra."

If you are having difficulty, try it with actual coordinates. Draw a picture.

15. If A and B are points, the vector from the origin to $A + B$ is one diagonal of the parallelogram whose sides are $\overrightarrow{OA}$ and $\overrightarrow{OB}$. Show that the other diagonal is (as a vector) equivalent to the vector from O to $B - A$.

16. In the picture below, the head of $\overrightarrow{OA}$ is $(1, 1)$ and the head of $\overrightarrow{OB}$ is $(1, 0)$.

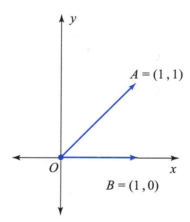

Find x and y so that $x\overrightarrow{OA} + y\overrightarrow{OB}$ equals

a. $(2, 1)$.

b. $(4, 3)$.

c. $(11, 1)$.

d. $(5, 6)$.

e. $(0, 8)$.

f. $(1, -10)$.

Suggestion: Draw the two vectors and the points on a piece of graph paper.

Then think about how you would arrive at each of the points if you could only travel in the directions allowed by the two vectors.

17. **Checkpoint** Use "vector algebra" to locate the given points in the picture below.

   a. $2A + 3B$

   b. $3A - B$

   c. $-2A - \frac{1}{2}B$

   d. $0.7A + 3.6B$

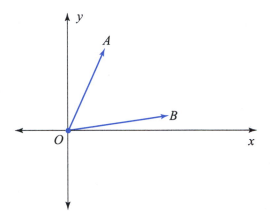

ACTIVITY 2

Head Minus Tail

Theorem 5.9 says that $\vec{AB}$ is equivalent to the vector from O to $B - A$. Some people think of this as "moving $\vec{AB}$ to the origin." They say, "To move $\vec{AB}$ to the origin, I just have to make the vector from O to $B - A$."

Head minus tail.
Easy enough.

Other people say, "To move a vector to the origin, just subtract the tail from the head." The next few problems show you how this technique can be applied to the geometry of vectors.

18. **a.** To move a vector around, first move the vector to the origin. Then move it to the new spot. In the picture below, you can move $\vec{AB}$ to the origin simply by calculating "head minus tail." What does that give you?

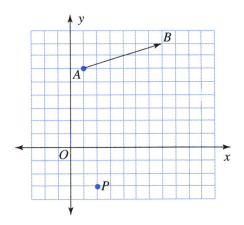

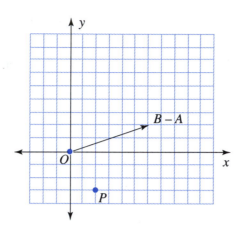

Move it to the origin ("head minus tail") and then move it back out by translating.

b. Next, take the vector anchored at the origin and move it to P by adding P to the head and to the tail. But adding P to the tail (which is just O) produces P. And adding P to the head produces the head of the vector you want. What is it?

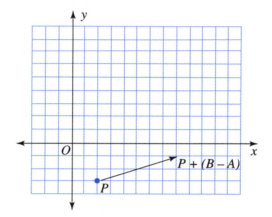

19. Find and draw a vector equivalent to $\overrightarrow{AB}$ that

   a. starts at $C = (8, 3)$.

   b. starts at $J = (-8, 3)$.

   c. starts at $K = (0, 3)$.

   d. ends at O.

   e. ends at $C = (8, 3)$.

   f. starts at B.

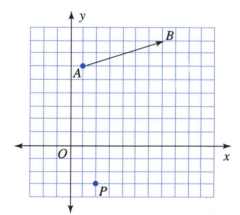

20. Suppose $A = (3, 4)$, $B = (9, 0)$, $C = (-1, 2)$, and $D = (5, -2)$. Show that $\overrightarrow{AB}$ is equivalent to $\overrightarrow{CD}$ by moving by moving both vectors to the origin. Show also that $\overrightarrow{AC}$ is equivalent to $\overrightarrow{BD}$.

21. This "move it to the origin" technique is useful in situations that do not even seem to involve vectors. For example, a quadrilateral can be moved to the origin by subtracting any one of its vertices from all the rest. There are, then, four ways to do it. In each of the figures below, find the coordinates of the quadrilaterals at the origin. You could do this by counting squares (except some of the figures run off the graph paper), but it is easier to do the arithmetic with the vertices of *ABCD*.

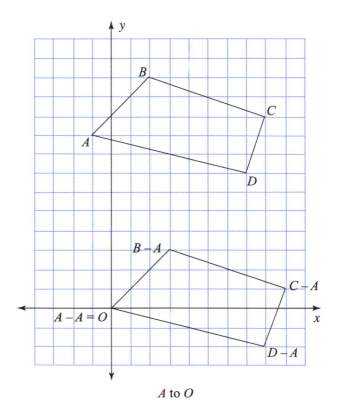

A to *O*

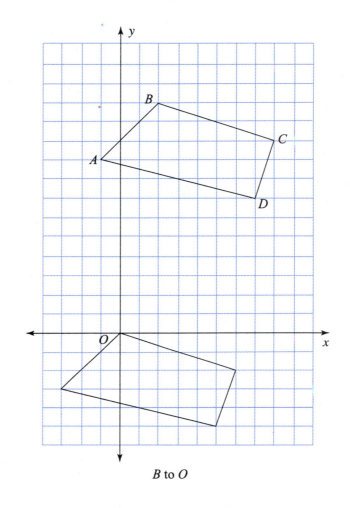

B to *O*

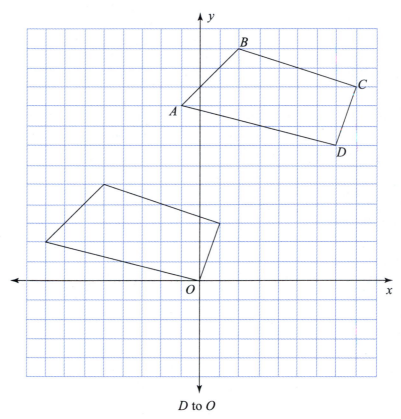

D to *O*

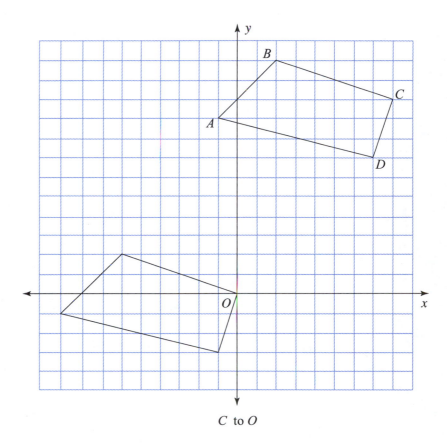

C to O

Since it is so easy to do, many people use the move-it-to-the-origin technique as a way to test vectors for equivalence rather than calculating heading and length. The strategy is the following: Suppose you have two vectors and you want to see if they are equivalent. Move each of them to the origin (by finding "head minus tail" for each of them), and see if you get the same thing.

> *Head minus tail test.* Two vectors are equivalent if and only if head minus tail on one vector gives you the same result as head minus tail on the other. In symbols: $\overrightarrow{AB}$ is equivalent to $\overrightarrow{CD}$ if and only if
>
> $$B - A = D - C.$$

22. Prove the following theorem:

Theorem 5.10

Adding the same point to the tail and the head of a vector produces an equivalent vector.

Draw a picture. If you are having difficulty, try it with actual coordinates.

Calculating head minus tail, which is the same as moving a vector to the origin, can be used to compare vectors for other things besides equivalence. The next problem asks you to extend the head minus tail test to cover more general relations.

23. Below are some vectors.

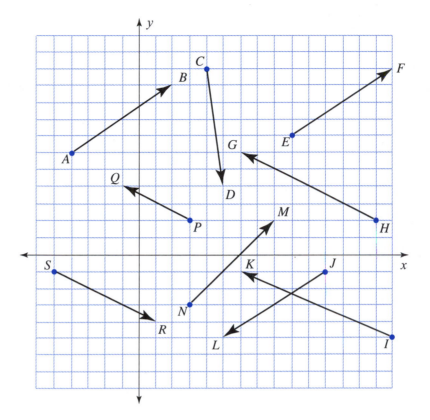

   a. Which vectors are parallel?

   b. Which are parallel in the same direction?

   c. Which are parallel in the opposite direction?

   d. Which are equivalent?

   e. Come up with "head minus tail" tests for two vectors to be

      i. parallel in the same direction.

      ii. parallel in the opposite direction.

All this can be summarized in one "megatheorem":

Theorem 5.11

Two vectors $\overrightarrow{AB}$ and $\overrightarrow{CD}$ are parallel if and only if there is a number k so that

$$B - A = k(D - C).$$

If $k > 0$, the vectors have the same direction. If $k < 0$, the vectors have opposite directions. If $k = 1$, the vectors are equivalent.

---

24. **Checkpoint** Find and label four points A, B, C, and D so that the vector from A to B is equivalent to the vector from C to D. Without labeling any other points, find another pair of equivalent vectors. Explain why what you say is true.

"Not equivalent" is not a sufficient answer to this problem.

1. If $A = (3, 5)$, $B = (7, -2)$, $C = (9, 1)$, and $D = (5, 8)$, is the vector from A to B equivalent to the vector from C to D? If so, why? If not, what *would* you call them?

2. If $A = (3, 5)$, $B = (7, -2)$, $C = (9, 1)$, and $D = (17, -13)$, what word describes the relationship between the vector from A to B and the vector from C to D?

It is amazing how vector methods can simplify proofs of what are otherwise complicated theorems. For example, do you remember the Midline Theorem?

> *The line segment that connects the midpoints of two sides of a triangle is parallel to the third side and half as long.*

A vector proof might go like this:

> Take two sides of the triangle to be $\overrightarrow{OA}$ and $\overrightarrow{OB}$. Then the midpoints are labeled as in the picture below.

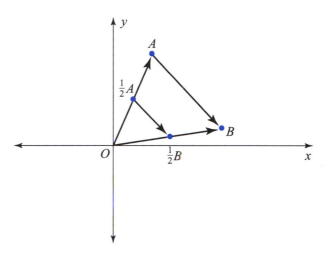

Next, translate $\overrightarrow{AB}$ and the midline to the origin.

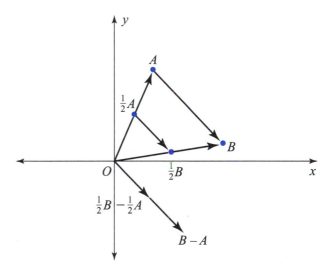

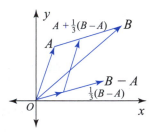
3. Complete the proof of the Midline Theorem.

4. Show that the point

$$\frac{2}{3}A + \frac{1}{3}B$$

is on $\overline{AB}$, $\frac{1}{3}$ of the way from A to B.

5. What point is on $\overline{AB}$, $\frac{1}{4}$ of the way from A to B?

6. Suppose $A = (3, 5)$ and $B = (7, 1)$. Find C if B is the midpoint of $\overline{AC}$.

7. Suppose $A = (0, 8)$, $B = (3, 12)$, $C = (12, -1)$, and $D = (6, -5)$. Show that the quadrilateral formed by joining the midpoints of $ABCD$ is a parallelogram.

8. Suppose A, B, C, and D are arbitrary points that are the vertices of a quadrilateral.

   a. Express the midpoints of the sides of the quadrilateral in terms of A, B, C, and D.

   b. Show that the quadrilateral formed by joining the midpoints of $ABCD$ is a parallelogram.

9. A British friend of the authors posed the following problem:

   Draw an arbitrary quadrilateral $ABCD$. Place a point, labeled P, anywhere you like, as in the figure below. Construct a segment through A with P as an endpoint and A as its midpoint. Begin at the endpoint of this new segment and construct another segment, this time with B as the midpoint. Continue doing this until you construct a segment with D as the midpoint. The finishing point of this whole journey around $ABCD$ is labeled Q in the figure.

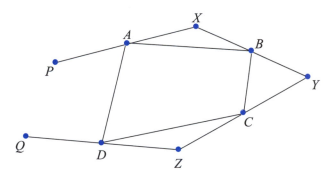

This is an ideal experiment to try with geometry software.

   a. Draw $\overrightarrow{QP}$. If you change P's location, $\overrightarrow{QP}$ changes location, but does not change direction or length. That is, all possible vectors $\overrightarrow{QP}$ are equivalent. Prove this. (Hint: Show that $P - Q$ can be written solely in terms of A, B, C, and D.)

   b. How can you make $\overrightarrow{QP}$ have length 0? That is, when does $P = Q$?

10. Below is a picture of a polygon whose sides are vectors.

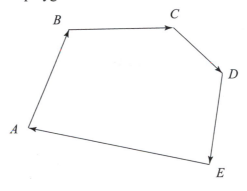

Move each side of the polygon (each vector) to the origin, to get five vectors, each with tail O.

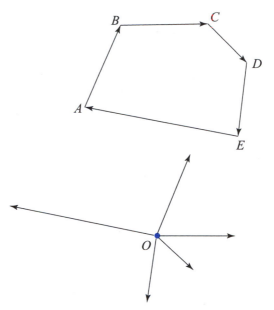

a. Label the heads of each of these vectors.

b. Show that the sum of these vectors is O.

11. Use vectors to show that the diagonals of a parallelogram bisect each other. (Hint: You can assume that one vertex of the parallelogram is O. If A and B are the vertices adjacent to O, the fourth vertex is $A + B$).

12. Show that in a parallelogram, the sum of the squares of the lengths of the diagonals is the same as the sum of the squares of the lengths of the sides.

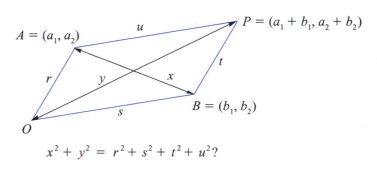

$$x^2 + y^2 = r^2 + s^2 + t^2 + u^2?$$

Draw a picture. If you are
having difficulty, try it
with actual coordinates.

13. Suppose A and B are points. Use vectors to explain how to locate each point below.

a. $\frac{1}{3}A + \frac{2}{3}B$ **d.** $\frac{3}{4}A + \frac{1}{4}B$

b. $\frac{2}{3}A + \frac{1}{3}B$ **e.** $\frac{3}{5}A + \frac{2}{5}B$

c. $\frac{1}{4}A + \frac{3}{4}B$ **f.** $kA + (1 - k)B$ (here, $0 \le k \le 1$)

14. Let A, B, and C be points, and let

Hint: $\dfrac{A + B + C}{3} =$
$\dfrac{1}{3}B + \dfrac{2}{3}\left(\dfrac{A + C}{2}\right)$

$$P = \frac{1}{3}(A + B + C).$$

Show that in $\triangle ABC$, P is $\frac{2}{3}$ of the way from any vertex to the midpoint of the opposite side.

15. George has a way to find the population center for three cities, each of the same size. He puts the cities on a coordinate system, adds the coordinates of the three cities, and scales by $\frac{1}{3}$. In what sense is George's point the "population center"?

16. Martha extends George's method by allowing for cities of different sizes. She first coordinates the map, scales the coordinates of each city by its population, adds the results, and then divides the resulting point by the sum of the populations of all three cities. In what sense is Martha's point the "population center"?

17. Why does a pantograph work? That is, suppose you build a collection of linkages like the one below. The rods are pivoted at D, E, B, and F. Why, if you nail O to a table and trace out a figure with point A, does point B trace out a smaller but similar figure?

Points E and F are
midpoints of $\overline{DA}$ and $\overline{DO}$.

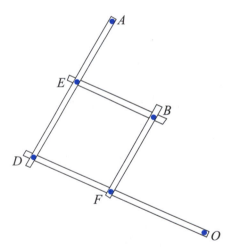

Hint: Does the picture below suggest any explanations to you?

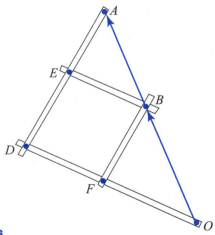

Slope and Equations of Lines

In this lesson, you will relate ideas about vectors to slope and equations of lines.

There are lots of words you probably remember from previous mathematics classes: words like *slope, y-intercept, point-slope equation* and a host of others that had to do with finding equations of lines. In this lesson, you will revisit these terms and connect them with vectors and the algebra of points.

Suppose $A = (3, 1)$ and $B = (8, -2)$. Points A and B determine a line, conveniently called $\overleftrightarrow{AB}$.

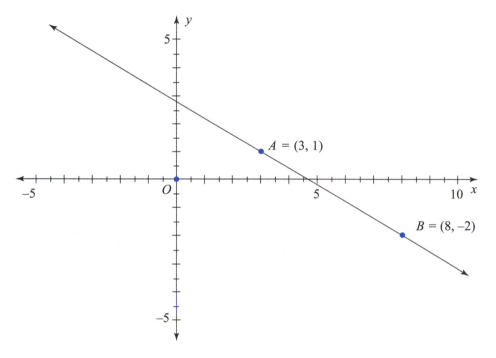

Not every point in the plane is on $\overleftrightarrow{AB}$. So the question is, given a point $X = (x, y)$, how can you tell if it is on or off the line? Of course, for some points, like $(0, 100)$, you can just eyeball it. But for others, you cannot be so sure. For example, is $X = (1, 2)$ on $\overleftrightarrow{AB}$? That is something to consider.

Explore and Discuss

Either X is on $\overleftrightarrow{AB}$ or it isn't. Vectors can help you tell. Think of the vectors $\overrightarrow{AX}$ and $\overrightarrow{BA}$. When you move them to the origin, you get vectors from O to $X - A$ and from O to $A - B$. If X is on the line, $X - A$ will be a multiple of $A - B$. If it isn't on the line, it won't be a multiple *of* $A - B$.

Why is this? That is, explain the statement "If X is on the line, $X - A$ will be a multiple of $A - B$. If it isn't on the line, it won't be a multiple of $A - B$."

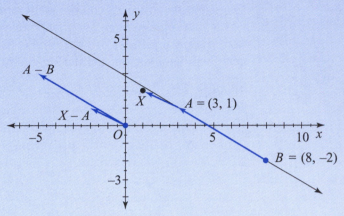

X not on $\overleftrightarrow{BA}$

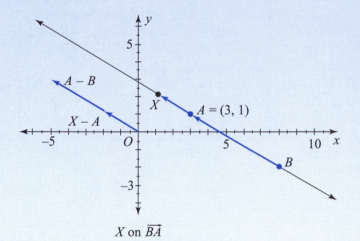

X on $\overleftrightarrow{BA}$

Notice that in this case, $A = (3, 1)$, $B = (8, -2)$, and $X = (1, 2)$. So,

$$X - A = (-2, 1) \quad \text{and}$$

$$A - B = (-5, 3).$$

If $X - A$ were a multiple of $A - B$, that multiple would have to be $\frac{1}{3}$ (look at the last coordinates). So, it would have to be true that

$$(-2, 1) = \frac{1}{3}(-5, 3)$$

and that's just not true. It is *close* (that's why you couldn't eyeball it), but it is not a true equation. So, X isn't on $\overleftrightarrow{AB}$.

On the Line?

The following problems ask you to test lots of points to see if they are on a given line. The goal is to build up to a *generic* calculation.

1. a. Suppose $A = (3, 1)$ and $B = (8, -2)$. Which of the following points are on $\overleftrightarrow{AB}$?

$(13, -5)$ $(-7, -5)$

$(14, -6)$ $\left(4, -\frac{2}{5}\right)$

$(-2, 3)$ $\left(4, -\frac{3}{5}\right)$

$(-2, 4)$ $\left(0, 2\frac{4}{5}\right)$

b. What must be true about x and y if $X = (x, y)$ is on $\overleftrightarrow{AB}$?

Ways to Think About It

$X = (x, y)$ is on $\overleftrightarrow{AB}$ if and only if $X - A$ is a multiple of $A - B$. But

$$X - A = (x, y) - (3, 1) = (x - 3, y - 1) \quad \text{and}$$

$$A - B = (3, 1) - (8, -2) = (-5, 3).$$

So $X = (x, y)$ is on $\overleftrightarrow{AB}$ if and only if $(x - 3, y - 1)$ is a multiple of $(-5, 3)$.

That is, $X = (x, y)$ is on $\overleftrightarrow{AB}$ if and only if

$$(x - 3, y - 1) = k(-5, 3)$$

for some number k. You can write this as two equations (look at each coordinate): $X = (x, y)$ is on $\overleftrightarrow{AB}$ if and only if

$$x - 3 = -5k \quad \text{and}$$

$$y - 1 = 3k.$$

You may never have seen two equations like this before; there is this pesky k in there. But you can get rid of the k by *dividing* the two equations:

$$\frac{y - 1}{x - 3} = -\frac{3k}{5k} = -\frac{3}{5}.$$

The k is gone. You have a nice test to see if (x, y) is on $\overleftrightarrow{AB}$. You just calculate $\frac{y-1}{x-3}$ and see if you get $-\frac{3}{5}$. If you do, (x, y) is on the line. If you don't, (x, y) is not on the line.

Put another way, you have an *equation* for $\overleftrightarrow{AB}$:

$$\frac{y - 1}{x - 3} = -\frac{3}{5}.$$

You can use this equation to test points for being on $\overleftrightarrow{AB}$.

For example, to test the point $(9, -2)$, you plug $x = 9$ and $y = -2$ into the equation and see if it works:

$$\frac{-2 - 1}{9 - 3} \overset{?}{=} -\frac{3}{5}.$$

The left side is $-\frac{3}{6} = -\frac{1}{2}$ which is not the same as $-\frac{3}{5}$. So the point $(9, -2)$ is not on the line.

In order to divide the equations, you have to assume $x \neq 3$. That's ok. You already know how to tell if $X = (x, y)$ is on the line when $x = 3$: y has to be 1.

2. What would y have to be if $(8, y)$ were on the line?

3. Suppose $A = (-5, 12)$ and $B = (5, -4)$.

    a. Find b if $(0, b)$ is on $\overleftrightarrow{AB}$.

    b. Find c if $(c, -14)$ is on $\overleftrightarrow{AB}$.

    c. Find the point on $\overleftrightarrow{AB}$ whose x- and y-coordinates differ by 48.

4. Find an equation of the line passing through the given two points. Give three points that satisfy each equation, three that don't, and draw a picture of the whole setup.

    a. $A = (3, 5)$, $B = (-1, -5)$

    b. $P = (8, -1)$, $Q = (1, 7)$

    c. $R = (6, 7)$, $S = (2, -3)$

    d. $T = (3, 5)$, $U = (11, 11)$

5. **Checkpoint** The equation of a line is

$$\frac{y - 1}{x - 7} = -\frac{2}{5}.$$

    a. What is y when $x = 3$?

    b. What is x when $y = 2$?

    c. Graph the line.

    d. Find y when $x = 0$ (this is called the y-intercept).

    e. Find x when $y = 0$ (this is called the x-intercept).

    f. Find y when $x = 7$.

ACTIVITY 2 ▸ ## Slope

Let's go back to the situation in Problem **1:** Let $A = (3, 1)$ and $B = (8, -2)$. To test $X = (x, y)$ to see if it's on $\overleftrightarrow{AB}$, you built the equation

$$\frac{y - 1}{x - 3} = -\frac{3}{5}.$$

The left side is a "change in y over change in x" from point $X = (x, y)$ to point $A = (3, 1)$. The right side is the "change in y over change in x" from point $A = (3, 1)$ to point $B = (8, -2)$. This number, change in y over change in x can be calculated for *any* two points. It is called the **slope** between the two points.

So, m(A, B) is the change in y over change in x from A to B.

Definition

If $A = (a_1, a_2)$ and $B = (b_1, b_2)$, the *slope* from A to B, is the number $m(A, B)$ given by the formula

$$m(A, B) = \frac{a_2 - b_2}{a_1 - b_1}$$

provided $a_1 \neq b_1$.

For example, if $P = (8, -12)$ and $Q = (-100, 45)$, then

$$m(P, Q) = \frac{-12 - 45}{8 - (-100)} = -\frac{57}{108}$$

It turns out that slope is a convenient and useful idea when working with lines. The following problems lead up to the reasons why.

6. For each pair of points, plot the points, draw the line segment between the points, and find the slope.

    a. $A = (4, 5)$, $B = (9, -1)$

    b. $A = (4, 5)$, $C = (5, 7)$

    c. $X = (14, -7)$, $A = (4, 5)$

    d. $P = (0, 3)$, $Q = (5, -3)$

7. What is the relation between $m(A, B)$ and $B - A$?

8. Suppose $A = (-2, 3)$ and $B = (12, 5)$.

    a. Where does $m(A, B)$ show up in the equation for $\overleftrightarrow{AB}$?

    b. Write four points X that are on $\overleftrightarrow{AB}$. Show that for each of your points, $m(X, A) = m(A, B)$.

Why do you have to rule out vertical lines?

9. Suppose A, B, and C are three points that are not on a vertical line. Show that A, B, and C are collinear if and only if $m(A, B) = m(A, C)$.

For Discussion

People say that the *slope of a line* is the slope between any two points on the line. Why does it not make a difference which points they pick?

10. Suppose $A = (6, -2)$. Graph the line that passes through A and has the given slope.

    a. 1   **d.** -3

    b. 0.5   **e.** 0

    c. 2   **f.** $-\frac{2}{3}$

11. Establish the following properties of slope.

    a. $m(A, B) = m(B, A)$

    b. $m(A, B) = m(A + C, B + C)$

    c. $m(A, B) = m(O, B - A)$

    d. $m(A, B) = m(tA, tB)$

    e. $m(P, P + 3A) = m(O, A)$

    f. $m(P, P + tA) = m(O, A)$ for all numbers $t \neq 0$

12. Show that two nonvertical lines are parallel if and only if they have the same slope. What can you say about their slopes if they are perpendicular?

13. A line has the property that if you calculate the slope between any two points, you always get the same number. Is this true about any other figures?

14. Checkpoint

   a. What can you say about a line if its slope is

   i. positive?

   ii. negative?

   iii. zero ?

   b. What kinds of lines have no slope?

ACTIVITY 3

More About Equations of Lines

Recall that when $A = (3, 1)$ and $B = (8, -2)$, you found an equation for $\overleftrightarrow{AB}$ to be:

$$\frac{y - 1}{x - 3} = -\frac{3}{5}.$$

Ways to Think About It

Some people describe this equation by saying: $X = (x, y)$ is on $\overleftrightarrow{AB}$ if and only if

$$m(X, A) = m(A, B).$$

They say, "X is on $\overleftrightarrow{AB}$ if and only if the slope from X to A is the same as the slope from A to B."

This is a good way to think about equations of lines. Suppose you want to find the equation of the line between $P = (7, 1)$ and $Q = (-3, 5)$. Ask yourself, "How do I know if $X = (x, y)$ is on $\overleftrightarrow{PQ}$?" Answer: "If the slope from X to P is the same as the slope from P to Q." But

$$m(X, P) = \frac{y - 1}{x - 7} \quad \text{and}$$

$$m(P, Q) = \frac{1 - 5}{7 - (-3)} = -\frac{4}{10} = -\frac{2}{5}.$$

So, the equation of the line is

$$\frac{y - 1}{x - 7} = -\frac{2}{5}$$

as long as $x \neq 7$. If you multiply both sides by $x - 7$, you can leave out the pesky condition "as long as $x \neq 7$." You get

$$y - 1 = -\frac{2}{5}(x - 7)$$

Remember, the line does go through P = (7, 1).

Now you *can* let $x = 7$. If you do, you get $y = 1$, which is right. Many people simplify the equation even more.

$$y - 1 = -\frac{2}{5}(x - 7)$$

$$5(y - 1) = -2(x - 7) \quad \text{(from multiplying both sides by 5)}$$

$$5y - 5 = -2x + 14$$

$$2x + 5y = 19$$

This is easy to use, although some people prefer

$$5y = -2x + 19$$

or even

$$y = -\frac{2}{5}x + \frac{19}{5}$$

because these let you easily find y if you know x.

15. Find equations of the following lines. Graph the lines and find three points on each line.

   a. The line through $A = (3, -1)$ and $B = (9, 12)$.

   b. The line through $A = (3, -1)$ and O.

   c. The line through $P = (7, 14)$ with a slope of 2.

   d. The line through $P = (7, 14)$ with a slope of $-\frac{1}{2}$.

   e. The line through $Q = (-3, 4)$ parallel to the line whose equation is $3x + 2y = 7$.

16. Find the equations of the lines that contain the sides of the square in the figure below.

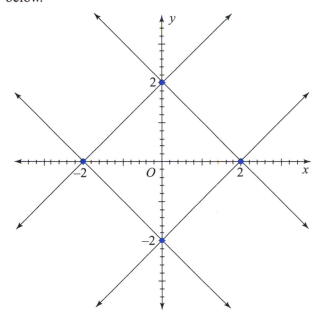

That's why these are called "linear equations."

17. Show that the graph of any equation in the form $y = ax + b$ is a line. What is its slope?

18. There is more than one way to do things. Suppose $A = (4, 2)$ and $P = (7, -1)$. Show that the set of all points X that satisfy

$$X = P + tA$$

for some number t is a line. What is its slope? What is its "ordinary" equation?

This is called the vector equation of the line.

19. Checkpoint Consider these problems.

    a. Is the point $(1, 3)$ on the line with equation

        i. $y = 4x - 1$?

        ii. $y = 3x + 1$?

        iii. $y = \frac{2}{3}x + \frac{7}{3}$?

        iv. $\frac{y-3}{x-1} = \frac{2}{3}$?

        v. $\frac{y+1}{x+3} = \frac{1}{3}$?

    b. Write the equation of one other line that the point $(1, 3)$ is on.

On Your Own

1. If $A = (3, 2)$ and $B = (5, 7)$, determine whether or not the points below are on the line containing A and B. Describe how you decided.

    a. $(4, 4)$

    b. $(1, -3)$

    c. $(2, 0)$

2. Use the following pairs of points to find

    a. the equation of the line that passes through these points.

    b. three other points that satisfy this equation.

    c. three points that do not satisfy this equation.

        i. $(1, -1)$ and $(4, 1)$

        ii. $(0, 2)$ and $(3, 4)$

        iii. $(4, -1)$ and $(7, 1)$

        iv. $(-2, 3)$ and $(1, 5)$

3. Find the equation of the line that passes through $(2, 3)$ and $(7, 3)$.

4. Find the equation of the line that passes through $(2, 3)$ and $(2, 5)$.

5. Use the following pairs of points to

   a. find the slope between the points.

   b. plot the points.

   c. draw the line between the points.

   i. $(6, 4)$ and $(-3, 4)$

   ii. $(6, 4)$ and $(6, -3)$

   iii. $(3, 1)$ and $(5, -2)$

   iv. $(5, -2)$ and $(7, -5)$

6. On one set of coordinate axes graph the lines through $(7, -2)$ with the given slopes.

   a. -4

   b. -1

   c. 0

   d. 0.2

   e. 0.4

   f. -0.6

   g. 1

   h. 4

7. Carpenters use *pitch* when they describe how steep a roof is. A pitch of 6 means the roof rises 6 inches for every 12 inches.

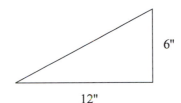

6"

12"

What is the slope of a roof whose pitch is

   a. 6?

   b. 8?

   c. 4?

   d. 10?

8. In New England roofs are very steep.

   a. What is the pitch of this roof?

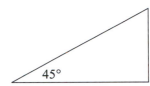

45°

   b. What is its slope?

9. Triangle ABC has vertices $A = (-2, 1)$, $B = (3, 4)$, and $C = (0, 9)$

   a. Draw $\triangle ABC$ on a coordinate plane.

   b. Find equations for the sides of the triangle.

   c. Find the lengths of the sides of the triangle.

   d. Verify that this is a right triangle.

   e. How do the slopes of the perpendicular sides compare?

10. **a.** Find the equation of the line through $A = (3, 2)$ and $C = (9, 4)$.

b. Find the equation of the line through $B = (4, 8)$ and $D = (8, -2)$.

c. Show that $ABCD$ is a parallelogram.

d. Find the point of intersection of $\overline{AC}$ and $\overline{BD}$.

Unit 5 Review

1. Draw coordinate axes. On the axes, plot and label the following points.

   A: (3, 1)

   B: (1, −3)

   C: One more vertex of a square with $\overline{AB}$ as one side. Give the coordinates of *C*.

   D: The fourth vertex of a square with $\overline{AB}$ as one side. Give the coordinates of *D*.

2. Use the distance formula to confirm that opposite sides of the square you made in Problem **1** have the same length.

3. Find the midpoints of the sides of your square from Problem **1.**

4. What are the coordinates of the points shown below?

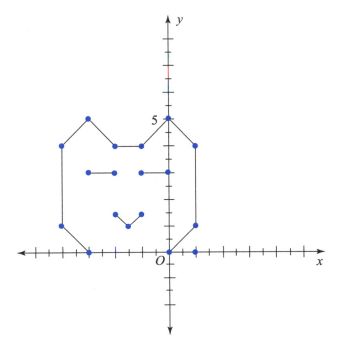

5. **a.** Sketch all the points whose *y*-coordinate is −2.

   b. Sketch all the points whose *x*-coordinate is 5.

6. Draw a vertical line through the point with coordinates (−2, 1). Identify six other points on your line. What can you say about every point on your line?

7. Name and plot seven points whose first coordinate is the same as the second coordinate. Graph all the points whose first coordinate is the same as the second coordinate.

8. Graph all the points whose first coordinate is the negative of the second coordinate.

9. Find and identify eight points that are 5 units away from the origin. Sketch the picture of all the points that are 5 units from the origin.

10. Name six points in three dimensions, whose z-coordinates are not zero, that are 5 units away from the origin.

11. Suppose $P = (12, -5, 3)$. How far is P from the origin? From $(15, -9, -4)$?

12. Describe a method for finding the distance between two points if you know their coordinates.

13. a. Suppose $A = (8, -12, -12)$ and O is the origin. What are the coordinates of the midpoint of $\overline{AO}$?

    b. Suppose $C = (3, 5, 8)$ and $D = (-9, 10, 3)$. If D is the midpoint of $\overline{CE}$, find the coordinates of E.

14. Describe a method for finding

    a. the coordinates of a segment's midpoint if you know the coordinates of its endpoints.

    b. an unknown endpoint if you know the coordinates of a segment's midpoint and one endpoint.

15. Write and Reflect Explain how the distance formula is related to the Pythagorean Theorem. Be sure to include a picture.

16. Describe how to scale a figure by operations on the coordinates of the points. Draw a sketch to demonstrate the technique.

17. What happens when you multiply the coordinates of the points on a figure by a negative number?

18. Describe how to translate a figure by operations on the coordinates of the points. Draw a sketch to demonstrate the technique.

19. Apply each rule below to the coordinates of the points on a figure. For each rule, decide if it produces a figure congruent to the original, similar to the original, or neither.

    a. $(x, y) \mapsto (x + 2, y + 2)$

    b. $(x, y) \mapsto (2x, 2y)$

    c. $(x, y) \mapsto (x + 3, y - 5)$

    d. $(x, y) \mapsto (3x, -5y)$

    e. $(x, y) \mapsto (3x - 1, 3y + 1)$

    f. $(x, y) \mapsto (\frac{1}{2}x, -\frac{1}{2}y)$

    g. $(x, y) \mapsto (y, x)$

    h. $(x, y) \mapsto (-x, y + 4)$

20. Group the following rules according to which are *equivalent*.

Suppose you perform these rules on the same figure. Which rules give the same result?

    a. $(x, y) \mapsto \left(\frac{1}{2}(2x + 2) + 3, 3y - 12\right)$

    b. $(x, y) \mapsto \left(\frac{1}{2}(2x + 8), 3(y - 12)\right)$

    c. $(x, y) \mapsto (x + 4, 3(y - 4))$

    d. $(x, y) \mapsto \left(\frac{1}{2}(4x + 8), 3(y - 2) - 6\right)$

21. Make up your own rule. Draw a sketch to illustrate what it does to a figure. Write your rule using $(x, y) \mapsto$ notation.

22. When you scale a point A by a number c, where is cA in relation to A and the origin?

23. Plot a point P on a coordinate system. Then plot every multiple of P.

24. Describe two ways to add two points. Illustrate each way with a sketch.

25. Copy the picture below. Then draw vectors equivalent to $\overrightarrow{AB}$ starting from O, P, and R.

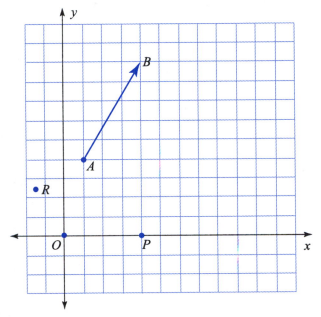

26. How can you tell if two vectors are

   a. parallel?

   b. equivalent?

27. **a.** Give the equations of two lines that you know are parallel.

   b. Give the equations of two lines that you know are perpendicular.

UNIT 6

Optimization
A Geometric Approach

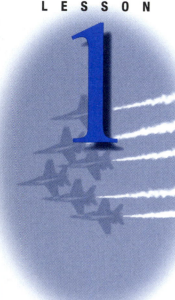

1

Making the Least of a Situation

In this lesson, you will find methods to minimize paths.

The words *optimize* and *optimization* are relatively new to the English language.

> **optimize** *tr. v.* 1. To make as perfect or effective as possible. 2. To make the most of. (*American Heritage Dictionary of the English Language,* Third Edition, 1992.)

> **optimization** *n.* The procedure or procedures used to make a system or design as effective or functional as possible, especially the mathematical techniques involved. (*American Heritage Dictionary of the English Language,* Third Edition, 1992.)

People create mathematics to help find the best solutions to practical as well as theoretical problems. It is no wonder, then, that the theme of optimization shows up repeatedly in mathematics. Much mathematics has been developed in attempts to answer optimization questions like these:

- What's the fastest way?
- What's the shortest way?
- What's the smallest that ... ?
- What's fairest?
- What's the cheapest way?
- What's the longest way?
- What's the biggest that ... ?

Optimization often involves *minimizing* something—making it as small as possible. One might minimize rate, time, or the number of ways something can happen. In geometry, the things to minimize tend to be measures like distance, length, area, volume, or angle measure. Minimization problems occur in many fields, such as architecture, computer programming, economics, engineering, and medicine.

The **distance** between two fixed points is the length of the shortest path between them. It cannot be minimized or maximized; it is just what it is. A **path** between two points can be as long as one wants but cannot be shorter than the distance between those two points.

Humble as the subject seems, there is a lot to think about here. For example, you know that there is only one shortest path between points in a plane and that the length of the path is the segment connecting the points. In Unit 1, you thought about the shortest path between two points on the surface of a ball and found that this path was along a great circle. You also found that, for some points, there were infinitely many "shortest paths" on the surface of a sphere. You could ask more questions, such as: "What about shortest paths between two points on a cylinder?"

Explore and Discuss

Finding the shortest path between two points might seem to be the simplest minimization problem in geometry, but sometimes you cannot take the shortest path because things are in the way.

Suppose you and a friend are at a corner of a parking lot and it is raining fiercely. All the spaces have been taken, and you want to get to your car in the shortest way possible. You can walk or run either around the cars or between them.

a What route minimizes the length of your path to the car?

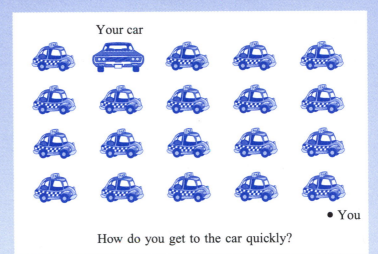

How do you get to the car quickly?

You are lounging on the beach at *L*. You want to run to the shoreline and swim out to your friends on a raft at *K*.

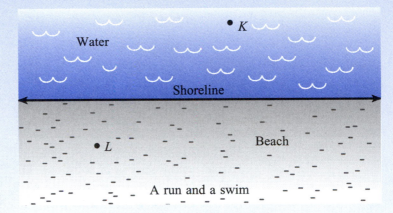

A run and a swim

b You want the swim to be as short as possible. To what point on the shoreline should you run to minimize the path you swim? Why?

c You want to reach *K* with the least amount of running possible. At what point on the shoreline should you enter the water now? Why?

d You want to reach *K* in the least possible total distance. Now where should you hit the shoreline? Why?

Is there more than one best route to the car?

Does it matter which way the cars are facing?

This problem will come up again. We will call it the run-and-swim problem.

Make sketches to show where you should hit the shoreline.

Minimizing Distance

Sometimes you cannot take the shortest path because things are in the way; sometimes you do not take the shortest path because you care about the means by which you get yourself to your destination. Still other times you do not take the shortest path from one point to another because there is a third place you want to visit first.

Imagine, for example, that you are motorboating on a river and the boat is very short on fuel. You *must* drop off a passenger on one riverbank first, and then you can go to refuel at a station on the other riverbank.

Below is a picture of the situation. The boat is at *A*. You will drop off the passenger on the lower bank (call that not-yet-chosen place *P*), and then you will refuel on the opposite bank at *B*. *You* choose the location for *P* (you can drop off the passenger anywhere on that bank), but, of course, you want to minimize total distance because you are low on fuel.

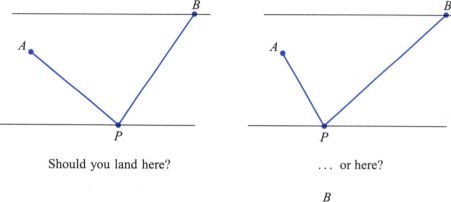

Should you land here?   ... or here?

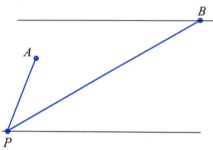

... or maybe here?

1. Find a way to locate the best spot for *P* that *doesn't* involve trial and error. Explain and justify your solutions. The following questions may help you. Be sure that you keep records as you work. Write your conjectures even if you later reject them. You might look at what happens to *AP* and *PB* separately.

    a. Study what happens to the sum *AP* + *PB* as *P* moves along the riverbank. Find the optimal place for *P* by experimentation.

    b. How do the positions of *A* and *B*, relative to the lower bank, affect the best spot for *P*? For example, if *A* and *B* are exactly the same distance away from the lower riverbank, where is the best location for *P*? If *A* is half as far as *B* is from the lower bank, where is the best location for *P*?

    c. If you hold *A* constant but widen the river to move *B* farther away, does the best location for *P* move nearer to *A* or farther from *A*?

Ways to Think About It

When you look into a mirror, your reflection appears to be the same distance from the mirror as you are, but on the opposite side of the mirror from you. This makes the images and the real scenes that cast the images symmetric with respect to the mirror.

Suppose there is a mirror along the lower bank. The path from the reflection of *A* to *P* to *B* is the same as the path from *A* to *P* to *B*. Why?

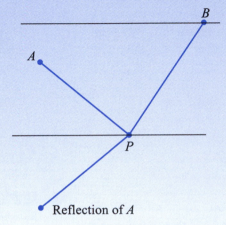

You can make the trip from the reflection of *A* to *B* as short as possible by connecting them with a straight line segment. So that same spot minimizes the trip from *A* to *B* with a stop on the riverbank.

This problem will come up again. Let's call it the burning tent problem.

2. Returning from a hike while on a camping trip, you see that your tent is on fire. Luckily, you are holding a bucket, and you are near a river. At what point along the river should you get the water to minimize the total distance you have to travel to get back to the tent? Justify your answer.

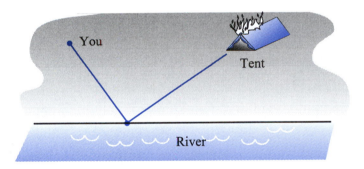

Where should you stop to fill the bucket?

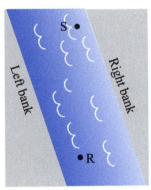

Back and forth . . .

3. In the picture at the left, a canoe is in a river at *R*. It must first let off a passenger on the left bank, then pick up a passenger who will meet the canoe on the right bank, and finally deliver that passenger to an island at *S*. Explain how to find the drop-off and pickup points that minimize the total distance traveled. Check your solution using a compass and ruler, string, or geometry software.

4. You are at an arbitrary point M in an oddly-shaped swimming pool with many sides. How can you find the shortest way out of the pool? Will you go to a corner or to a side?

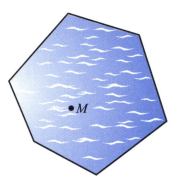

How would you describe all the points that are a given *distance away from* your spot in the pool?

5. Can you draw a pool or a position for M in such a way that a person would choose to swim to a corner as the shortest way out?

6. **Checkpoint** You are in a rectangular swimming pool at K, out of reach of the sides of the pool. Before swimming to L, you want to swim to a side of the pool to put down your sunglasses. Explain how to find the place to put your sunglasses that minimizes the length of the path you have to swim.

Does your method work for any polygonal pool?

What does polygonal mean?

Don't lose your sunglasses.

On Your Own

1. Define these words.

optimization
minimum
maximum
perpendicular
bisect
polygon
distance
length
path

2. What is the shortest way to get from one point to another?

3. What is the shortest way to get from a point to a line?

4. How would you find the shortest path to get from *A* to

    a. the *line BC*?

    b. the *segment BC*?

Explain your reasoning.

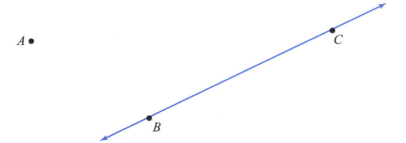

5. Describe your general *strategy* for minimizing the total path from a point to a line and then to another point not on that line.

6. Write what you know about perpendicular bisectors. Describe how you can use them to help you solve an optimization problem.

7. In Unit 1, Lesson 1, you learned about the Triangle Inequality.

    a. Restate the Triangle Inequality.

    b. What does the Triangle Inequality have to do with the run-and-swim problem or with any of the other problems in this lesson?

8. Is the statement below true or false? Justify your answer.

> *In any quadrilateral, the sum of the lengths of the diagonals is smaller than the perimeter of the quadrilateral and bigger than half its perimeter.*

9. A hungry spider sees a bug on the ceiling. The picture below shows the spider on the left wall of the room. What is the shortest path along the walls between the spider and the bug? (No air travel is allowed.) Make a sketch to show what you think is the shortest path, but make sure to describe how to find it as well.

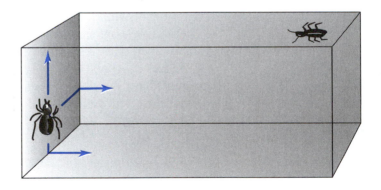

Which way?

10. In the following figure, $\overline{NL}$ and $\overline{QM}$ are perpendicular to $\overleftrightarrow{LM}$.

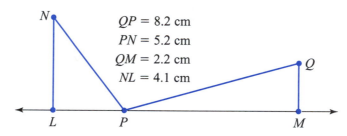

$QP = 8.2$ cm
$PN = 5.2$ cm
$QM = 2.2$ cm
$NL = 4.1$ cm

  a. What is the measure of $\angle NLP$?

  b. What is the distance from Q to $\overleftrightarrow{LM}$?

  c. What is the length of $\overline{LP}$?

  d. Is P equidistant from N and Q? Explain.

  e. Is $\overline{NL}$ parallel to $\overline{QM}$? Explain how you know.

  f. If P were free to move to the right along $\overleftrightarrow{LM}$, what would happen to NP?

11. Assume that in Problem **10**, P is a point that is free to move along $\overleftrightarrow{LM}$. Tell whether the following statements are true or false. Correct any false statements.

  a. No matter where P moves on $\overline{LM}$, $NP + PQ < NQ$.

  b. No matter where P moves on $\overline{LM}$, $NP + PQ$ will be the same.

  c. If P is at the midpoint of $\overline{LM}$, then $LP = MP$.

  d. If P is at the midpoint of $\overline{LM}$, then $NP = PQ$.

  e. Based upon the current location of P, $NL + LP = NP$.

  f. If $\angle NPQ$ is a right angle, then $NP = PQ$.

  g. If N' is the reflection of N across $\overleftrightarrow{LM}$, then $m\angle NPL = m\angle N'PL$.

12. Swimming lessons are being given at a local pool this summer. To get the swimmer's badge, a student must swim the length of the pool underwater. The last test is to dive into the pool and swim underwater until you get to the other end of the pool. You must touch the bottom of the pool once during the swim.

Of course, the swimmers want to swim the shortest possible route, especially the ones who are not sure how long they can hold their breath. Carol, who is a careful thinker, takes a starting position and then dives in. The pool is 50 feet long, 20 feet wide, and 8 feet deep. Draw a diagram of the pool showing where Carol must have stood to start and the best path underwater. (Assume a constant speed for the entire path.) Explain why this is the best path.

13. Given any $\triangle ABC$, let M_1 be the midpoint of $\overline{AB}$ and M_2 be the midpoint of $\overline{BC}$. Find the shortest path from M_1 to $\overline{AC}$ to M_2. Label the point at which the path hits $\overline{AC}$ as X. Will X always, sometimes, or never be the midpoint of $\overline{AC}$? Explain your answer.

14. Points *S* and *T* are the endpoints of a diameter of circle *O*. What are the possible values for *m∠SPT*? Where should you put *P* on the circle to minimize *SP* × *PT*? To maximize *SP* × *PT*?

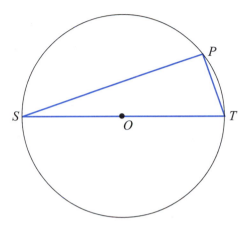

15. Assume that points *S* and *T* are fixed in the positions shown below and that *P* is free to roam around the circle. Angle *SPT* is called an *inscribed angle* in the circle. What are the possible values for *m∠SPT*? Where should you put *P* to minimize *SP* × *PT*? To maximize *SP* × *PT*?

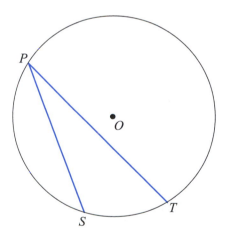

16. Again, points S and T are fixed, but this time P is free to move off the circle. Suppose that P starts inside the circle and gradually moves to be outside the circle. What happens to the size of $\angle SPT$ in the process? Why?

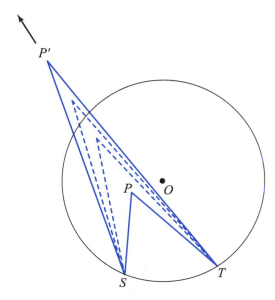

17. Electricians, computer scientists, and others look for optimal ways to connect various points to make networks.

   a. Using the rectangular set of points below, can you create a network whose connecting horizontal and vertical paths have a total length less than the one shown?

   b. What if the restriction of horizontal and vertical segments is removed? Find another network that uses segments that are not necessarily horizontal or vertical and that has a total length less than the one shown.

How many paths are there (if it matters where you start) so that ABCD is different from DCBA?

Perspective

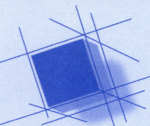

Student Research Experience

Several problems in the previous lesson are related to some of the questions Tonya McLean investigated at a summer mathematics research project for undergraduates. Tonya sent us two of those problems, along with the two at the end of the following essay, as an introduction to the minimal networks topic that she and her colleagues researched.

The "SMALL" Undergraduate Research Project (named after participating professors) is probably the largest in the country. In this summer program, undergraduates spend nine or ten weeks working on problems of current research interest, proving theorems, and writing up results, often for publication and presentation. Approximately seven or eight groups of five or six students each work in such areas as geometry, dynamical systems, knot theory, group theory, and quantum mechanics (a new physics component). Funding from the National Science Foundation's REU (Research Experiences for Undergraduates) program, the NECUSE (New England Consortium for Undergraduate Science Education), individual colleges, universities, and other sources provides money for student stipends and expenses. Contact the National Science Foundation for a list of REU sites if you might want to apply.

The students worked on a lot of different problems to gain insight and experience into their main challenge: finding algorithms that would help them create truly minimal solutions after getting *approximate* solutions (such as from a computer) for a minimal network of connected points. A solution might involve minimizing distances, lengths, or cost. Below is Tonya's description of her experiences in finding and participating in the research project.

"Throughout my first two years in college, I only rarely thought about how I would use my education as a math major. My main choices were entering the job market or continuing my studies with graduate school. I knew I might have a lot of job opportunities with my undergraduate degree, even if I wasn't sure what they might be, yet somewhere in the back of my mind sat the question of graduate school. As my junior year progressed, however, I found myself continually trying to decide whether I should even consider going to graduate school—I wasn't sure it would be interesting or rewarding. No matter how much I thought about it, nothing ever seemed to convince me of a decision. Luckily, I soon stumbled upon an opportunity to do summer research at the undergraduate level; that was just what I needed to complete my decision-making process.

"During each of the three summers prior to my junior year, I had worked at the grocery store in my hometown. Now I was determined to find a better job, a job where I could use my mind and finally get some reward for my three years of higher education. What can a math major possibly do for summer employment? Plenty! Sadly though, I did not know this yet. I didn't even know where to begin looking for a new summer job.

"After asking around, I learned that the campus Career Development Office had a binder with listings of summer opportunities available for math majors. I found listings of colleges which had programs offering undergraduates an experience in conducting research in mathematics. Each of the abstracts which described the research to be conducted was intriguing, and most could be understood without looking up any of the terms in a textbook. How perfect! Since conducting research is a key ingredient in attending graduate school, not only would this experience provide me with a challenging and rewarding summer, it would also help me decide what path to pursue after college.

"I applied to several of the research programs and was accepted at one called SMALL, located at a school in Massachusetts. I was assigned to work with two research teams. Each team consisted of five undergraduates and a faculty advisor. One undergraduate for each team was selected to become the group leader, whose responsibility was to facilitate communication among group members and ensure that the group's research was moving along in the desired direction. Each group met daily to discuss any progress that individual members had made since the last meeting and to determine what the next step would be to bring the group closer to a solution. When we weren't meeting as a group, we worked alone on the problems, but if at any point a member needed help, he or she would consult other members of the group for guidance.

"I was a member of both the Minimal Networks Group and the Symmetry Group, and I was also the group leader for the Symmetry Group. Alice Underwood was the faculty advisor for the Minimal Networks Group. We spent the first week of the program working on a few introductory problems and deciding on a general direction for the group. The central problem we addressed was finding a formula that would provide the minimal solution of connecting a network, given an approximate solution for that network. At the end of the program, it was exhilarating to have actually solved a problem which we had seen for the first time only weeks earlier.

See the end of this Perspective for a short description of Symmetric Groups.

"Robert Mizner was the faculty advisor for the Symmetry Group. The project for this group was built around the book by B. Sagan, *The Symmetric Group: Representations, Combinatorial Algorithms, and Symmetric Functions*. This was a graduate level algebra text that we were reading. We read and discussed the text extensively and compiled a paper that was designed to be an introduction to symmetric groups—a guide for students unfamiliar with the topic. We also added solutions to relevant problems in order to aid the reader in solidifying concepts.

"Prior to participating in SMALL, if anyone had asked me what mathematics research was like, I would have been speechless. Similarly today, when discussing what I do, I often encounter people who are dumbfounded when they learn that there are academic and business professionals who are currently conducting new research in mathematics and making new discoveries. Some feel that mathematics is a lost subject that doesn't help us with anything but balancing our checkbooks. Nothing could be more wrong. Mathematics is just as important as other sciences, such as chemistry and physics, and can be applied to solve most any problem.

"My experience at SMALL helped me decide to attend graduate school. I am currently enrolled in a Master of Arts program. As a teaching fellow at my college, I have encountered undergraduates who tell me they enjoy their mathematics courses, yet they don't choose to major in mathematics because they "can't do anything but teach" with a math major. It is wonderful to see some relief in their faces when I tell them that mathematics majors can do anything and enter into any fields; that the academic and professional worlds deeply respect the kind of training, curiosity, and mental discipline that mathematics majors develop. (Law schools, for example, actively recruit mathematics majors because of their training and the ways of thinking they have developed.) I urge those who enjoy mathematics to speak to mathematics professors, other professionals, and/or a career counselor or two before deciding on a college or a job path, for I have found that mathematics does nothing but open doors for one's career."

1. Assume that only horizontal and vertical paths may be used to connect the set of points contained in the network below. Can you come up with a *better* solution (a shorter path) than the one depicted? Can you find a minimal solution? If so, be sure to explain how you know it is minimal.

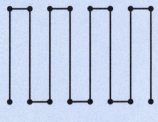

Connected points

2. Information may be transmitted only along the edges of the box below. The box has a square base and a height that's half the length of either of the other two dimensions. The manufacturer of this device wants to connect all of the vertices. Information sent along the path used should "visit" each vertex once and only once. The solution shown here in solid lines could be written as *ABCHGFED*.

   a. Find a better solution—one that has a shorter total path—than the one shown.

   b. What would be a worse solution—one that has a longer total path—than the one shown?

   c. Find a best (minimal) solution. Explain why it is best.

   d. If the connections between the vertices don't have to follow a continuous path from one vertex to the next, without retracing some part of the path, is there a solution that is shorter in total length than the one you found in part **c**? Explain.

   e. How do your answers change if the box has a nonsquare rectangular base?

Here, information can travel in any direction along the edges.

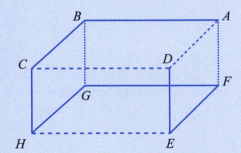

What Is a Symmetric Group? Mathematics is often concerned with finding a common underlying structure in things that may appear quite different.

For example, consider the four arithmetic operations on integers: addition, subtraction, multiplication, and division. At one level, they're all the same; they are arithmetic operations. At another level, they are all different. There are a couple of useful ways to classify the operations as pairs, but there is something that applies to exactly *three* of them that is a particularly important mathematical idea. That property is called *closure.* For integers, closure means that applying the operation to two integers always results in another integer. The set of integers is "closed" with respect to three of the operations because you don't have to look outside the set for the *result* whenever you apply one of them. To which three operations do you think closure applies?

When a set (like the integers) with one operation (like multiplication) has this closure property and a few other properties, mathematicians call it a *group.* The *symmetric groups* are an important class of groups that involve permutations—rules or functions that arrange and rearrange objects. These groups arise naturally in algebra, as well as in geometry (in crystallography) and in physics.

There are many ways to shuffle a set of cards. For example, when you have three cards, every shuffle produces one of six possible arrangements of the cards. (What are they?) Two shuffles can be "composed" by doing the first one and then doing the second one. The mathematical system whose elements are the card shuffles, and whose operation is this kind of composition, is an example of the *symmetric group on three symbols*. How many distinct shuffles are there on four cards? On five cards? On *n* cards?

Two shuffles are considered the same if they produce the same arrangement of cards.

LESSON

Making the Most of a Situation

In this lesson, you will find methods to maximize the areas of given figures with fixed perimeters.

Just as you sometimes want to minimize things, there are times you want to *maximize* things—to make things as large as possible. In general, people want to minimize things that cost money, require a lot of boring work, waste time, or make them uncomfortable. People often want to *maximize* things that make food or money, involve a lot of interesting work, save time, or make them more comfortable.

Explore and Discuss

Perhaps the most common thing geometers try to maximize is area.

a Describe situations in which maximizing area would help in making food or money, saving time, or providing more interesting work or more comfort.

The most basic area-maximization problem is this:

For all shapes with the same perimeter, which one has the greater area?

b Make some conjectures about the area-maximation problem. Give explanations for your conjectures. Don't restrict your ideas only to polygons because "shapes" means figures formed by curves as well.

ACTIVITY

Some Maximization Problems

These problems will help you develop some maximization strategies.

1. Triangles of many different sizes may have two sides of lengths 5 and 6. Which of these triangles encloses the most area?

2. Of all the parallelograms whose sidelengths are 20, 30, 20, and 30, which one encloses the most area?

3. Suppose you want to build a house with some sort of rectangular base. The most expensive part of framing the house is the walls. Given your budget, you decide you can afford a house whose floor has a total perimeter of 128 feet. What dimensions should you choose for the floor of the house if you want to maximize the floor area?

This is one of those problems in which you have to read a proof and understand it, so read carefully.

4. A friend came up with the following argument which shows that a square with sidelength 32 feet is the best solution to Problem **3:**

> I'll show that a 32×32 square is best by demonstrating that any other rectangle with a perimeter of 128 has a smaller area than the square. I'll do this by showing that I can cut up such a rectangle and make it fit inside the 32×32 square with room to spare.

Suppose, for example, that I had a 40 × 24 rectangle. First, I would cut it like this:

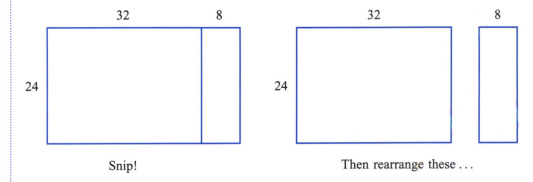

Snip!   Then rearrange these . . .

Then, I would take the small strip off the side, turn it, and put it on top of the 24 × 32 rectangle:

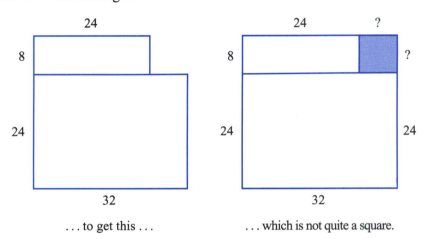

. . . to get this . . .   . . . which is not quite a square.

See? The operation *rearranged* the area, but I didn't add or lose any area. Meanwhile, my two pieces cover *some* of the 32 × 32 square but not all of it. The shaded part isn't covered, so the square has more area! Therefore, the area of the 32 × 32 square is *bigger* than the area of the 40 × 24 rectangle.

a. Does the cutting argument for the 40 × 24 rectangle really work? Rewrite it in your own words. Explain each step.

b. Try using a similar argument for a rectangle with different dimensions but with the same perimeter of 128 feet.

c. Use a cutting argument to show that *any* nonsquare rectangle with a perimeter of 128 feet has a smaller area than a square with the same perimeter.

5. Use the cutting argument to show that an $a × b$ nonsquare rectangle has a smaller area than a square with the same perimeter.

6. Suppose you have 32 meters of fencing to build a rectangular pen. You can create a bigger pen by building it against your barn, which is 25 meters long, because fencing will be needed for only three of the sides. What dimensions maximize the area of this pen?

*Geometry software is
an ideal tool for
experimenting with this
problem.*

7. Suppose you have 32 meters of fencing and you want to build a rectangular pen against a not-so-long barn, one that is only 13 meters long. Again, you plan to use the barn as one side of the pen. What size rectangle maximizes the area of the pen now?

8. To score in soccer, one has to kick a ball between two goal posts. Suppose you are running straight down the field toward the goal, but off to the side because of the other team's defense. As you run, you have various openings on the goal posts. From what position should you shoot if you want the widest angle for getting the ball between the goal posts?

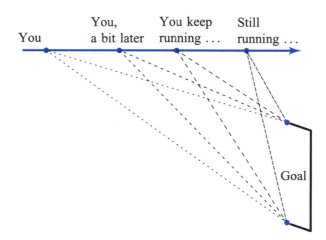

9. What if you are playing soccer and you are moving the ball along some straight line that is not necessarily perpendicular to the goal line? Describe how to locate the spot that maximizes the angle in the picture.

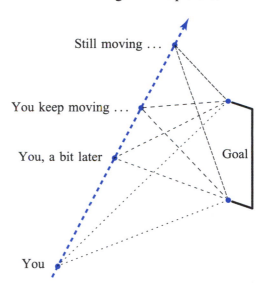

10. **Checkpoint** Of all the rectangles with a given perimeter, which one has the greatest area? Explain your answer.

On Your Own

1. Define these words.

   rectangle
   square
   perimeter
   angle
   polygon
   inscribe

What about tracing, cutting up the polygons, and then comparing the parts? What else might you do?

2. Which of these two polygons has the greater area? Find a way to convince your teacher or someone else that your answer is correct.

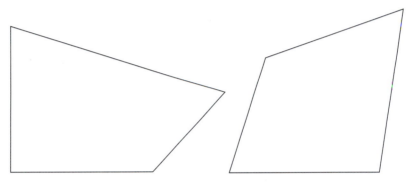

At least two angles are important—the bottom-to-top angle and the side-to-side angle—and they maximize in different places.

3. You are in a gallery looking at a picture on the wall in front of you. Your eye level is 5 feet, the bottom of the picture is 5.5 feet above the ground, and the picture is 4 feet tall and 6 feet wide. How far from the wall should you stand for the maximum viewing angle? Is this likely to be the best place to stand?

4. Which of the two marked angles below is larger? Why do you think so?

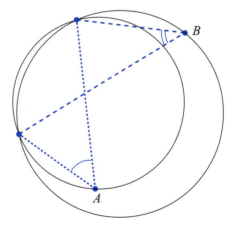

5. Explain why the hypotenuse is the longest side of a right triangle.

6. Make a list of all the strategies you have used so far for solving optimization problems. Illustrate each strategy with an example of a problem that can be solved with it.

7. Describe a way to create a polygon with the same sidelengths as the polygon below but with greater area.

8. A farmer owns a square farm and wishes to buy adjacent farms for her children. Because of highways and natural borders, such as rivers, the adjacent farms are triangles. (To keep the examination of the situation simple, assume the triangular farms are equilateral and congruent.) Look at the diagram below. Answer the questions that follow.

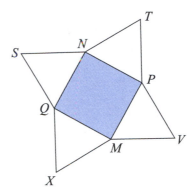

   a. If another square were drawn adjacent to the first farm so that it shared side $\overline{NP}$, would point T be inside, outside, or on the border of the square?

   b. Assume that the perimeter of the first farm $NPMQ$ is 8 kilometers.

      i. What is the area of the first farm?

      ii. What would be the perimeter and area of each child's farm?

      iii. What would be the total perimeter and area of the land owned by the entire family?

      iv. What is the maximum land area the farmer could acquire if she could buy as much land surrounding her farm as possible, given the restriction of a total perimeter no greater than the one just calculated in part **iii?** Assume there are no restrictions on the shape of her children's farms.

      v. What would be the shape of the family farm then?

9. Drink boxes are a fairly new type of packaging. A typical boxed juice contains about 250 milliliters (ml) and measures 10.5 cm × 6.5 cm × 4 cm. What is the volume of this box in cubic centimeters?

a. Is this the best box size? Suppose you are the manufacturer and you want to maximize your profit by reducing packaging costs. What size rectangular drink box will hold 250 ml and use the least amount of cardboard?

b. Design a drink container of any shape that would hold 250 ml of juice, using the least amount of cardboard. Be sure that the container could be held comfortably by a child or an adult and could be packaged, shipped, and stacked on a store shelf. Explain your choice with enough information to convince a packaging engineer that this is the best shape.

10. Cut wood is often sold by the *cord* and the *half-cord*. A cord is a stack that measures 4 feet × 4 feet × 8 feet. The wood can be bought in logs which are 8 feet long ("log lengths"), cut into smaller logs, or cut and split into easy-to-manage pieces. For firewood, many people buy a cord or half-cord of cut and split logs. Paper and lumber mills, however, often purchase wood in log lengths.

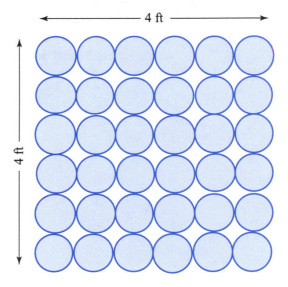

This picture shows one way
to stack a cord of log lengths.

Suppose that, when Jamal orders a cord stacked like this, he may choose the diameter of the logs. What diameter should he pick if he wants to maximize the amount of wood in the stack and minimize the amount of air? Of course, wood doesn't come this way; there are many different diameters in a pile. But this is one way to start thinking about the problem. Should Jamal buy one big log? Many small logs? Does it matter that small logs burn faster than large ones?

Theorem 6.1

Of all the polygons having a given perimeter and a given number of sides, the regular *polygon has the greatest area.*

Try proving this for a special case: Suppose the given perimeter is 120 ft and the given number of sides is 6.

You might want to make use of Theorem 6.1 again.

11. Solve the following generalization of Problem **6** on page 480 using Theorem 6.1.

Suppose you have 840 feet of fencing and you want to build a four-sided pen (not necessarily a rectangle) against a very long stone wall, using the wall as one side of the pen and your fencing for the other three sides. What shape and dimensions maximize the area of the pen?

12. Suppose you have 840 feet of fencing and you want to build a *five*-sided pen against the stone wall, using the wall as one side of the pen and your fencing for the other four sides. What shape and dimensions maximize the area of the pen?

13. Compare the area of the pen you found in Problem **12** to the area of the pen you found in Problem **11**. State your conjectures.

14. In the figure below, imagine that A and B are fixed but C can move anywhere as long as $AC + BC$ stays the same. As you move C, $\triangle ABC$ changes shape. For what position(s) of C does the triangle have the greatest area? Why?

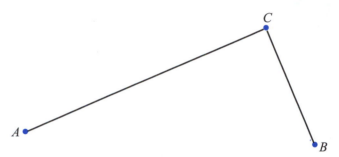

One way to model this situation is to take a piece of string—its length representing $AC + BC$—and pin one of its ends at each of points A and B. Pull out the slack so that the string describes two line segments. The intersection of these two segments is one possible location for C.

You can do this as a thought experiment if there is no string available.

Pull to remove slack

15. In Problem **14,** what kind of a curve do you get if you drag C around while keeping $AC + BC$ constant?

16. Of all triangles of perimeter 24, which one has the greatest area? Justify your answer. Does your finding relate to any conclusions drawn in Problems **11–13?**

How is this a "related problem?"

17. In Problem 4 on pages 479–480, you saw that of all the rectangles with the same perimeter, the square has the greatest area. Here is a related problem: Of all the rectangles of area 64, which one has the *smallest* perimeter?

18. If a rectangle has sides a and b,

   a. what is the area of the rectangle?

   b. what is the side of a square with the same area?

   c. which has the greatest perimeter, a nonsquare rectangle or a square, if the areas are the same? Prove it.

Contour Lines and Contour Plots

In this lesson, you will explore functions on the plane and use contour lines to find minimal values for these functions along a path.

Have you ever seen a topographic map? A topographic map usually depicts land elevations and sea depths by curves or lines that represent points of about the same elevation.

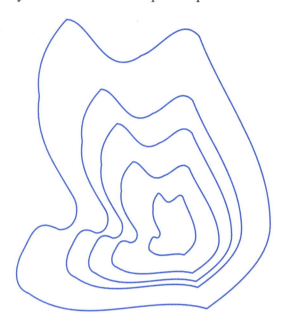

All the points on the same curve are about the same distance above or below sea level. Is this a map of a mountain? A lake?

A topographic map is one example of a contour plot.

What does topography mean?

If you like being outdoors and you are interested in interpreting maps, why not give this a try?

Although this diagram shows only closed curves, other maps may include open (partial) curves, depending on the size and scale of the map.

Contour plots are used extensively in mapmaking. The U.S. Geological Survey began looking for mapping volunteers in the summer of 1994 and may continue to need help updating the maps of your area. Volunteers can sign up for a specific map or part of a map. Volunteers of any age and background are accepted; in particular, the USGS is looking for people with some hiking ability, basic mathematics skills, and good vision. It is also helpful if you can learn how to interpret contour lines.

Maps need to be updated because many changes can happen to the topography around you, sometimes because of such things as the construction of roads or buildings, changes in the course of a stream or river, the digging or filling in of quarries or mines, and the destruction caused by earthquakes or floods. In addition to the need for updated information on changes, many of the existing features on current maps must be verified. You can sign up as an individual or with a group of friends or relatives. Check your library for recent address and telephone information for the U.S. Geological Survey.

Explore and Discuss

Make a contour plot of one of the following:

- Your gym (with all the bleachers pulled out)

- Your football stadium (including stands, if there are any)

- A local sports stadium (including the stands)

- An outdoor amphitheater (like Blossom in Ohio or Great Woods in Massachusetts)

- An auditorium or music hall that you have visited

ACTIVITY 1 ▶ **Interpreting Contour Plots**

Contour plots provide a new way to visualize functions, and they can often be used to solve optimization problems.

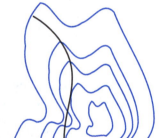

1. Suppose the picture at the left is the topographic map of a mountain and there is an increase of 1000 feet between contour lines.

   a. If the lowest part of the mountain is 100 feet above sea level, sketch the places on the mountain that are about 1600 feet above sea level.

   b. The exact peak is unmarked here, but given what you know about these contour lines, what is the maximum elevation that it could be?

   c. What is the steepest part of the mountain? Explain.

2. If the dark curve on the contour plot shows the path of some hikers on the mountain, what is the highest elevation they reach on their hike?

3. The same picture could be a map of a pond, in which the outer contour line represents the edge of the pond and each inner contour line represents an increase in depth of 10 feet. Imagine that a camp owns the property bounded by the quadrilateral shown on the map below. The swim director wants to rope off a children's swimming area with water no deeper than 5 feet. Trace the map on another piece of paper. Sketch a proposal for such a roped-off area.

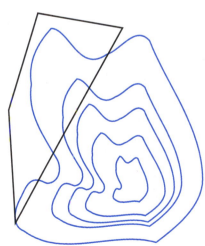

Why would people use the words *contour* and *level* to describe these paths?

The picture looked like this:

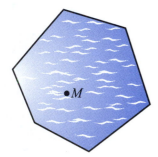

The individual curves in a contour plot are called *contour lines* or *level curves*. A more precise definition of a contour line will come up soon, but for now you can think of a contour line as a curve that shows where a particular feature of a situation is invariant. When the contour lines make shapes that you recognize (circles or polygons, for example), you may be able to use the geometry of those shapes to solve optimization problems. For example, recall Problem **4** in Lesson 1:

> You are at an arbitrary point *M* in an oddly-shaped swimming pool with many sides. How can you find the shortest way out of the pool? Will you go to a corner or to a side?

One student, Pam, thought about doing the problem this way:

> Wait until the water in the pool gets very calm. Then take your fist and pound the water, right by your waist. Ripples will go out from the place you hit the water, and they will be circles. Swim to the first place that a ripple touches a side of the pool.

4. a. Draw a picture that shows what Pam is talking about. Why does her method work?

b. In Pam's method, the ripples are contour lines. Exactly what is constant for all the points on the same ripple?

c. Each ripple is an expanding circle. Even after each circle has expanded so much that it touches a side, it continues to expand. Draw a sketch to show what a ripple looks like a short time after touching a side.

d. The ripple you have just drawn should intersect the side in two places. Why are those two points of intersection not the best places to which to swim?

5. **Checkpoint** You are in a circular pool. You want to get out as quickly as possible. If you are at point *K* in each picture below, which point on the edge of the pool should you swim to? Describe your method for deciding.

a.

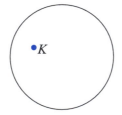

b.

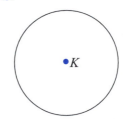

A Contour Plot for Soccer

Remember the previous soccer problems? Suppose you are on a team and your coach wants you to learn how to find the best kicking spots for scoring a goal. This is a complicated skill to learn, so first you look at some simpler cases.

6. The coach draws a line straight across the field. Then she lines up the whole team on that line.

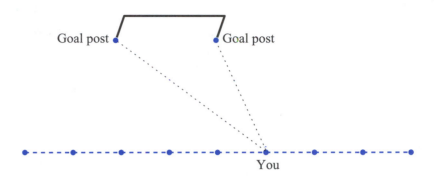

Goal post   Goal post

You

a. Which player has the maximum kicking angle?

b. How does the angle vary as you cross the field?

c. How many players on the line have a kicking angle that is 5° less than the maximum? Approximately where are they?

d. Are there any players on the line with a kicking angle of 1°? Approximately where are they?

7. At the next drill, suppose everyone agrees that the player in the center has the best angle (suppose it is 40°). The coach asks, "Are there any other places on the field from which you can kick from a 40° angle?" Place yourself in one of those positions. Sketch the locations the players can stand. Your sketch will show a *contour line for a kicking angle* of 40°.

8. The team developed a contour plot for the kicking angles over the whole field.

a. Make a sketch to show what this plot might have looked like. Show both goal posts and several contour lines.

b. Are the contours straight, curved, irregular, or circular? Explain.

c. Label the contour lines on your plot with the measures of the kicking angles. Make sure to include the 90° contour line.

d. How do the numbers behave as you move away from the goal in any direction?

e. If you add a contour line between two existing ones, what can you say about the number attached to it?

f. Suppose a player with the ball is crossing the field in a straight line but not parallel to the goal line. Draw such a line on your contour plot. Explain how the plot can be used to determine the spot on the line from which the player has the largest kicking angle.

In a real game of soccer, you wouldn't have to run in a straight line, nor would you care about the *best* angle, as long as it was good enough. Depending on your strength and kicking accuracy, you might be willing to sacrifice the width of the angle a bit if it gave you a more open shot. But the size of the angle can be one important consideration.

9. Checkpoint Invent another problem that uses some of the same mathematics as the soccer problem.

The player's position is the vertex of the "kicking angle," and each goal post is on a side of the angle.

This problem is more realistic because players can kick from anywhere on the field.

As you move along any contour line, the kicking angle stays constant. Does the contour plot tell you what that constant angle is?

Contour Lines and Functions

So far, contour plots have been used like this:

> The nature of the problem assigns a number to each point on the plane. The number might represent temperature, barometric pressure, height above sea level, depth of a pond, distance from a swimmer in a pool, measure of the kicking angle on a soccer field, or a value of many other types.

A *contour line* is a set of points—a curve of some type—whose assigned values are all the same. A *contour plot* is a collection of contour lines with enough information to let you reason about the values of points in the regions between the lines.

Ways to Think About It

The process of assigning a number to each point on the plane is an example of what is known as a mathematical **function.** One way to think about a function in this new context is to imagine a point *P* moving around on the plane carrying a little calculator on its back. The calculator is programmed to always do the same calculation (to measure the kicking angle, for example), and that calculation depends solely on *P*'s current location. Sometimes there are many different points which cause the calculator to produce the same number or *value.* A contour line for the function is just the set of all points for which the calculator produces the same value.

10. Using geometry software, define three points *A*, *B*, and *P*. Consider the function *f* that calculates the sum of the distances *PA* + *PB*. Use your software to measure *PA* + *PB*. Then, as you drag *P* around, the measure will change, showing you the particular sum—the *value* of your function. On a transparency taped over the screen, mark enough points to let you draw a few contour lines for *f*.

You can name a function; call it, for example, *g*. Then the value of *g* for any point *P* is usually written $g(P)$.

The function $g(P)$ is pronounced "*g* of *P*" and means "the *value* of *g* at *P*." If you think of *g* as a calculating machine, $g(P)$ is the number produced by the machine when you "feed it" point *P*. Or it is the *output* of machine *g* that you get when point *P* is the input.

One might sometimes ask, "What number does *g* assign to this particular position for *P*?" One can also ask, "At what positions for *P* does *g* produce this particular number?" The set of all positions for *P* that cause a particular value for *g*, such as 5 or 8, is a *contour line* for *g* corresponding to that number. A *contour plot* for *g* is a collection of *g*'s contour lines, each corresponding to different numbers.

11. Suppose *C* is some fixed point on the plane and *g* is the function that takes a point *P* in the plane and calculates its distance from *C*; that is, $g(P) = PC$.

> *You might want to use your geometry software or a piece of string.*

   a. Make a contour plot for *g*.

   b. What shape is each contour line? Provide a proof.

   c. This contour plot would give more information if each contour line were labeled with a number that told the value of *g*. Label each contour line with the appropriate number.

12. Suppose ℓ is some fixed line on the plane and f is the function that calculates a point's distance from ℓ.

  a. Make a contour plot for f.

  b. What shape is each contour line? Provide a proof.

  c. Label each contour line with an appropriate number.

This is an extension of Problem 10.

  d. What would the contours for f look like if you were looking for all points in *space* (not just in the plane)? That is, what set of points in space is a fixed distance from a given line ℓ?

13. Put two points, A and B, on the plane. Let f be the function that assigns $XA + XB$ (a number) to X (any point). Make a contour plot for f.

To find the value that f assigns to a point X, measure the distance from X to A and the distance from X to B. Then add.

  a. Have you seen curves like these contour lines before? Describe them as precisely as possible.

  b. Label each contour line with an appropriate number.

  c. Of course, your contour plot cannot show *every* contour line. Counting your smallest contour line as "the first," imagine drawing a new contour line between your second and third contour lines. What are the possible numbers that could belong to that new contour line?

  d. Draw a straight line across your contour plot. Explain how to locate the point on that line for which f produces the smallest value.

  e. On your contour plot, draw a circle that contains A and B in its interior. Explain how to locate the point or points on the circle for which f produces the smallest value.

A gadget like this might help.

  f. Can any number you pick be a value for *some* contour line? If so, explain why. If not, give an example—a number that couldn't possibly be a value for f—and explain why not.

14. Suppose E and F are fixed points on the plane and f is a function defined so that $f(Z) = ZE + 2ZF$ for any point Z.

  a. Make a contour plot for f. Label it appropriately.

  b. Draw a straight line on your contour plot. Explain how to locate the point on the line that minimizes the value of f.

Try to invent a gadget using string that will help you draw the contour lines for j.

15. Suppose A, B, and C are three fixed points on the plane and j is a function defined to measure the sum of the distances from a point P to A, B, and C:

$$j(P) = PA + PB + PC.$$

Make a properly-labeled contour plot for j.

16. **Checkpoint** Consider these problems.

  a. Explain how to draw a contour plot for a function that is defined on the points in a plane. Illustrate with examples.

  b. Explain how a contour plot can sometimes help you solve an optimization problem. Use an example in your explanation.

Revisiting the Burning Tent Problem

Recall the burning tent problem from Lesson 1:

> Returning from a hike while on a camping trip, you see that your tent is on fire. Luckily, you are holding a bucket and you are near a river. Where along the river should you fill the bucket with water to minimize the total distance you travel to get back to the tent?

You should already have a method to find the best spot, but you can also get insight into the problem by looking at some "wrong" answers. Pick a non-optimal spot along the line, like the one pictured below.

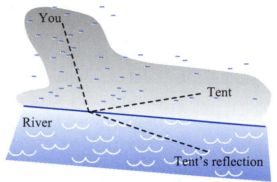

Why is this not the best spot?

Surely, there are other spots P (not necessarily on the river's edge) that are exactly as bad as that spot—no better and no worse.

17. What about those spots *not* along the river that are just as bad (an equal total distance from You and the Tent) as some particular non-optimal point along the river? That is, what is the set of points some constant *total* distance from You and the Tent? Pick a total distance and perform an experiment to determine the contour line for that value.

18. Below is a construction geometry software technique that allows you to make contour lines like the one in Problem **17**:

- Locate two points, A and B.

- Make a segment (other than $\overline{AB}$) that has your desired total distance.

- Put a point P on the segment so that it is partitioned into two parts, d_1 and d_2. You will need to construct these two parts as separate segments.

- With a center at A, create a circle with radius d_1. With center B, create a circle with radius d_2.

- Mark where the circles intersect and trace those points. As P moves along your segment, you should get a contour line.

You might want to hide the circles to neaten your sketch.

  a. If the two circles do not intersect, what is wrong with your sketch? How can you fix it?

  b. This sketch gives you one contour line. Alter the sketch to give you several different contour lines at the same time.

The contour lines you have been drawing are called **ellipses.**

Definition

Begin with two points *A* and *B* and a positive number *k*. The set of all points *P* for which $PA + PB = k$ is called an *ellipse*. Points *A* and *B* are called the *foci* of the ellipse. (*A* is one *focus*, and *B* is the other *focus*.)

Or you can define an ellipse as a contour line.

Definition

An *ellipse* with foci *A* and *B* is a contour line for the function *f* that is defined by the calculation

$$f(P) = PA + PB.$$

Two definitions are equivalent if they define the same object.

19. Are these two definitions of an ellipse equivalent? Explain.

When builders need to construct elliptical shapes, they use a method like the one illustrated in the picture below. They drive nails in at the two foci of an ellipse and then tie each end of a piece of string around the nails. With a pencil, they trace around the nails, keeping the string taut.

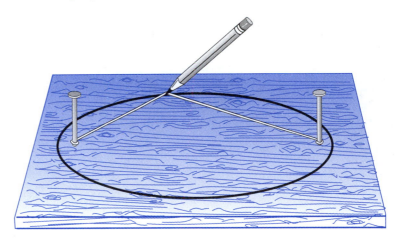

20. Explain how this setup guarantees that all the points drawn by the pencil are the same total distance from the two nails.

Below is a set of drawings for the burning tent problem. Each one shows some path from You to the River to the Tent and a curve containing all the other points on the plane (not just on the river) that are the same total distance from the Tent and You.

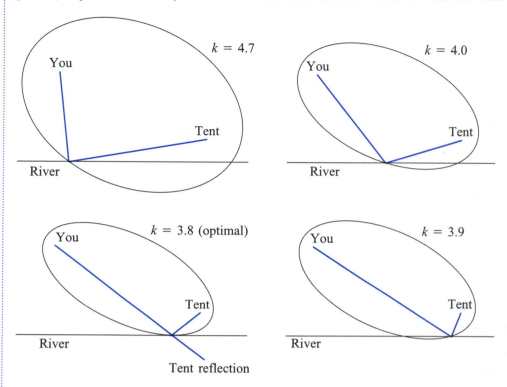

Here is a picture with many of the contour lines superimposed—a contour plot for the burning tent problem.

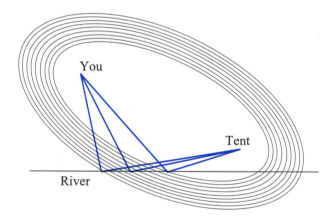

In mathematics, relief maps are called surface plots. Surface plots are supposed to be 3-dimensional; what you see in this lesson are 2-dimensional drawings of surface plots.

21. Notice how these contour lines relate to each other and to the optimal spot on the riverbank. Explain how the contour lines can be used to solve the burning tent problem.

22. a. Investigate topographic maps and relief maps. Describe how they are different.

   b. Suppose O is a fixed point and f is defined so that $f(X) = OX$. What would a relief map for f look like?

   c. Let A and B be fixed and g be defined so that $g(X) = AX + BX$. Describe g's relief map.

d. Below is a depiction of a 3-dimensional relief map of a function. What might the contour plot look like?

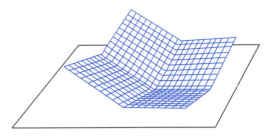

23. **Checkpoint** Draw a circle and a line tangent to the circle.

    a. Which point on the tangent line is closest to the center of the circle?

    b. When you draw a radius from the point of contact to the center of the circle, how are the radius and the tangent line related?

On Your Own

1. Write and Reflect Find at least two ways in which contour plots are used by weather forecasters. You may have to contact a local weather service for help.

2. Define these terms.

contour line	ellipse
contour plot	focus
function	tangent

3. Imagine you are driving through the country and the views are so beautiful that you hardly mind the relentless straightness of the road. A building to your left, which is quite plain in appearance from the rear, has a beautiful front facade that you are eager to photograph. You figure that you are now about as *close* to the building as you will ever be on this road, but because the building is not set parallel to the road, perhaps you will be at a better *angle* further along. What *is* the best viewing position along the road?

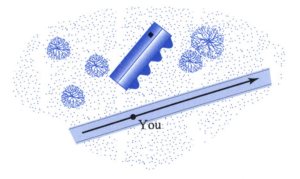

You

Look at Problem 14 in Activity 3.

4. Suppose A and B are fixed points and P is free to roam anywhere in the plane (on both sides of $\overleftrightarrow{AB}$). Suppose h is a function defined as

$$h(P) = m\angle APB.$$

You could use color-coding to explain certain aspects of your map.

   a. Make a properly-labeled contour plot for h.

   b. Are there any points on the plane that seem to be on *every* contour line in your plot? Are there any points that couldn't be on *any* contour line in *any* contour plot for h? That is, are there any points on the plane for which h produces *no* number? Explain.

5. In Problem **4** above, what is the contour line that P traces if P is above $\overleftrightarrow{AB}$ and $m\angle APB = 90°$? Describe how $m\angle APB$ behaves if P is inside this contour line and above $\overleftrightarrow{AB}$. If P is outside?

6. Ellipses are interesting curves that have many wonderful properties. For example, the tangent property for the ellipse says the following.

   Suppose you have an ellipse with foci A and B and you draw tangent $\overline{ST}$ to the ellipse at P. Then $m\angle SPA = m\angle TPB$.

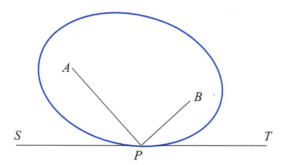

Think of A as You and B as the burning Tent.

This is also why the focus is called a "focus."

   Suppose you place a ball at each focus of an elliptical billiard table and shoot one of the balls in *any* direction. The tangent property implies that, if you hit the ball hard enough, it will hit the other ball after just one bounce off the elliptical cushion, even if you don't aim! Prove that the tangent property stated above is true for any ellipse.

7. Natasha is in a circular swimming pool at K. She wants to swim to L, but first she wants to swim to the edge of the pool to leave her sunglasses there. Explain how to find the best place to put the sunglasses to minimize the total amount of swimming.

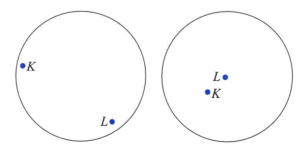

   Here are two possible arrangements for this problem.

8. Suppose you are on another camping trip and your tent is, once again, on fire. Luckily, you have a bucket, and you are near a river. Where should you get the water to minimize the *time* it takes to get back to your tent? Justify your answer.

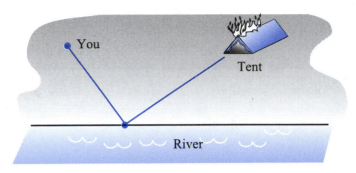

9. Take an acute triangle and *inscribe* another triangle in it. Below is one example.

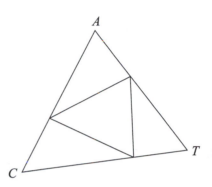

Of all the possible triangles inscribed in △*CAT*, which has the smallest perimeter?

This is a type of minimal path problem.

10. In △*ABC*, *P* is a point on $\overline{AB}$, $\overline{PD} \perp \overline{AC}$, and $\overline{PE} \perp \overline{BC}$. Where should *P* be located on $\overline{AB}$ to make the length of $\overline{DE}$ as small as possible?

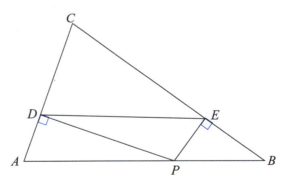

11. In △ARF, P is a point on $\overline{AR}$, $\overline{PD} \parallel \overline{RF}$, and $\overline{PE} \parallel \overline{AF}$. Where should P be located on $\overline{AR}$ to make $\overline{DE}$ as small as possible?

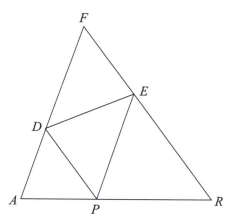

12. Recall Problem **6** in Lesson 1.

> You are in a rectangular swimming pool at K, out of reach of the sides of the pool. Before swimming to L, you want to swim to a side of the pool to put down your sunglasses.

Explain how to use contour plots to find the place to put down your sunglasses in order to minimize your swimming.

13. In studying the burning tent problem, many students make this false conjecture: "It doesn't matter where you stop at the river; the distance AP + PB will always be the same."

Write a description or proof explaining why this is false. How can you be sure that the total distance (AP + PB) will *not* be the same everywhere along the river and that, in fact, there *is* some point P which minimizes the total travel AP + PB? Your explanation must use logic and mathematics. Simply measuring some number of points is not a sufficient response, although you may want to use the information you gain from doing so in your explanation.

Some people can swim
faster than they can run.

B •

Wildlife
Refuge

A •

This is a 3-dimensional
version of Problem 11 in
Activity 3.

14. Rafael is lounging on the beach at L, thinking about how to minimize the total time it will take him to run to the shore and swim out to a raft at K. If he can run twice as fast as he can swim, where should Rafael hit the shore to minimize the *time* it takes to get to the raft?

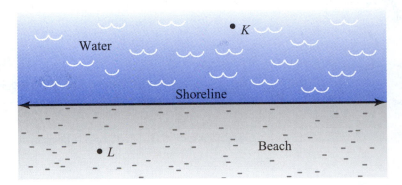

15. Below is a problem and a proposed answer. Read both carefully. Then critique the problem and support or refute the answer. (Note: There are some ambiguities in the problem and some weaknesses in the answer.)

> *Problem:* Two cities A and B, separated by a large wildlife refuge, decide to pool their resources to build a recreation center. By agreement, the new center cannot be located in the refuge, nor can the traffic generated by the center cross the refuge. New roads must be built. City officials want to find a location that minimizes that cost. Where should the recreation center be placed?

> *Answer:* Clearly, to do the *least* amount of road building to connect the two cities at the same point, we would have to build the roads along the line segment AB. But we cannot, because that would cross the refuge. So the total road building must be some length greater than AB. All the locations with that very same total distance from A and B lie along an ellipse. So, the best places to locate the recreation center are anywhere on the smallest ellipse that has A and B as foci and does not cross the refuge.

16. Suppose ℓ is a fixed line, E is a fixed point not on the line, and f is a function that assigns a numerical value to each point P this way:

- f finds the distance from P to E;

- f also finds the distance from P to ℓ;

- f adds the results and assigns *that* number to P.

Make a contour plot for f. Label your contour lines with the appropriate values calculated by f.

17. Suppose you have a function g that calculates the distance between a point P in 3-dimensional space and some fixed point O. What would a picture analogous to a contour plot look like for such a function?

Rich's Function

In this lesson, you will use reasoning by continuity to solve problems.

Some years ago, a student of one of the authors faced this problem on a standardized, multiple-choice achievement test:

Problem

Given an equilateral triangle of sidelength 10 and a point *D* inside the triangle, what is the sum *DR* + *DQ* + *DP*?

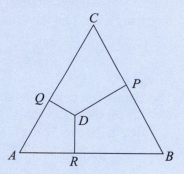

A. 10

B. $5\sqrt{3}$

C. 30

D. 8.6

The student's name was Rich. From then on, in the spirit of professional mathematicians, the students in that class referred to the function that calculates the sum of the distances from an interior point to the sides of an equilateral triangle as "Rich's Function."

Explore and Discuss

Rich had no ruler and no computer. See what you can do with the problem without those tools.

Before reading further, think about the problem above and see if you can find a strategy for solving it. Remember that the distance from a point to a line is the distance along the *shortest* path. In the diagram, this means that $\overline{DR}$ is perpendicular to $\overline{AB}$, $\overline{DQ}$ is perpendicular to $\overline{AC}$, and $\overline{DP}$ is perpendicular to $\overline{BC}$.

Reasoning by Continuity

Perhaps the writers of the test question intended to find out if the students knew a particular theorem. As it turns out, Rich didn't know such a theorem. But he either outwitted the test or demonstrated something the testers may have been looking for: the ability to explore and solve a problem through a combination of deduction, experimentation, and reasoning by continuity. Here is how Rich went about it.

1. First, Rich decided that *one* of the four numbers listed had to be right. Why is that a logical conclusion?

Then he tried to imagine placing point D at several different spots inside the triangle in order to compare the sums of the distances. Yet there were no tools with which to measure anything during the test. That's when Rich realized that the location of D within the triangle must not matter; in fact, the problem said nothing specific about where inside the triangle D was located. This was crucial: *No matter where D is placed* inside the equilateral triangle, the sum of the distances from D to the sides of the triangle must be the same.

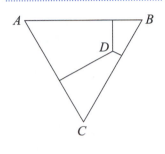

Point D could go almost anywhere.

By drawing this conclusion, Rich got another good idea. He imagined moving point D close to a vertex to see what that would do to the three distances.

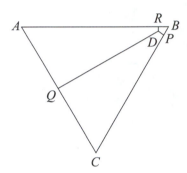

2. a. With D close to vertex B as pictured, which two lengths are very small?

    b. Can you make those two lengths even smaller? How?

    c. How small can you make them?

    d. In Unit 4 you used $\rightarrow$ notation to show something getting very close to a value. As $D \rightarrow B$, fill in the following:

    i. $DP \rightarrow$ _____

    ii. $DR \rightarrow$ _____

3. As DP and DR disappear, what happens to DQ? What part of the triangle does it get closer to?

4. **Write and Reflect** Teachers who looked at this clever solution said that the student used "reasoning by continuity." What do you think that means?

Rich came back to class after the test with the following conjecture:

The sum of the distances from any point inside an equilateral triangle to the sides of the triangle is equal to the length of an altitude of the triangle.

These facts will help: The triangle has sides of length 10, and the altitude of an equilateral triangle bisects the base.

5. **Checkpoint** If you have not already found the sum of the distances to the sides of the triangle presented at the beginning of this lesson, do so now by using Rich's reasoning, what you know about equilateral triangles, and the Pythagorean Theorem. See if Rich's reasoning leads to one of the answers given on the multiple-choice test.

Proving Rich's Function Is Constant

Rich made the conjecture about the altitude of the triangle during the test, under pressure, and didn't have time then to prove whether or not the conjecture was true. But, as it turned out, experimenting with the position for point D and looking at unusual spots or extreme cases helped solve the problem. The class spent the next couple of days discussing Rich's insight and the kind of mathematical thinking that led to it. Students wondered whether they could *prove* the conjecture and turn it into a theorem.

Here is one way to prove the conjecture. In this proof, the area of $\triangle ABC$ is calculated in two different ways: as one triangle and as the sum of the areas of three small triangles. Read through the proof. Important keys to understanding this proof are the two area formulations:

A common geometric habit of mind: Calculate area in more than one way.

$$\text{Area}(\triangle ABC) = \frac{1}{2}bh$$

and

$$\text{Area}(\triangle ABC) = \text{Area}(\triangle ADC) + \text{Area}(\triangle CDB) + \text{Area}(\triangle BDA).$$

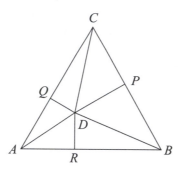
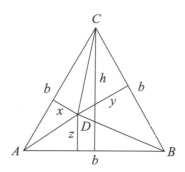

Here is the proof, using the diagrams above:

$$\frac{1}{2}bh = \text{Area}(\triangle ABC)$$

$$= \text{Area}(\triangle ADC) + \text{Area}(\triangle CDB) + \text{Area}(\triangle BDA)$$

$$= \frac{1}{2}bx + \frac{1}{2}by + \frac{1}{2}bz$$

$$= \frac{1}{2}b(x + y + z)$$

To divide both sides of an equation by b, you must be sure that b ≠ 0. If b were zero, you wouldn't have a triangle.

Since $\frac{1}{2}bh = \frac{1}{2}b(x + y + z)$, $b \neq 0$, then $h = x + y + z$.

6. Referring to this text as little as possible, rewrite the preceding proof in your own words.

A popular section of Mathematics Magazine (published by the Mathematical Association of America) is called "Proof Without Words."

There are other ways to prove Rich's conjecture. "Proofs without words" are designed to prove something or convince you of something using only pictures. On the next page is a proof without words of the conjecture which appeared on the cover of *Mathematics Teacher* (Vol. 85, No. 4, April 1992). Study the pictures and try to make sense of what is going on from one step to the next. After you have looked at the pictures, try to write the words that justify each step.

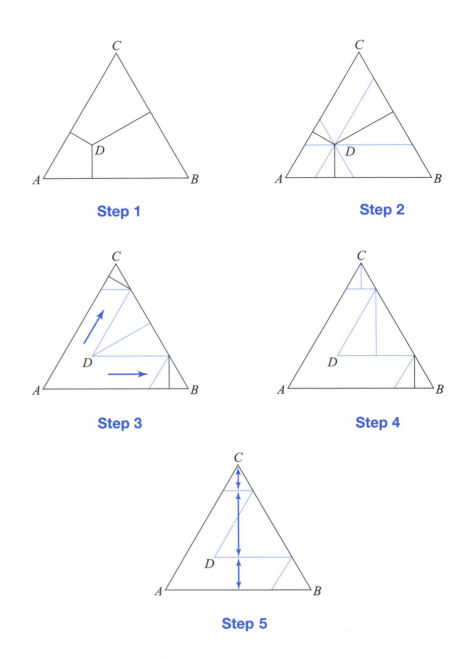

Step 1

Step 2

Step 3

Step 4

Step 5

7. For each step starting with Step 2, write exactly what changed between that step and the previous one. Make sure to supply reasons that explain why each transformation is valid. For Step 1 you might write, "*ABC* is an equilateral triangle with point *D* in the interior. The perpendiculars have been drawn from *D* in order to mark the distances from *D* to the sides of the triangle."

Now, rather than just a conjecture, the class finally had a theorem:

Theorem 6.2 Rich's Theorem

The sum of the distances from any point inside an equilateral triangle to the sides of the triangle is equal to the height of the triangle.

Theorem 6.2 can be used to establish inequalities. The next problem gives an example of its application.

Hint: Recall the arithmetic mean and the geometric mean from Unit 4. Which is bigger? When are they equal?

8. In the picture below, $\triangle EFG$ is equilateral, and perpendiculars from D to the sides of the triangle are drawn, going to points A, B, and C. Point W is inside the triangle. Show that

$$WA + WB + WC > DA + DB + DC.$$

(Hint: Draw perpendiculars to the sides from W, too.)

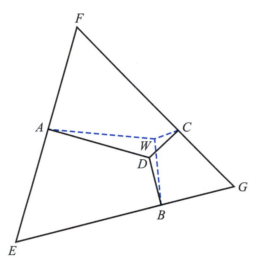

9. Checkpoint Look back at the two proofs of Rich's function. For each proof, describe where you use the fact that the triangle is equilateral.

ACTIVITY 3 › What if the Triangle Isn't Equilateral?

Rich's Theorem says that from any point inside an equilateral triangle, the sum of the distances to the sides of the triangle is always equal to the height of the triangle. Can this theorem be generalized for any triangle?

There is a way that you can figure this out without any measuring at all.

10. If the triangle isn't equilateral, will the sum of the distances to the sides still be equal to the height of the triangle?

In the previous activity, you looked at two proofs that showed that the sum of the distances to the sides of an equilateral triangle equals the height of the triangle. One proof used algebra and calculated the area in two different ways:

$$\frac{1}{2}bh = Area(\triangle ABC)$$

$$= \frac{1}{2}bx + \frac{1}{2}by + \frac{1}{2}bz$$

$$= \frac{1}{2}b(x + y + z)$$

Thus, $h = x + y + z$.

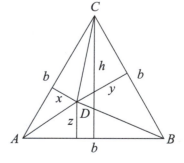

The other proof used geometry and the transformations of smaller equilateral triangles and their altitudes:

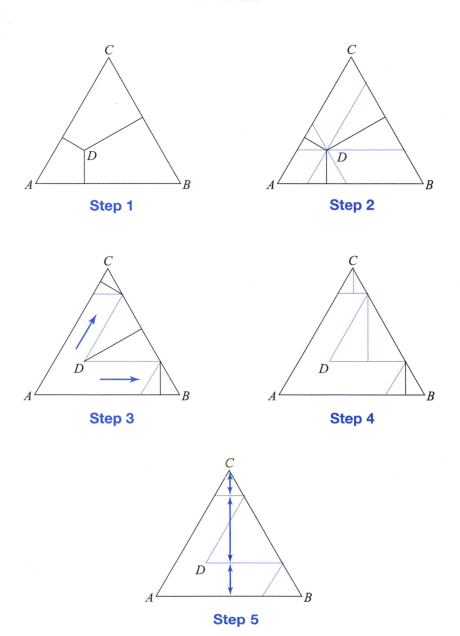

Step 1

Step 2

Step 3

Step 4

Step 5

11. Write and Reflect What goes wrong with each of these proofs when you no longer assume the triangle is equilateral?

What you have done in Problem **11** is another use of proof. By studying what goes wrong with a proof when you change what is given, you can often find new results. You can develop a conjecture for this problem by experimenting, and that is a good thing to do. But by looking at the proofs and what goes wrong with them, you may actually see a way to prove your conjecture.

So, in the case of function R, if the triangle isn't equilateral, the function is no longer constant on the triangle's interior. You will have to call the function by some other name, such as, S (for the sum of the distances to the sides of a nonequilateral triangle).

12. Checkpoint Consider these problems.

   a. Where is this new function, S, smallest?

   b. Where is it largest?

   c. What is that minimum *value* for the function S on the *interior* of the triangle?

   d. What is the maximum value?

1. Given an equilateral triangle of sidelength 8 centimeters and a point *G* inside the triangle, what is the sum of the distances from *G* to the sides of the triangle?

2. Define these words.

   translation
   rotation
   function
   conjecture
   theorem

3. Triangle *FGH* is isosceles, with $GF = HF$; $\overline{GK}$ and $\overline{HJ}$ are medians.

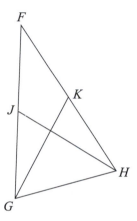

Choose the correct symbol to compare the relative sizes or values of each pair of objects named. You may use =, >, <, or ng (not enough information given).

   a. *GJ* _____ *FJ*

   b. *GJ* _____ *GH*

   c. *FK* _____ *FJ*

   d. *FJ + JH* _____ *FK + KH*

   e. Area($\triangle GJH$) _____ Area($\triangle GHK$)

   f. $\angle KHG$ _____ $\angle JGK$

   g. Area($\triangle FKG$) _____ Area($\triangle HKG$)

   h. Distance from *G* to *K* _____ Distance from *G* to $\overline{FH}$

   i. *JK* _____ *GH*

4. What if $\triangle FGH$ were scalene and $\overline{GK}$ and $\overline{HJ}$ were still medians? Which of the relationships in Problem **3** would stay the same, and which would change? For example, $GJ = FJ$ in Problem **3a.** Is this still true if $\triangle FGH$ is not isosceles? For each relationship in Problem **3,** write *same* if the relationship stays the same, or choose the correct new symbol if the relationship changes. Try to do this exercise without drawing. Instead, try to visualize the scalene triangle. When you have finished, you can draw by hand or use geometry software to construct a scalene triangle to see if you are right.

5. Define these terms.

equilateral triangle	acute triangle
isosceles triangle	parallel
scalene triangle	angle bisector
obtuse triangle	altitude of a triangle

6. Make a list of everything you know that is true about equilateral triangles. Once you have a complete list, decide which characteristics are true for all triangles; which are true for some, but not all, triangles; and which are true only for equilateral triangles. Find a way to present this information so that other students can read and understand it easily. The format is up to you.

7. The term *function* has appeared many times in this lesson ("the function R," "constant function," "a function that calculates the sum of the distances," and so on). Explain what a function is. Since the term has not been officially defined for you in this lesson, write what you think it means. Give some new examples if you can, and write any confusions or questions that you have about the term.

8. Pictured below is a circle with center O, diameter $\overline{AB}$, a point P on the circle, chords $\overline{AP}$ and $\overline{PB}$, and AO = 2 cm. Assume P can move around the circle but the circle itself is a fixed size. As P moves to each new location on the circle, some things change, and some are invariant. For example, as P moves around the circle, the length of $\overline{AP}$ changes, at times increasing in length and at times decreasing. As P moves around the circle clockwise from A to B, which measures change, and which are constant?

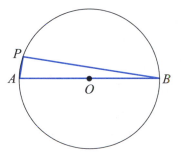 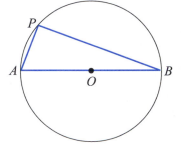

a. For each of the measures listed below, indicate whether the measure is constant or changing.

   i. $m\angle APB$

   ii. the distance from P to O

   iii. the perimeter of $\triangle APB$

   iv. the area of $\triangle APB$

   v. the ratio of the circumference of the circle to the diameter of the circle

   vi. the sum of the distances AP + PB

   vii. the ratio of AP to BA

   viii. the ratio of AP to PB

   ix. the distance from M to N, where M is the midpoint of $\overline{AP}$ and N is the midpoint of $\overline{PB}$

b. For each of the measures in part **a** that are constant, state what that constant value is or how it can be found.

c. For each of the measures in part **a** that are changing, use a graph or a clear written explanation to explain how they change.

9. In Problem **12** in Activity 3, you looked for the minimum value and the maximum value for the function on the interior of a nonequilateral triangle. In what kinds of triangles are these two values equal? For these triangles, what does that say about the function?

10. Here is a different function:

Point D is moving only along $\overline{AB}$ of $\triangle ABC$. The perpendiculars are drawn so that, at each position, the function calculates the sum of the distances to the other two sides of the triangle. What position for D minimizes the sum? What is the minimum *value* for this function? What position for D maximizes the sum, and what is that maximum value? For what kinds of triangles is the function constant along $\overline{AB}$? Prove or support all your conjectures.

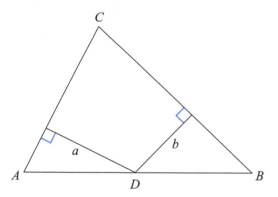

11. Here is yet another function:

Again, point D moves only along $\overline{AB}$ of $\triangle ABC$. The perpendiculars are drawn to the other sides to calculate the distance, but this time the value of the function is given not by the sum of the distances but by c, the length of $\overline{QR}$.

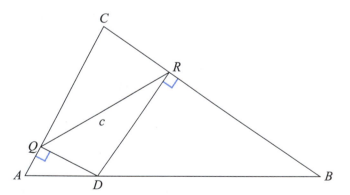

a. What position for D minimizes the value of this function. What is that minimum value?

b. What position for D maximizes the value of this function. What is its maximum value?

Take It Further

12. Use the strategy of calculating area in more than one way to find the length of $\overline{BN}$ if $AB = 5$ and $BC = 12$.

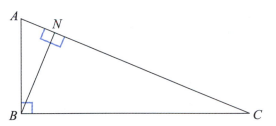

Do you know another way to find the length of $\overline{ST}$?

13. In the triangle below, $AB = 5$, $BC = 12$, and $SB = 4$. Find the length of $\overline{ST}$ by calculating area in more than one way. (Recall that the area of a trapezoid is half the product of the height and the sum of the lengths of the bases.)

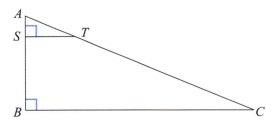

14. Draw a "proof without words" to show that the sum of the lengths of the perpendiculars to the sides of an equilateral triangle from a point anywhere on the triangle is equal to the height of the triangle. (Do not trivialize the problem by using one of the vertices.)

LESSON 5

The Airport Problem

In this lesson, you will explore possible "best spots" for placing an airport and learn about continuous variation.

Three neighboring cities, all about the same size, decide to share the cost of building a new airport. City officials hire your group as consultants to find possible locations for the airport.

In the following activities, you will develop the mathematics to determine some theoretically best spots for the airport (ignoring things like lakes, mountains, waste dumps, and so on). After you have developed your conjectures and proved your theorems, you will write a final report to the city council members explaining how you have chosen the airport location and why it is best. When you write that final report, you should show where some *theoretically* best spots are located and also include practical concerns or reasons (if any) for *not* recommending these mathematically-determined spots.

Why bother with a theory that ignores such realities as topography, politics, and land use? In fact, such a theory is essential. One *must* have a way of saying, "If all else were equal, the best spot would be ... ." Then and only then can one have a standard from which to evaluate plans that (almost inevitably) depart from the model.

Explore and Discuss

a On a map of your state, find three cities of about the same size. Make them the cities in your airport project. In your group, discuss where you think the airport should be located and why. Then write a short, preliminary report describing your conclusions.

In locating the best spot for the airport, you have to determine what the people in the cities care about. For example, they might want the location of the airport to be *fair* in the sense that each city will be approximately the same distance from the airport.

b Come up with some other possible meanings or definitions of *best*. Discuss the advantages and disadvantages of each definition.

c Which of the definitions of *best* would you recommend? Why? Are any of them the same? Which would be acceptable to a city council?

d Explain how to locate the best spot for the airport if the city officials agree to the "fairness solution" suggested above.

e Give an example that shows how the *fair* location for an airport for three cities seems just plain silly.

510 UNIT 6 OPTIMIZATION

The Environmental Solution

Why only "if the cities make roughly equal use ..."?

The environment is probably best served by minimizing the total distance traveled by all users. If the cities make roughly equal use of the airport, this environmental solution is also the most economic, as it minimizes the amount of roadway that must be constructed. Not only can road building costs be substantial at the outset, but maintenance costs can be high since the highway department charges for every mile of roadway that must be maintained. The shorter the roads, the lower the total cost for building and operating the airport.

This environmentally- and economically-best spot turns out to be more difficult to locate than the *fair* spot, but it is far more practical. To locate it, you must also develop and use some beautiful mathematics. So, you have another *optimization* problem: the airport problem. You want to minimize the sum of the distances from the airport to the three cities. If you label the cities as vertices of a triangle, then the problem can be stated in the following way.

The Airport Problem Where should you put point D if you want to make $DA + DB + DC$ as small as possible?

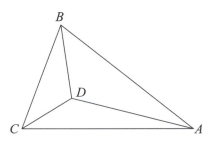

Should it go here?

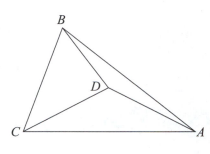

Here?

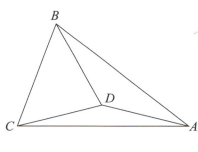

Here?

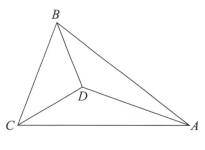

Here?

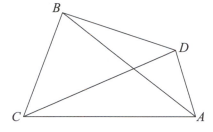

Or maybe even here?

No ... definitely not there.

*For example, what is
DA + DB + DC if D is at
A? At B? At the foot of
the perpendicular from C
to $\overline{AB}$?*

1. How is the "environmental solution" different from the "fairness solution"? Would they ever be the same? If so, under what conditions?

2. Sometimes you can get a sense of a system by investigating its behavior in special cases. Draw △ABC on a piece of paper so that $AB = 9$, $AC = 6$, and $BC = 8$. Place D at several points and calculate the sum of the distances to the vertices. Of all the possible positions for D that you try, which one makes $DA + DB + DC$ the smallest?

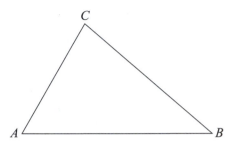

A 6–8–9 triangle

3. **Checkpoint** Describe some differences between the airport problem and the model here in terms of finding the distances to the vertices of a triangle. What are some problems you might have even if you know the mathematically best spot?

ACTIVITY 2 Special Cases and Models

Finding the point inside a triangle that minimizes the sum of the distances to the vertices is not an easy problem. Just coming up with a reasonable conjecture may take some time.

Here are three ideas for directions you might take in working on the airport problem: 1) explore simpler problems and special cases, 2) make a mechanical model, or 3) make a computer simulation.

Idea 1: Look at Simpler Problems and Special Cases

4. What if there are only two cities? Where can the airport go?

• City B

City A •

5. What if there are three cities, but they do *not* form a triangle? That is, what if they are collinear?

a. Make an argument for an airport site that minimizes the total distance.

b. Can you find a *fair* solution—a spot that is equally distant to each city?

• City B

• City C

City A •

Will the best spot move very far if you move one of the cities only a little bit?

6. What if the cities are *almost* collinear?

   a. Where (approximately) is the environmentally best spot?

   b. Where is the *fair* spot?

Looking for special and extreme cases is a very useful way to think about problems. Putting the cities in a line was a special case that made it easier to find a best location. Now try some special kinds of triangles with the hope that the special cases will help you make good guesses about the general case. Here are two ways that might simplify the three-city problem.

Small diagrams make it difficult to measure the effects of different choices for locating the airport. Use a whole sheet of paper and measure carefully. If you are using geometry software, use the entire screen for your diagram.

What is it that makes this a simpler case? Are there other triangular special cases that you would expect to yield insights?

7. What if one city is very far away from the other two? The sketch below doesn't necessarily show the best position for the airport. Where (roughly) would that position be? Explain your reasoning. How would the best position change if City C were twice as far away from $\overline{AB}$? This may be a good question to investigate using geometry software.

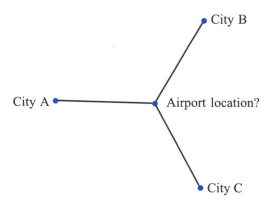

8. What if the cities form an equilateral triangle? The center of the triangle looks like a good spot for the airport. Test enough locations to see if the center spot really minimizes the total distance when the cities are equidistant from each other.

Here the cities form an equilateral triangle.

The guess-AND-CHECK method makes some strong assumptions about the nature of a problem.

You cannot check *every* point in a region, even a tiny region. Between any two points, no matter how close, there is always another point. Perhaps, tucked in somewhere among the very worst of locations will be one that is far better than the best spot you have found in a good-looking region. If small changes in the location of the airport *could* make such a big difference, then the process of checking your best guess with some nearby spots would not be reliable.

9. Experiment enough to get a sense for how the total distance changes with location. Give an argument to support this claim:

    "If I move the airport just a little bit, the sum of the distances to the cities won't change much."

Or, stated more formally:

    "The sum of the distances varies continuously with the location of the airport."

Maybe a contour plot would help here.

10. Could there be more than one spot for which the sum of the distances to the vertices is as small as possible? Defend your answer. After all, there were *many* best spots in the two-city airport problem.

Idea 2: Make a Mechanical Model

Sometimes a machine or physical model can help you experiment with a system. Three physical models for investigating the airport problem are described in this activity. If you already have a conjecture, a model can help you test it. If you do not have a conjecture, one of these three models may help you come up with one.

Take a flat piece of wood and pound nails into it, more or less in a circle.

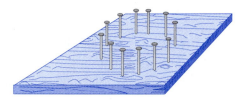

With enough nails in your circle, you can choose three to represent cities in a triangle of almost any shape.

You can also use push pins in wood or cardboard, but they are not quite as strong.

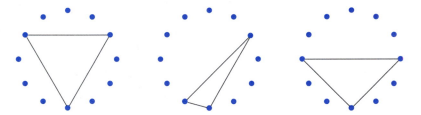

Pick three nails that give you the shape you want.

A model can be crude but still help you think about a problem.

The model is completed with a small metal ring to represent the airport and a strong, slippery thread or string (nylon works well) whose length represents the sum of the distances from the airport to the cities (actually twice the sum, because the thread must be doubled to loop around the nails). The sketch on the following page shows how the thread is passed through the ring and looped around the nails.

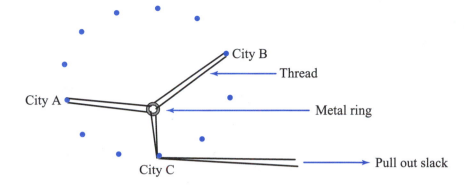

City B

Thread

City A

Metal ring

Pull out slack

City C

11. Explain how this device can be used to locate the best airport location for three cities. Why does it work?

12. Use the model for several arrangements of three cities. Record each experiment by placing a piece of paper inside the circle of nails and tracing the roads to the best airport location for each arrangement.

13. For a given set of three cities, could there be more than one spot where the sum of the distances to the cities is as small as possible? Defend your answer.

Ways to Think About It

The thing you want to minimize is the sum of the distances to the cities. This is a *number,* and you can calculate it for *any* point.

So, you have a new function. Call it the *airport function* and name it *a.* Call the variable point *U.* The cities are labeled *A, B,* and *C.* For any point *U,*

$$a(U) = UA + UB + UC.$$

The notation $a(U)$ is read "*a* of *U*" or "*a* evaluated at *U.*" In other words, $a(U)$ is the output of the function *a* when *U* is the input.

You can now restate the airport problem:

The Airport Problem What point *U* produces a minimum value for the airport function *a*?

14. Use the string-and-nails device to make a contour plot for the airport function *a.* Use your contour plot to defend your answer to Problem **13.**

15. Explain how the contour plot could be used to solve the problem of locating an airport for three cities where an obstacle, such as a power plant, prevents the use of the best spot.

A big part of doing
mathematics is learning
to ask the right kinds of
questions.

16. Experiment with the two-city airport problem using this model. What do the contour lines look like this time?

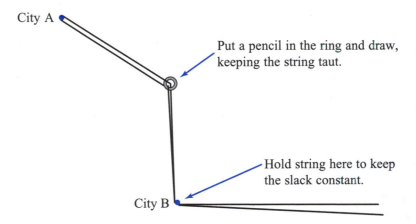

City A

Put a pencil in the ring and draw, keeping the string taut.

Hold string here to keep the slack constant.

City B

Idea 3: Make a Computer Simulation

You can also use geometry software to model the airport problem.

Build a dynagraph that lets you see the value of the airport function both numerically and geometrically. Shown below is such a dynagraph.

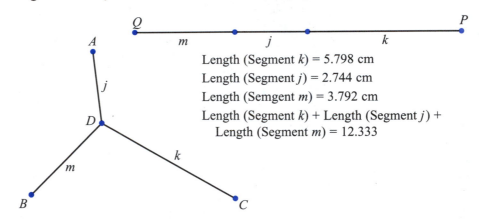

Length (Segment k) = 5.798 cm
Length (Segment j) = 2.744 cm
Length (Semgent m) = 3.792 cm
Length (Segment k) + Length (Segment j) +
   Length (Segment m) = 12.333

As you move point D around, the total distance increases and decreases. The length QP is the value of the airport function for point D. Because you are trying to minimize this length, have the software trace the locus of P. As you move D, what will happen to P when the total distance is minimized?

17. Use your geometry software to draw the contour lines for the airport function.

18. You also can use your computer to keep track of angles in the airport problem. Try it. The pattern(s) you find might support a conjecture or two.

19. Checkpoint State any conjectures you have so far about the airport problem. Provide evidence for your conjectures.

Establishing the Conjecture

The experiments in the previous activity make it seem reasonable to state the following conjecture:

> *The Airport Conjecture* If three cities are arranged in a triangle and no angle of that triangle is too big, then the best place for the airport is the spot at which the roads to the cities form 120° angles with each other.

What does "too big?" mean? If it means "bigger than 60°," then the conjecture is true but of very limited value.

A proof often requires a new insight, a new strategy, or a new way of looking at things. The process can seem quite mysterious from the outside, where the bizarre turn of thought that made the proof possible appears to come completely out of the blue. In fact, these ideas do not come out of the blue: No one invents a proof out of thin air. People walk into situations with a lifetime of experience and with a collection of habits of mind that develop over that lifetime. These experiences and habits include general principles, such as the following.

- If a system varies continuously and at one point its value is 5 and at another point its value is 10, then somewhere between these two points, its value must be 8.

- The shortest way to get from A to B is to travel along the segment from A to B.

Some people believe it helps to memorize things like this and then apply them according to some problem-solving strategy. The development of good habits of mind doesn't work that way. Habits are things you do routinely. You don't even have to think about them. You develop habits over time. Good habits of mind are essential ingredients but rarely the only ones. Proofs arise from a combination of *general* habits and *specific* experience with a problem.

Most people improve their ability to create proofs by studying the proofs of others.

Following someone else's proof to see how it works often provides a good picture of the inner workings of a complex problem. The proof outlined below was invented by the German mathematician J. E. Hoffman in 1929. As you will see, Hoffman essentially applied the strategy outlined in the burning tent problem but with a few more complications.

Recall the setup. You have a system of three fixed points and a moving point D. You want to minimize the sum $DA + DB + DC$. In the picture below (and in all the following ones), D is picked not as the *best* spot but just as *some* spot that will later be improved upon.

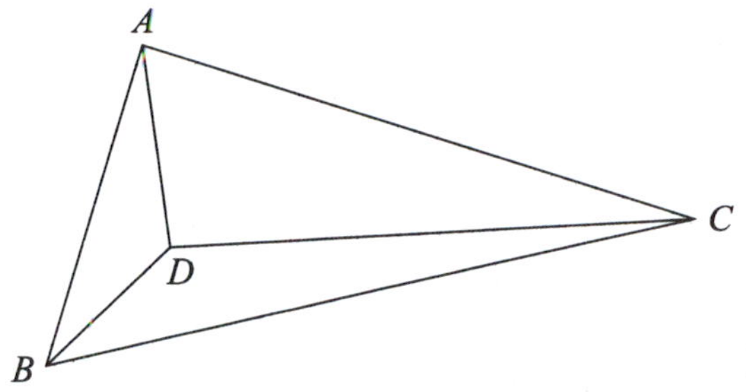

Step 1 Hoffman thought of a similar problem that is easier to solve using an equivalent system—a system of three lengths. In this new system, the best spot is simpler to locate. Hoffman's ingenious construction rotates the smallest triangle, *BDA*, 60° counterclockwise around *B*. It is not clear how Hoffman thought of this, but it is the trick that makes the proof work. Hoffman may have had the 120° conjecture in mind before doing the rotations. Maybe something special about the 120° point made the following construction look reasonable.

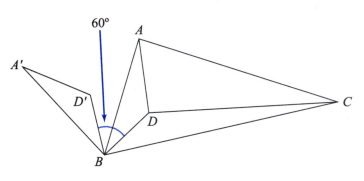

20. Does the position of A' depend on where you put D? Explain.

21. Below is a picture of the same situation, with an alternate location for D, marked D_2. Copy the picture. Rotate $\triangle BD_2A$ 60° about B. Mark the location of D_2'. What do you notice about A'?

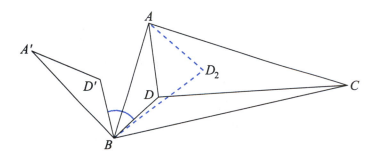

22. In the picture below, two dashed lines have been added to Hoffman's construction to complete $\triangle DBD'$ and $\triangle ABA'$.

Hint: They must be isosceles, so the base angles are equal. Why? By construction, m∠DBD' = m∠ABA' = 60°.

a. Prove $\triangle DBD'$ is equilateral.

b. Prove $\triangle ABA'$ is equilateral.

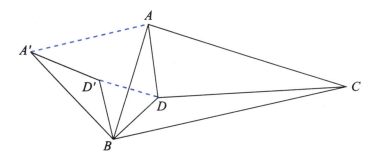

Triangles ABA' and DBD' are equilateral.

Triangle ADB is congruent to triangle A'D'B. Rotation doesn't change any of the original triangle's measurements.

Because $\triangle DBD'$ is equilateral, the distance from D to D' is the same as the distance from D to B. Look at these distances a couple of times, because the fact that they are equal makes the proof work: $DD' = DB$.

The last thing to notice is that $D'A' = DA$ because $D'A'$ is the rotated version of DA. So keep track of these two things:

$$DD' = DB \text{ and } D'A' = DA.$$

In the following picture, $\overline{CD}$, $\overline{DD'}$, and $\overline{D'A}$ are dashed. No matter where you put D, the construction ensures that $DD' = DB$ and $D'A' = DA$.

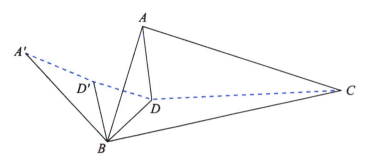

So, what does all this have to do with the distance from the cities to the airport? The sum of the distances from the airport to the cities is $DC + DB + DA$. But that is the same as $CD + DD' + D'A'$:

$$DC + DB + DA = CD + DD' + D'A'.$$

That means that the dashed line has the same total length as the sum of the distances to the airport. So, finding the *minimum* value for the airport function is the same as finding a place for D so that $CD + DD' + D'A'$ is as small as possible.

Step 2 A geometry software construction of the above setup helps show what to do next. Draw $\triangle ABC$. Create a point D. Rotate $\triangle BDA$ 60° about B. Then mark the segments as shown in the previous picture.

You can now move D about, looking for a place that makes $CD + DD' + D'A'$ as small as possible. But notice that $\overline{CD}$, $\overline{DD'}$, and $\overline{D'A'}$ form a path from C to A'. You must put D in a place that minimizes the total trip from C to A'.

Should I put D here?

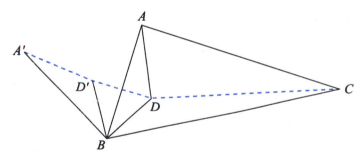

23. What is the shortest path from A' to C? Do you think it is possible to position D so that the three dashed segments form that path? Explain.

... or here?

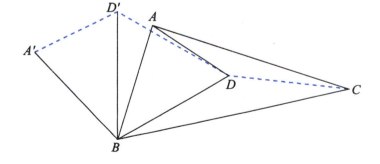

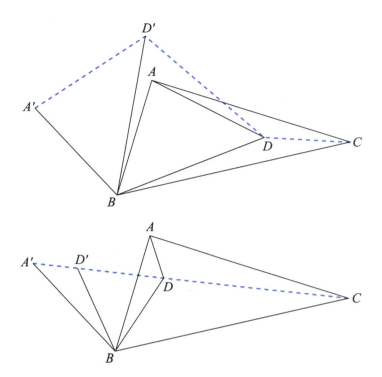

No, ... definitely not there.

What about here?

That will do it. The shortest path from C to A' is along a straight line. So if you can line things up in such a way that $\overline{CD}$, $\overline{DD'}$, and $\overline{D'A'}$ all lie along a straight line, then you have minimized the sum $CD + DD' + D'A'$. That means that $CD + BD + DA$ is also as small as possible.

Remember, the construction is set up to keep CD + DD' + D'A' = DC + DB + DA.

Think about the angle made by B, D, and D'. That is inside the little equilateral triangle you made by rotating $\triangle BDA$. If $\triangle BDD'$ is equilateral, $m\angle BDD'$ has to be 60°.

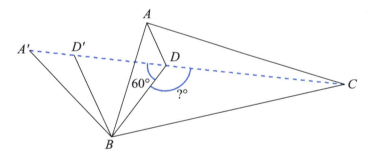

24. If A', D', D, and C are all collinear, what is the measure of $\angle CDB$?

25. Does this prove what you want to prove? In particular, is the measure of $\angle ADB$ also 120°?

26. Copy the figure above. Build another equilateral triangle using a *different* side. Connect its outer vertex to the opposite vertex of $\triangle ABC$. The proof says that the best location for the airport is somewhere on that connecting line. If the proof is valid, that line will go through point D already created by the proof. Check it out.

27. How do you think Hoffman might have thought of the main idea for his proof?

Here is one way to state the theorem you have just proved:

> ### Theorem 6.3
>
> *The best place to put the airport is the point at which the roads make 120° angles with each other, unless there is no such place inside the triangle.*

28. Improve the statement of the theorem by putting it into your own words to make it more precise; that is, tell exactly when there is no such point and what to do in those cases.

Having the proof of something can make you feel good. It can make you think you *really* understand something; it lets you know *why* something is true. But proofs are used in mathematics for more than making people happy. Coming up with a proof can be a useful research technique in the sense that the proof helps you find new facts and do and learn new things. In the case of the airport problem, Hoffman's work was designed to prove that the best airport location within the triangle will have 120° angles between the roads around it if there is such a place. But the proof does more than that. It also provides a method for *finding* that point without hunting around for it.

29. Invent a construction for the best airport location.

30. Go back to some of your earlier diagrams, and compare this construction's *best* place to the *best* place you found earlier.

31. One of the author's students proposed a construction to locate the 120° spot. The construction takes two steps: (1) Build equilateral triangles on two sides of the original triangle, and (2) connect the new vertices of the equilateral triangles to the opposite vertices of the original triangle. According to the student, the intersection, point *D* of those two connecting lines, is the 120° spot. That is, ∠*ADB*, ∠*ADC*, and ∠*CDB* are each 120°.

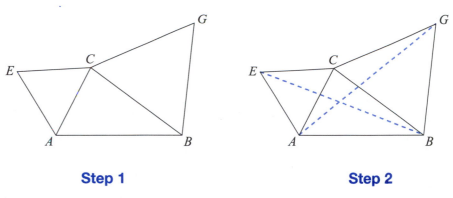

Step 1   **Step 2**

Does this method work? Why or why not?

32. Review the construction in Problem **31.** Use the Hoffman proof to explain why this construction produces a point such that all the angles around it measure 120°.

33. Hoffman's proof also explains why the 120° conjecture won't work for triangles with very big angles. The picture below shows what happens when one angle made by the cities is too big.

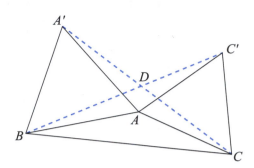

a. Why would *D* be a silly location for the airport in this case?

b. What *is* the best spot for the airport in this case?

c. For what kinds of triangles will the Fermat point lie outside the triangle made by the location of the three cities?

Compare the construction in Problem **31** to the following, similar construction usually attributed to the Italian mathematician Toricelli, who gave the first known simple geometric solution to the problem.

Toricelli proposed this solution before 1640, but it was not published until 1659—by Viviani, one of Toricelli's students.

Step 1 Toricelli constructed equilateral triangles on the outside of each of the sides of the given △*ABC*.

Step 2 Next, Toricelli circumscribed circles around each of these equilateral triangles. The three circles intersect at the desired point.

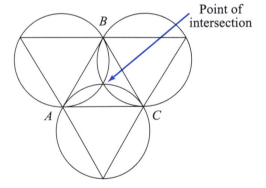

Point of intersection

This is the way the airport theorem is stated in this book.

The book, *Exercitationes Geometricae,* published in 1647, showed that, when you construct lines from any two of the given points to the Fermat point, the angle formed has a measure of 120°.

34. Write and Reflect How might Toricelli have *come up with* the construction in Steps 1 and 2? That is, what *reasonable* things might have led to this approach?

35. Show that Toricelli's circle construction really does produce the Fermat point. That is, show that the circles really meet at *one* point and that that point is the 120° spot.

36. Checkpoint Go back to your original three cities. Find the Fermat point for your three cities. Decide whether or not that point is really the best spot to put an airport. If it is not, explain why and use techniques from this unit to pick a good spot. Write up your work as a report from your consulting group.

The Airport Problem Revisited

Lemma *comes from Greek for* proposition *or* statement. *The German word for theorem is* satz. *The word for lemma is* hilfsatz, *a "helper-theorem."*

One thing leads to another. In mathematics, when one result is used to prove another, the first result may be called a *lemma*. This activity presents a chain of ideas that uses a theorem you studied earlier (Rich's Theorem) as a lemma to prove the 120° result for the airport problem. You will have to fill in a few links in the chain.

But why *another* proof? Hasn't Hoffman's proof already elevated the airport conjecture into the airport theorem? Isn't it convincing enough?

Yes and no. Yes, the airport theorem is established. But a proof is more than a convincing argument. In fact, many theorems are proved long after people are convinced of them. One of the most important and useful functions of proof in mathematics is that proofs establish logical *connections* between results. These connections can be used to gain insights into established facts and to invent new conjectures and facts.

One way to connect Rich's function to the airport problem is to use the same point with different triangles. Start with the airport problem, with cities *A*, *B*, and *C*. Then, at each city, construct a perpendicular to the road that goes to the airport. The sum of the distances to the *vertices* of △*ABC* is the same as the sum of the distances to the *sides* of △*EFG*. If △*EFG* were equilateral, you could make a connection to Rich's function.

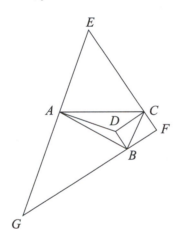

37. Show that it is sometimes possible to make △*EFG* equilateral by moving point *D* around. Is it *always* possible?

38. Show that if the outside triangle *is* equilateral, then *D* must be the Fermat point—the alleged best spot for the airport—for the small triangle.

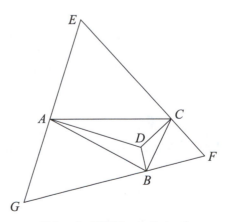

Triangle *EFG* is equilateral.

So, when D is at the Fermat point, the big triangle is equilateral. How can you use that fact to prove the airport conjecture?

One way is to use a kind of indirect reasoning. According to the airport conjecture, the sum of the distances from any point *other than* D to the cities is greater than the sum of the distances from D to the cities. Let W be such a point.

Indirect reasoning is another important strategy in mathematics. If you can show that every other point is worse than the 120° point, then the 120° point is best.

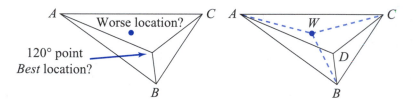

Is W worse than D?

You want to show that the sum of the distances from D to the three cities is less than the sum of the distances from W to the cities. In symbols, you want to show that

$$WA + WB + WC > DA + DB + DC.$$

39. Reconstructing the big equilateral triangle by sketching perpendiculars to the roads to D at the cities, you have this situation:

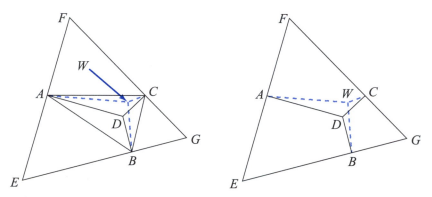

The figure on the right doesn't include the triangle connecting the cities.

Complete the argument to show that $WA + WB + WC > DA + DB + DC$.

40. Make a diagram with cities in a triangle that has one angle greater than 120°. Investigate the above construction in this case. It might be useful to use geometry software here. Is Theorem 6.2 on page 503 any more helpful than Hoffman's proof of the 120° conjecture in deciding where to put the airport?

41. Checkpoint Put together the pieces of the puzzle. Write a proof, based on Rich's function, that the 120° spot—the Fermat point—minimizes the total distance to the vertices for most triangles.

On Your Own

1. Give an algorithm for locating the "fair solution," where the airport will be equally distant from each city in any triangle.

2. The following different kinds of *centers* of triangles are also called *points of concurrency.*

a. Define each of the following centers. Provide a simple sketch to show what each one is. You may need to refer to a math textbook or dictionary.

centroid
circumcenter
incenter
orthocenter
Fermat point

b. Which of the points of concurrency listed above would you use in the following situations?

i. Find the location for an airport so that it will be the same distance from the three cities.

ii. Find a location for an airport so that it will be the same distance from already-existing highways that connect each of the cities to the other two.

c. Write your own question, whether about the airport situation or some other triangular configuration, for which the correct answer is *centroid*.

3. Triangle *ABC* below is equilateral. Points *D*, *E*, and *F* are midpoints. Point *G* is the centroid.

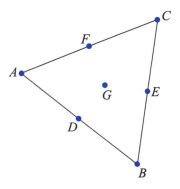

a. Which other points of concurrency does *G* represent? (See Problem **3** in Lesson 1.)

b. Listed below are some objects that might be found in △*ABC* by connecting any of the given points, *A* through *G*. Indicate which of the items can be found by connecting various points. How many of each can you find?

median
equilateral triangle
isosceles triangle
trapezoid
rhombus
kite
other convex, 4-sided polygons
convex, 5-sided polygons
concave, 5-sided polygons
regular polygon
congruent right triangle

4. Suppose $\triangle ABC$ from Problem **3** is drawn on an *x-y* coordinate grid, with A at the origin and B at (4, 0).

   a. What is the slope of $\overline{AC}$?

   b. What is the slope of $\overline{BD}$?

   c. Write equations for $\overline{BC}$ and $\overline{AF}$.

   d. What are the coordinates of points C, D, and G?

   e. What is the length of $\overline{AF}$?

5. You have performed the Hoffman construction on *two* sides of an original triangle (Problem **26**). Predict what would happen if you did the construction for the third side. Give a reason for your prediction. Then check it by performing the construction.

6. *What makes a proof good or bad? Correct or incorrect? Elegant or awkward?* These are questions students and mathematicians think about. Write about one or more of the following questions. Be sure to include examples.

   a. What makes a good proof?

   b. What are some things that are illegal in a proof?

   c. Several proofs of different types are presented in this lesson. Choose one that you particularly liked or understood well. Write about what makes it a good proof.

   d. You may have heard someone say, "That is a really elegant proof." What makes a proof "elegant"?

   e. Take a statement that was proved in this book, and prove it in another way. (Check with your teacher about your choice of statement.)

7. Define these terms.

   rotation
   center of rotation
   angle of rotation

8. The figure below shows a 60° rotation of $\triangle LPM$.

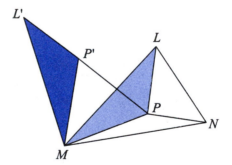

   a. What appears to be the center of rotation?

   b. You were given the angle of rotation (60°). Name all the angles in the sketch that must be 60°. Be sure to explain why each must be 60°.

c. Given that rotating a triangle does not change the lengths of its sides or the size of its angles, decide which of the following statements are true. If a statement is false, correct it by changing the measures, points, or segments that make it false.

$LP = L'P'$

$MP = MP'$

$\triangle MPP'$ is isosceles

$m\angle MPP' = m\angle MP'P = 60$

$MP + PN = PP' + PN$

$L'P' + PP' + PN = L'N$

9. The following figure also shows a rotation. Here, $m\angle DBD' = 45°$ and D is the midpoint of $\overline{AC}$.

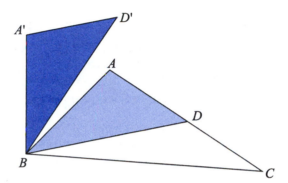

a. What has been rotated in the figure above?

b. What is the center of rotation?

c. What is the angle of rotation?

d. Name seven or eight pairs of congruent objects in the figure. For example, because D is the midpoint of $\overline{AC}$, $\overline{AD} \cong \overline{AC}$.

e. Suppose that $m\angle ABD = 35°$, $m\angle DBC = 20°$, and $m\angle ADB = 50°$. What other angle measures can be determined? Find as many as you can.

f. What if D is the center of rotation and $\triangle ABD$ is rotated 180°? Draw and describe what the picture will look like.

10. Here is another variation on the drawing above. Could this be a picture of a rotation? If it is, what is the angle of rotation and center of rotation? What is being rotated? If it is not, explain why not.

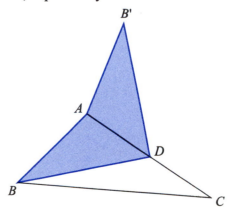

11. Relate a strategy used in the Hoffman proof (see Activity 3) to a strategy that might be used to solve the burning tent problem in Lesson 1. Explain what is similar and what is different in the use of that strategy for each of the two problems.

12. Now you know how to locate the point that minimizes the airport function. Find a simple way to describe the actual *value* of the airport function at the best spot. That is, find a simple way to describe the actual sum of the distances to the cities from the Fermat point.

Take It Further

13. Explain how the contour lines for the airport function might be thought of as generalized ellipses. How would you define an ellipse with *four* foci? How could you tie some string to draw one?

14. Suppose the airport for three cities has to be placed on a specific line. Assume there's an existing highway that must be close to the airport. Now the problem is to locate, among the points *on the line*, the one that minimizes the airport function. Propose and explain a method for finding that spot.

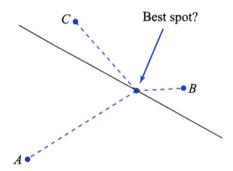

15. What does the airport function's surface plot look like?

Perspective

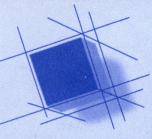

The Research Continues Today

For the last 300 years, dozens of mathematicians have looked at the airport problem and have attempted to glean new insights from it. Fermat, Steiner, Gauss, Toricelli, and Fesbender are just a few of the people who have played important roles in the history of the problem. Along with many others, these mathematicians have produced results that are both beautiful and practical.

The first recorded version of the airport problem occurs in the writing of Pierre de Fermat. In an essay on optimization, he appears to have posed the problem as an application of his work. He worded it in this way: "Let [they] who do not approve of my method attempt the solution of the following problem: Given three points in the plane, find a fourth point such that the sum of its distances to the three given points is a minimum!"

Frank Hwang is a mathematician working at Bell Laboratories. He was born in Shanghai on August 24, 1940. Two months before his 20th birthday, he earned a B.A. degree in modern languages. In 1962, he passed the TOEFL (Test of English as a Foreign Language) and came to the United States to study. The only problem was that he didn't know what to study. His uncertainty was reflected in one year of International Relations at City College of New York and then two years earning an M.B.A. at the Baruch School. While earning his M.B.A., he was required to take a statistics course. This became his next major, and he worked toward a Ph.D. at North Carolina State University.

Dr. Hwang has worked at Bell Labs since 1967. His mathematical interests changed, gradually shifting from statistics to discrete mathematics. He has published about 300 research papers (one-tenth in Chinese), written three books (one in Chinese), and edited two books. His major extracurricular activity is the card game bridge. He has been on the Republic of China's national bridge team several times, the first time when he was seventeen. The pinnacle of his bridge achievements took place in 1969; he was the runner-up at the World Bridge Team Championship.

Dr. Hwang's work at Bell Labs has often put him in contact with the airport problem. In fact, he co-authored a book (*The Steiner Tree Problem* by F. K. Hwang; D. S. Richards; and P. Winter; New York: North-Holland, 1992) which gives some of the modern applications of the result and its generalizations. The authors asked Dr. Hwang to give them a few examples. He sent them the following summary.

Here are some modern developments associated with the Steiner problem:

In 1934, two Czechs, Jarnik and Kossler, first studied the Steiner problem in its full generality (with *n* points). Unfortunately, their work was published only in a Czech journal and did not receive the attention it deserved; it was not even referenced in the Courant and Robbins book.

Kuhn, H. W., "Steiner's Problem Revisited," Studies in Optimization, Dantzig, G. B., and B. C. Eaves, eds: Washington, D.C., Mathematical Association of America, 1993.

Since 1960, the Steiner problem has picked up full steam, not only as a plane geometry problem, but also for higher dimensions and for other metrics. One metric of special importance is the rectilinear metric, sometimes called the Manhattan metric, which superimposes a rectangular grid through the given points and measures the distance between two points by their vertical and horizontal distances on the grid. This is the metric used in routing the VLSI (Very Large Scale Integrated circuit) chips. The wires interconnecting a given set of electrically common points on the chip are allowed to bend or to meet only in angles of 90°. The two sets of wires, vertical and horizontal, are usually routed in different layers of the chip so that two wires in the same layer maintain a constant distance (wires getting too close can cause electrical problems). An interconnection with minimum length indicates in general an interconnection realized in a small area of the chip, which is desirable in our never-ending effort to shrink the chip.

Sometimes the cost of connecting two points is not proportional to the distance between them. Then we need a graphical model to represent the more general cost structure. For this version of the Steiner problem, the set of given points is connected by various *edges* (line segments), and each edge in the graph can be assigned an arbitrary length which represents the cost of connecting the two endpoints. The absence of an edge between two points indicates either that it is not feasible to connect them physically or that the cost would be exorbitant. Again, you want to minimize the total length of such an interconnection.

The Steiner problem also unexpectedly rears its head in biology. For more than 100 years, biologists have attempted to infer *evolutionary trees*. These trees provide information on basic classifications, on when divergence occurred, and on how far apart one species is from another. These questions can be represented by a Steiner problem. The input is a set of species, together with information about each species and relations between the species. The output is a tree (a form of interconnection) that best fits this information. The species are at the leaves of the tree. The internal nodes of the tree, corresponding to Steiner points, are inferred ancestral species.

Unit 6 Review

1. Draw two points on your paper. What is the shortest path between your two points?

2. What is the shortest path from a point to a line?

3. Describe the burning tent problem and its solution using reflection.

4. **a.** Of all rectangles with perimeter 40 meters, which has the largest area?

   b. Prove it is the largest.

5. **a.** A square and an equilateral triangle have the same perimeter. Which has the larger area?

   b. A square and a circle have the same perimeter (circumference). Which has the larger area?

6. What is a contour line?

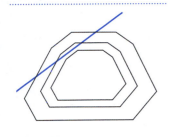

7. At the left is a contour plot for a function in a plane. If the value increases for each contour line,

   a. where is the minimum of the function along the line? Explain.

   b. can you find a maximum for the function along the line? Explain.

8. Draw a contour plot for each of the following functions in the plane. Describe what shape the contour lines form and how you know.

   a. $f(P)$ = distance of P from $B = (-2, 3)$

   b. $g(P) = AP + BP$, where $A = (0, 0)$ and $B = (-2, 3)$

   c. $h(P) = m\angle APB$, where $A = (0, 0)$ and $B = (-2, 3)$

9. What is the shortest path from A to $\overline{BC}$? Use reasoning by continuity to justify your answer.

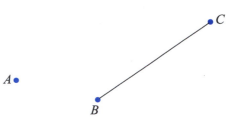

10. What is the difference between a conjecture and a theorem?

11. What point on the interior of a triangle minimizes the distance to the three sides? Does it matter what kind of triangle is being considered?

12. What point on the interior of a triangle minimizes the distance to the three vertices? Does it matter what kind of triangle is being considered?

Index